Essentials of Semiconductor Physics

Essentials of Semiconductor Physics

W. Tom Wenckebach

Department of Applied Physics
Delft University of Technology

JOHN WILEY & SONS, LTD
Chichester · New York · Weinheim · Brisbane · Singapore · Toronto

Copyright © 1999 by John Wiley & Sons, Ltd,
Baffins Lane, Chichester,
West Sussex PO19 1UD, England

National 01243 779777
International (+44) 1243 779777
e-mail (for orders and customer service enquiries):
cs-book@wiley.co.uk
Visit our Home Page on http://www.wiley.co.uk
or http://www.wiley.com

All rights reserved. No part of this publication may be reproduced, stored in a retrieval system,
or transmitted, in any form or by any means, electronic, mechanical, photocopying, recording, scanning
or otherwise, except under the terms of the Copyright, Designs and Patents Act 1988 or under the terms
of a licence issued by the Copyright Licensing Agency, 90 Tottenham Court Road, London,
UK W1P 9HE, without the permission in writing of the Publisher.

Other Wiley Editorial Offices

John Wiley & Sons, Inc., 605 Third Avenue,
New York, NY 10158-0012, USA

WILEY-VCH Verlag GmbH, Pappelallee 3,
D-69469 Weinheim, Germany

Jacaranda Wiley Ltd, 33 Park Road, Milton,
Queensland 4064, Australia

John Wiley & Sons (Asia) Pte Ltd, 2 Clementi Loop #02-01,
Jin Xing Distripark, Singapore 129809

John Wiley & Sons (Canada) Ltd, 22 Worcester Road,
Rexdale, Ontario M9W 1L1, Canada

Library of Congress Cataloging-in-Publication Data
Wenckebach, W. Tom.
 Essentials of semiconductor physics / W. Tom Wenckebach.
  p. cm.
 Includes bibliographical references and index.
 ISBN 0-471-96539-1 (hbk. : alk. paper). – ISBN 0-471-96540-5
(pbk. : alk. paper)
 1. Semiconductors. I. Title.
QC611.W39 1999   99-23420
537.6'22–dc21   CIP

British Library Cataloging-in-Publication Data
A catalogue record for this book is available from the British Library

ISBN 0-471-96540-5 (Paperback)
    0-471-96539-1 (Hardback)

Typeset in 10/12 pt Times by C.K.M. Typesetting, Salisbury, Wiltshire.
Printed and bound in Great Britain by Bookcraft Ltd, Midsomer Norton.
This book is printed on acid-free paper responsibly manufactured from sustainable forestry,
in which at least two trees are planted for each one used for paper production.

Contents

PREFACE		ix
NOTATION CONVENTIONS		xi
1	**INTRODUCTION**	**1**
	1.1 Metals, Insulators and Semiconductors	1
	1.2 Scope of the Book	4
2	**ELECTRONS, NUCLEI AND HAMILTONIANS**	**9**
	2.1 The Chemical Model	9
	2.1.1 The Silicon Atom	9
	2.1.2 Homopolar Binding in Silicon	14
	2.1.3 Other Semiconductors and Impurities	18
	2.2 The Single Electron Hamiltonian	19
	2.2.1 The Electron Hamiltonian of a Crystal	19
	2.2.2 Self-consistent Methods	20
	2.2.3 Core and Valence Electrons	22
	2.3 The Crystal Lattice	23
	2.3.1 Nuclear Motion	23
	2.3.2 Phonon Modes	27
	2.3.3 Electron–Phonon Coupling	30
3	**BAND STRUCTURE**	**35**
	3.1 Translational Symmetry	35
	3.1.1 Primitive Cells and Space Lattices	35
	3.1.2 Reciprocal Space and Brillouin Zones	38
	3.1.3 Transforming from Normal to Reciprocal Space	40
	3.1.4 Functions obeying Translational Symmetry	44
	3.1.5 Discrete Functions in Finite Crystals	47
	3.2 Bloch Functions and Band Structure	50
	3.2.1 The Single Electron Hamiltonian in Reciprocal Space	50
	3.2.2 Free Electrons	51

		3.2.3 Electrons in a Periodic Potential	53
		3.2.4 Band Structure of Typical Semiconductors	57
		3.2.5 Electrons and Holes	63
		3.2.6 Superlattices	66

4 THE k·p-APPROXIMATION — 69

4.1 General Formalism — 69
    4.1.1 The k·p-Hamiltonian — 69
    4.1.2 Basis States and Matrix Elements — 71
    4.1.3 Non-Degenerate Energy Bands — 75
    4.1.4 Degenerate Energy Bands — 76
4.2 Band Structure of Semiconductors — 78
    4.2.1 The Eigenstates at $\mathbf{k} = \mathbf{0}$ — 79
    4.2.2 k·p Treatment of the Conduction Band — 82
    4.2.3 k·p Treatment of the Valence Band — 84
    4.2.4 Spin-Orbit Coupling — 84
    4.2.5 The Valence Band near $\mathbf{k} = \mathbf{0}$ — 87
    4.2.6 Further Refinements — 90

5 EFFECTIVE MASS THEORY — 95

5.1 Externally Applied Fields — 95
    5.1.1 Introduction — 95
    5.1.2 The Hamiltonian with Externally Applied Fields — 97
    5.1.3 Transitions Induced by Externally Applied Fields — 100
5.2 The Effective Hamiltonian — 103
    5.2.1 Envelope Functions — 103
    5.2.2 The $\mathbf{P}\cdot\pi$ Method — 105
    5.2.3 Effective Schrödinger Equations — 107
    5.2.4 Plane Waves and Bloch Functions — 112
5.3 Quantization in Effective Mass Theory — 113
    5.3.1 Landau Levels — 114
    5.3.2 Hydrogenic Impurities — 119
    5.3.3 Quantum Wells — 123

6 THE CRYSTAL LATTICE — 129

6.1 The Lattice Hamiltonian — 129
    6.1.1 Reciprocal Space and Plane Standing Waves — 130
    6.1.2 The Lattice Hamiltonian in Reciprocal Space — 133
6.2 The Phonon Spectrum of Semiconductors — 136
    6.2.1 Optical and Acoustical Phonon Modes — 136
    6.2.2 Transverse and Longitudinal Modes — 143
    6.2.3 Phonon Dispersion Curves — 147

7 ELECTRON–PHONON COUPLING — 154

7.1 Electron–Phonon Coupling Mechanisms — 155
    7.1.1 Electron–Phonon Coupling in Reciprocal Space — 155
    7.1.2 k·p-Treatment of Electron–Phonon Coupling — 161

CONTENTS vii

		7.1.3	Normal Coordinates	163
		7.1.4	Matrix Elements between Phonon States	166
		7.1.5	Electron–Phonon Coupling in Effective Mass Theory	170
		7.1.6	Polar Semiconductors	172
	7.2	Electron–Phonon Coupling near the Γ-point		179
		7.2.1	Conduction Band Electron–Phonon Coupling	179
		7.2.2	Hole–Phonon Coupling	182

8 CHARGE TRANSPORT — 186

8.1 Quasi-Classical Motion — 186
    8.1.1 Elements of Charge Transport — 186
    8.1.2 The Classical Limit — 188
8.2 Carrier Scattering — 192
    8.2.1 Phonon Scattering Processes — 192
    8.2.2 Selection Rules — 194
    8.2.3 Density of States Functions — 197
    8.2.4 Density of States in Effective Mass Theory — 200
    8.2.5 Phonon Scattering Rates — 202
    8.2.6 Impurity Scattering — 209
    8.2.7 Screening — 213

9 OPTICAL TRANSITIONS — 217

9.1 Band Electrons in an Optical Field — 217
    9.1.1 The Optical Field — 218
    9.1.2 Bloch Electrons in an Optical Field — 220
    9.1.3 $\mathbf{k}\cdot\mathbf{p}$ Treatment of Transition Matrix Elements — 223
    9.1.4 Effective Mass Treatment of Optical Coupling — 226
    9.1.5 Optical Coupling Near the Γ-point — 228
9.2 Optical Absorption — 230
    9.2.1 Direct Semiconductors — 232
    9.2.2 Indirect Semiconductors — 240
    9.2.3 Quantum Wells — 249

A THE HYDROGEN ATOM — 254

B THE HARMONIC OSCILLATOR — 259

C PERTURBATION THEORY — 262

C.1 Time-independent Perturbation Theory — 262
    C.1.1 Non-degenerate Unperturbed States — 262
    C.1.2 Degenerate Unperturbed States — 265
C.2 Time-dependent Perturbation Theory — 269
    C.2.1 Time-independent Perturbations — 269
    C.2.2 Time-dependent Perturbations — 275

D TENSORS IN CUBIC CRYSTALS — 279

E	**THE CLASSICAL LIMIT**	**282**
	E.1 The Correspondence Principle	282
	E.2 The Velocity of a Particle	284
	E.3 The Force on a Particle	285
F	**SOME FOURIER TRANSFORMS**	**289**
	F.1 The Coulomb Potential	289
	F.2 The Potential of an Electric Dipole	291

EXERCISES	**293**
BIBLIOGRAPHY	**312**
INDEX	**314**

Preface

When first preparing a course on semiconductor physics at the University of Leiden in the eighties, I was surprised at the difficulty in finding a textbook covering the material I had in mind to treat. Being trained as a spectroscopist I was naturally inclined towards a quantum mechanical view, wanting to stress band structure, impurity levels, energy levels in quantum structures, etc., as well as the transitions between these bands and levels. Thus, the $\mathbf{k}\cdot\mathbf{p}$-approximation, effective mass theory and energy levels resulting from the latter theory would be central in my course. However, at that time most textbooks concentrated on transport rather than spectroscopy. As an afterthought this seems natural, transport properties of semiconductors being decisive for their use in devices.

On the other hand as a solid state physicist one looks for more. In particular, in a course one would like to bridge the gap between basic subjects such as quantum mechanics and Maxwells' equations on the one hand and the fundamental processes determining the behaviour of semiconductors on the other hand. Therefore, I kept to my original plans and extracted the necessary material from a variety of sources instead of using one of the existing textbooks.

Up to a large degree the present book is an extended version of the resulting lecture notes. Thus, I hope it will serve as a textbook for those who intend to study the quantum structure of semiconductors and who search for a basis for more advanced texts on quantum processes in these materials. In particular, instead of treating a great variety of subjects and effects, it concentrates on a reduced number of central issues like band structure, effective mass theory and electron–phonon coupling. As a result, the vast field of semiconductor heterostructures is largely left to future reading of one of the monographs that are entirely devoted to this subject. Still, I hope that the present text provides a solid basis for such future reading. To help the reader, a special section is included in chapter 1, where not only references for previous and future reading are discussed, but which is also intended to serve as a guide through the book for those who are interested in a specific subject.

As the present book is intended as an introduction to the subject rather than a monograph detailing research results, I have refrained from referencing original articles whenever an advanced textbook, monograph or review article was available.

Readers interested in the history of a given subject should consult those books or articles for references to original work. On the other hand, my efforts to maintain a constant level throughout the book, induced me to derive many results in my own unorthodox way, instead of reproducing those usually found in literature. For example, in the **k·p**-approximation I avoided the usual, but mathematically more involved variational technique and produced the results by directly applying second-order perturbation theory. In some cases this procedure resulted in some surprises as, e.g., in the case of the derivation of the effective Hamiltonian in section 5.2.3, where an extra term emerged which disturbed the beauty of a simple result. Still, most references are to 'orthodox' derivations, unless I have been able to trace a treatment similar to my own.

Chapters 1 to 5 are based on lecture notes for my course at the University of Leiden from 1985 to 1990 and at the Delft University of Technology from 1991 to 1993. Chapters 6 and 7 were written during a short sabbatical at the Centre d' Etudes Nucléaires de Saclay in the autumn of 1995. Chapters 8 and 9 were finished upon my return to Delft. The manuscript greatly profited from the reactions and comments of students who used it during my courses as well as of my colleagues in Delft and elsewhere. Among them I wish to thank S. F. J. Cox, Rutherford Appleton Laboratory, P. Harrison, University of Leeds, and C. A. M. Haarman, Department of Applied Physics, Delft University of Technology, in particular.

W. Tom Wenckebach
Delft, January 1999

Notation Conventions

Below follow some notation conventions that are adopted in the text. In the list Greek symbols follow latin symbols while the traditional sequence of the Greek alphabet is used. Other symbols follow at the end. Note that crystal structure is introduced in chapter 3, so chapter 2 uses a different convention for indentifying nuclei than the subsequent chapters.

$\mathbf{a}_\alpha$	basis vectors spanning primitive cell
$\mathbf{b}_\beta$	basis vectors spanning Brillouin zone
$f = (f_1, f_2, f_3)$	index identifying final **k**-vector after transition
$h = (h_1, h_2, h_3)$	index identifying reciprocal space lattice points, i.e. Brillouin zones (0 denotes the 1st Brillouin zone)
i, j	index identifying electrons (chapter 2)
$i = (i_1, i_2, i_3)$	index identifying initial **k**-vector before transition
j	index identifying lattice modes
$k = (k_1, k_2, k_3)$	index identifying **k**-vector
L_α	total number of primitive cells in the α direction
l	index identifying nuclei (chapter 2)
$l = (l_1, l_2, l_3)$	index identifying space lattice points, i.e. primitive cells
$m = (m_1, m_2, m_3)$	index identifying crystals when using periodic boundary conditions
N	total number of nuclei (chapter 2)
$N = L_1 L_2 L_3$	total number of primitive cells
n, m	indices numbering energy bands
$\lvert nk \rangle$	periodic part $u_k^n(\mathbf{r})$ of Bloch function as Dirac ket
$\langle nk \rvert, \langle _0 nk \rvert$	periodic part $u_k^n(\mathbf{r})$ of Bloch function as Dirac bra and integrating over the 1st primitive cell
$\langle _l nk \rvert$	periodic part $u_k^n(\mathbf{r})$ of Bloch function as Dirac bra and integrating over the lth primitive cell
$q = (q_1, q_2, q_3)$	index identifying **k**-vector when periodic boundary conditions apply

$\mathbf{R}_l = (X_l, Y_l, Z_l)$	position of lth nucleus (chapter 2)
$\mathbf{R}_l = (X_l, Y_l, Z_l)$	lth space lattice point
$\mathbf{r}_i$	position of ith electron
s	index identifying nuclei in primitive cell
V_B	volume of Brillouin zone
V_C	volume of primitive cell
(x_1, x_2, x_3)	coordinates in the frame of reference defined by the space lattice
(y_1, y_2, y_3)	coordinates in the frame of reference defined by the reciprocal space lattice
$\alpha = 1, 2, 3$	index denoting basis vectors of primitive cell
$\beta = 1, 2, 3$	index denoting basis vectors of Brillouin zone
$\mu, \nu = x, y, z$	indices denoting cartesian coordinates
$\Phi(\mathbf{k})$	(wave) function in reciprocal space
$\Psi(\mathbf{r})$	(wave) function in normal space
$\nabla_\mu = (\partial/\partial\mu)$	μ-component of nabla operator
$\int_V$	integration over volume V

1 Introduction

1.1 Metals, Insulators and Semiconductors

Semiconductors owe their dominant role in electronics to the special character of their electronic structure. As in all solids one approximates this structure by continuous energy bands separated by gaps. In the electronic ground state these bands are filled up to the so-called Fermi level and are empty above this level. As illustrated in Fig. 1.1 the position of the Fermi level with respect to the bands and the gaps determines the difference between metals, insulators and semiconductors.

In metals the Fermi level lies inside one of the energy bands, the so-called conduction band. The electrons in this partly filled band yield the high conductivity of metals. In insulators and semiconductors the Fermi energy level lies in a gap between two bands. In the ground state the band below the Fermi level, the so-called valence band, is completely filled, while the band above the Fermi level, again called the conduction band, is completely empty. Neither full nor empty bands contribute to conductivity. Hence, in the ground state, insulators and semiconductors are insulating. At finite temperature some electrons are thermally excited from the valence band into the conduction band, leading to so-called intrinsic conductivity. In this respect the distinction between insulators and semiconductors is merely a matter of the magnitude of the gap between the valence band and the conduction band. In insulators this gap is so large that intrinsic conductivity is negligible, while the gap of semiconductors is smaller, and conductivity is observed. In practice however, thermal excitation over the band gap is not the major source of conductivity of semiconductors. For example, in silicon the gap is equal to 1.17 eV corresponding to 13,600 K. At room temperature, 300 K, thermal excitation across this gap is still extremely small and intrinsic conductivity is negligible in comparison to other sources of conductivity.

A major characteristic of semiconductors is that they become conductive upon doping with certain impurities. In this respect they differ completely from insulators which remain insulators upon doping. The origin of this difference lies in the nature of their chemical binding. While most insulators are kept together by the electrostatic forces between the constituting ions or consist of molecules bound by Van der Waals forces, semiconductors are bound by homopolar binding, just as carbon atoms are bound in an organic molecule. Thus, the Si atoms in a silicon crystal can be

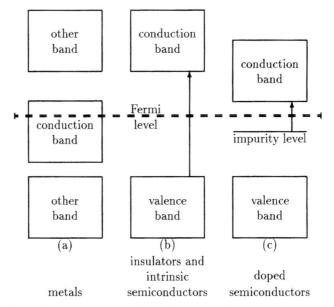

Figure 1.1 The band structure of metals, insulators and intrinsic semiconductors, and doped semiconductors. The arrows indicate excitations yielding conductivity.

considered to be split in Si^{4+} core ions and four valence electrons. In a 'balls and sticks' model like Fig. 1.2, the latter are the four 'legs' of the silicon atoms, that extend to other silicon atoms to bind them in a similar manner as carbon atoms are bound in an organic molecule. These legs can be interpreted as electron clouds consisting of two electrons each, that are positioned between the Si^{4+} core ions. One might consider a silicon crystal to be 'ionically bound' and built up of 'Si^{4+}- and '$(2e)^{2-}$-ions'. The valence band consists of all the quantum mechanical states forming these electron clouds. Furthermore, the conduction band contains excited states of the valence

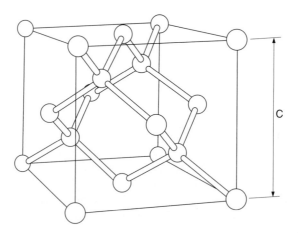

Figure 1.2 The crystal structure of silicon.

electrons, where they are ionized from these '$(2e)^{2-}$ ions' and move freely through the crystal.

To understand why an ionic crystal doped with an impurity remains an insulator, we consider the example of NaCl doped with oxygen. This crystal consists of Na^+ and Cl^- ions. Oxygen impurities form O^{2-} ions replacing Cl^- ions. As a simple replacement disturbs the neutrality of the crystal, NaCl reacts by creating a Cl^- vacancy or adding an interstitial Na^+ ion. Thus the doped crystal is just a 'new' type of ionic crystal which is again insulating.

If however, a homopolarly bound material like silicon is doped with an impurity, it behaves quite differently. We consider the case of a phosphorus atom replacing a silicon atom. The impurity atom differs from the host atoms by just one extra nuclear charge and one extra electron. When built into the crystal, it splits into a P^{5+} core ion and five electrons. The former replaces a Si^{4+} core ion, while four electrons are used for the homopolar bonds, i.e. to create '$(2e)^{2-}$ ions'. Then, however, the valence band is full and the fifth electron has no place to go except the conduction band. As a result this extra electron should contribute to the conductivity of our doped silicon crystal. Still, in the ground state it is insulating because this fifth electron is weakly bound to the phosphorus atom, which has an extra nuclear charge to attract it. On the other hand, the binding energy is only 46 meV corresponding to 535 K. So at room temperature this fifth electron is easily excited to a state in which it is completely free to move through the crystal and contribute to conductivity. The corresponding energy level scheme is given in Fig. 1.1c.

Phosphorus is called a donor impurity because it donates an electron to the conduction band. Other impurities called acceptors do the opposite by accepting electrons from the valence band. They induce conductivity via holes in this latter band. We distinguish n-type semiconductors conducting via electrons in the conduction band donated by donor impurities and p-type semiconductors conducting via holes in the valence band donated by acceptor impurities. Often semiconductors contain both types of impurities. They are said to be compensated. Energy is gained when an electron from a donor impurity falls from the conduction band to the valence band to fill a hole from an acceptor impurity. Hence such a compensated semiconductor shows the type of conductivity corresponding to the impurity with the highest concentration. As a result, all semiconductors can be clearly divided into n-type and p-type. This very important property of semiconductors allows the production of p–n junctions, where n-type and p-type material meet, and which are the basic building blocks of many semiconductor devices. Very importantly, these junctions are by nature microscopic, thus allowing extreme miniaturization of these devices.

Furthermore, the simultaneous existence of an energy gap and conductivity allows the use of semiconductors for opto-electronic devices. For example, in a semiconductor laser conductivity is used to transport electrons and holes to a well defined spot in the device where they annihilate each other and emit photons.

Elements and compounds that are held together by homopolar binding, are found in specific columns in the periodic table. As is seen in Table 1.1, homopolarly bound elements like silicon and germanium are found in the fourth column. Like carbon in organic molecules, the silicon atoms in a silicon crystal can be visualized to have four legs, each binding another silicon atom. As a result these elements crystallize with the

Table 1.1 The part of the periodic table containing the silicon-like semiconductors.

I	II	IIa	III	IV	V	VI	VII	0
H								He
Li	Be		B	C	N	O	F	Ne
Na	Mg		Al	Si	P	S	Cl	Ar
K	Ca	...Zn	Ga	Ge	As	Se	Br	Kr
Rb	Sr	...Cd	In	Sn	Sb	Te	I	Xe

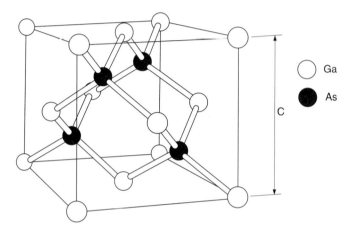

Figure 1.3 The crystal structure of GaAs.

diamond structure shown in Fig. 1.2. Binary compounds of elements of the third and fifth columns, the so-called III–V compounds like GaAs are bound in the same way as silicon. As a result, they crystallize with the very similar zinc blende structure shown in Fig. 1.3. Also some binary compounds of elements of the second and sixth columns, the so-called II–VI compounds, like ZnSe, crystallize with this latter structure. All these compounds, resembling silicon in chemical binding and crystal structure, may in principle be used in semiconductor devices.

On the contrary, other binary compounds of elements of the second and sixth column and of the first and seventh columns, are ionically bound and crystallize with a different structure. As a result these latter binary compounds are rarely considered for use in semiconductor devices.

1.2 Scope of the Book
PREVIOUS READING

The following chapters are intended to bridge the gap between basic quantum mechanics and advanced texts on quantum processes in semiconductors. The reader is supposed to have followed basic courses on formal quantum theory, so he or she understands the use of Hilbert space and Dirac's notation. To avoid the need to have a textbook on quantum mechanics at hand continuously, results for the harmonic

1.2 SCOPE OF THE BOOK

oscillator, the hydrogen atom, perturbation theory and the classical limit have been repeated in appendices A, B, C and E. In general textbooks on quantum mechanics, perturbation theory is often not treated up to the order needed here. Therefore, extensions to higher order have been added in appendix C. The same applies for the classical limit which is usually only given for the case of an isotropic mass. As we need the extension to anisotropic mass, this extension is given in appendix E. It is hard to advise on a choice for a textbook on quantum mechanics to use as a basis for the present book. Mostly the book of Cohen-Tanoudji *et al.* [11] is referred to, because it provides well worked out examples that are particularly useful as a basis for the present text. However, sometimes the more concise treatments of Dirac [15] or Landau and Lifschitz [23] were found to be more elegant.

Many of the more advanced texts on semiconductors make extensive use of group theory to characterize energy levels. While group theory is an extremely powerful tool for those who have had the opportunity to get acquainted with it, the use of group theoretical methods tends to look like black magic to the uninitiated. It proved to be quite easy to avoid the use of group theory altogether, provided the symmetry properties of a small number of tensors would be taken for granted. These symmetry properties were simply copied from Nye [29], summarized in appendix D and used whenever appropriate.

It might help when the reader has previous knowledge of solid state physics through introductory texts like Kittel [20] or Ashcroft and Mermin [1]. However, such previous knowledge is not required. Chapters 2 and 3 are intended to provide all the necessary material on general solid state physics in order for the reader to be able to follow all subsequent text, provided he or she has a good basic knowledge of quantum mechanics.

FURTHER READING

Except for the introductory chapter 1, each of the chapters could be conceived as an introduction to one or more specific advanced texts that the reader should be able to understand in detail after having finished that specific chapter. The list which follows is not meant to be exhaustive, however, it is intended to provide some idea of the books that could be used for further reading. Thus, after finishing chapter 2, the interested reader could continue studying *Electronic Structure and the Properties of Solids* by Harrison [17], which gives a detailed account of the physics of chemical bonding in solids and in particular of the type of homopolar bonding encountered in semiconductors.

From chapter 3 onwards, extensive use is made of crystal symmetry. Though group theory is avoided in the present work, it is a powerful tool for the study of all kinds of symmetry. Therefore, some readers may wish to learn about this subject after finishing the present book. In many cases the concise treatment in Yu and Cardona [43] will be sufficient to understand group theoretical arguments as they are often found in semiconductor literature. Those wishing a more elaborate treatment of group theoretical methods may read the classical text of Tinkham [39], which unfortunately treats point groups only, or the introduction to the subject by Cornwell [14], who does consider symmorphic space groups applying for composite semiconductors, but not the non-symmorphic space group of Si and Ge. For the latter the reader is referred to an early book by Cornwell [12] and in particular the work of Bir and Pikus [4] which specializes in semiconductors, but unfortunately these two books are out of print. Finally, Cornwell [13] provides a precise mathematical formulation of group theory and its application to solid state physics.

As mentioned above, chapters 2 and 3 are intended to provide the reader with the necessary introduction to solid state physics, so he or she will be able to follow the subsequent text. In case a deeper insight in general solid state physics is required the reader may continue with e.g. *Quantum Theory of Solids* of Kittel [21] or Harrison's *Solid State Theory* [16]. Cohen and Chelikowsky [10] provide a detailed account of band structure calculations for semiconductors. In their work, the interested reader will not only find an introduction to various methods for band structure calculation, but also extensive results for many semiconductor materials.

The $\mathbf{k}\cdot\mathbf{p}$-approximation treated in chapter 4 is of specific interest for semiconductors as it yields the band structure near the top of the valence band and the bottom of the conduction band. As this is where the action takes place, it can be found more extensively in a variety of texts. It is exhaustively treated in the book of Bir and Pikus [4], while less elaborate, but very useful treatments are found in an article by Kane [19] in the first edition of the *Handbook on Semiconductors* (but unfortunately not in the second edition) and in the book by Bastard [2].

Effective mass theory is an extremely powerful tool, because it provides a method to replace the complicated motion of a band electron in the Hartree–Fock potential of the crystal by that of a free electron with an adjusted mass, the effective mass. Chapter 5 treats effective mass theory and provides an entry to a variety of texts on this theory. The derivation of the effective Hamiltonian is a simplified version of the $\mathbf{P}\cdot\pi$ method, treated extensively by e.g. Zeiger and Pratt [44]. Once derived, numerous quantum effects can be treated. While amongst others, Ridley [32] treats Landau levels and impurity states in bulk semiconductors, Weisbuch and Vinter [40], Bastard [2] and Capasso [9] provide further reading on quantum effects in low dimensional systems.

The treatment of the phonon spectrum in chapter 6 derives directly from the classical book of Born and Huang [8]. Consequently, for further reading the reader is referred to that text. More important for the understanding of quantum processes in semiconductors is the subsequent treatment of electron-phonon coupling and its application to transport in chapters 7 and 8. Here, only the basic interactions and transition probabilities are given. However, once finished studying the present text, the reader should find no difficulty reading the books of Ridley [32] and Singh [36], who treat transport to the level where electron and hole mobilities can be fully understood. General textbooks, written from a more experimental point of view and with a large emphasis on transport are those of Seeger [35] and Böer [7]. A treatment of electron-phonon coupling in low-dimensional structures is found in ref. [33].

The texts of Ridley [32] and Singh [36] and in particular the book by Yu and Cardona [43] also provide further reading on optical transitions as introduced in chapter 9. Moreover, more of an experimentalist's view is offered by Klingshirn [22]. Again, let it be stressed that the texts mentioned here are just examples from the many available at present and that the omission of any book or article does not imply that it may not serve as an excellent text for further reading.

READERS GUIDE

Many readers will be interested in a specific subject only and may find it tedious to go through the whole book to find the answer for their specific problem. Also for teaching one might want to make a selection of the topics treated in the book. To help such readers let it first be noted that the book divides naturally into two parts. Chapters 2–5 treat the energy levels of electrons in semiconductors, while chapters 6–9 consider the quantum mechanical background of processes in these materials. Roughly, the former chapters concern solutions of the time-independent Schrödinger equation, while the latter chapters treat its time-dependent counterpart. By nature, quantum processes involve transitions between energy levels, so the first half of the book is indispensable for the second half. However, the opposite does not hold and chapters 2–5 may readily be used as the basis of an introductory course on semiconductors. Moreover, quantum processes not only involve the energy levels of electrons, but also phonons and/or optical fields and their mutual couplings. Hence, the treatment of such processes is inevitably more complicated and in a course one might prefer to make a selection of the material presented in chapters 6–9 rather than treat it all. To help making such selections the following short cuts are suggested.

The basic purpose of chapter 2 is to lower the entry level of the book. In particular section 2.1 may be skipped without much loss. Still, this section serves a particular purpose. There is a fundamental difference between the s- and p-type wave functions causing chemical binding and those serving as basis functions for the $\mathbf{k} \cdot \mathbf{p}$-approximation treated in chapter 4. Anyone wishing to avoid mixing up both types of wave functions should also read section 2.1. Furthermore, readers interested in band structure only may restrict themselves to section 2.2. On the other hand, those also interested in electron-phonon coupling need to study section 2.3 as well.

Reciprocal space as discussed in section 3.1 is basic for all subjects in the book. Though this subject is introduced in every textbook on solid state physics, one should be aware of differences in notation and normalization. The treatment of band structure in the rest of chapter 3 and of the $\mathbf{k} \cdot \mathbf{p}$-approximation in chapter 4 should also be considered as fundamental constituents of the text. An exception must be made for those few readers who are interested in the phonon spectrum only. They may skip these two subjects and restrict themselves to reading chapter 6 after having finished section 3.1.

Those interested in band structure only, will hopefully have found what they need once they have finished chapter 4. For further reading they are referred to specialized monographs. All others, studying transport, optical effects, etc., need to continue with section 5.1 introducing externally applied fields and section 5.2 providing a formal derivation of effective mass theory. At the beginning of section 5.3, the reader may choose to belong to three distinct groups. When mainly interested in semiconductor heterostructures or other quantization effects in effective mass theory, he should proceed with section 5.3. Subsequently, he is referred to specialized texts for further reading.

In the case where the reader intends to study transport, he may skip section 5.3 with the exception of the treatment of hydrogenic impurities in section 5.3.2. For a full understanding he should subsequently continue with chapters 6, 7 and 8, after which the reader is referred to specialized texts. However, if the results of chapters 6 and 7 are accepted, chapter 8 will prove to be quite self-contained. Even if the reader wishes to understand the details of the phonon spectrum and electron-phonon coupling, it may still prove useful to have a quick glance at chapter 8 first, as it provides the motivation for the more tedious chapters 6 and 7.

If readers are interested in optics only, most of section 5.3 may be skipped. However, the discussion of quantum wells in section 5.3.3 is needed to understand the treatment of optical transitions between subbands in section 9.2.3. Students of optics may then jump directly to chapter 9. They should then be capable of understanding the basic formalism treated in section 9.1 as well as direct optical absorption and intersubband transitions as discussed in sections 9.2.1 and 9.2.3. Unfortunately, phonons play an important role in indirect optical absorption treated in section 9.2.2. Again, if the results of chapters 6 and 7 are accepted, chapter 9 will prove to be quite self-contained. For a complete understanding of these subjects one cannot avoid studying chapters 6 and 7 however.

2 Electrons, Nuclei and Hamiltonians

2.1 The Chemical Model

In principle, the Hamiltonian of a solid is very complicated, containing the kinetic energy and mutual interactions of all constituting nuclei and electrons. To obtain insight in its structure it is instructive to start with a 'chemical' approach. One first splits the solid into separate atoms and determines their individual eigenstates. Then these atoms are split into cores—e.g. Si^{4+}-ions—and outer- or valence electrons yielding chemical binding. When combining the atoms into a crystal the cores are left unchanged. The wave functions of the valence electrons are constructed in the following simple way. From each one of two neighbouring atoms one takes an atomic wave function and subsequently one combines these two atomic wave functions into a linear combination. The result is called a bonding orbital. This chemical approach is especially useful to understand chemical binding, crystal structure and electron density in homopolarly bound materials such as silicon. However, because the wave functions are localized it fails to describe the origin of electrical conductivity. Therefore, in section 2.2 we take a different approach where we consider the wave functions of the valence electrons to extend over the whole crystal. Wherever possible we keep our discussion simple by restricting ourselves to silicon. At the end of section 2.1.2 the extension will be made to other semiconductors.

2.1.1 The Silicon Atom
THE ATOMIC HAMILTONIAN

The treatment of atomic structure can be found in many textbooks, e.g. ref. [41]. It is based on the wave functions of the hydrogen atom as summarized in appendix A. To simplify our discussion we assume the positions of the atomic nuclei to be fixed. This approximation is justified because the nuclear mass is large compared to the electron mass, so the centre of mass of any electron–nucleus pair is always very close to the nucleus. Moreover, the position of this centre of mass is well defined by the crystal structure. Thus, we replace the nucleus by a central force field at the origin of the coordinate system. As a second simplification we neglect interactions involving the electron spin. Then, the Hamiltonian of an atom consisting of a nucleus with a charge Ze and Z electrons is given by

$$\mathcal{H} = -\frac{\hbar^2}{2m_e}\sum_i^Z \Delta_i + \mathcal{V} \qquad (2.1)$$

Here, $\mathbf{r}_i = (x_i, y_i, z_i)$ is the position of the ith electron and Δ_i the Laplace operator:

$$\Delta_i = \nabla_i \cdot \nabla_i \quad \text{where} \quad \nabla_i = \left(\frac{\partial}{\partial x_i}, \frac{\partial}{\partial y_i}, \frac{\partial}{\partial z_i}\right)$$

$\mathcal{V}$ contains all electrostatic potentials:

$$\mathcal{V} = \sum_i^Z \mathcal{V}_i + \frac{1}{2}\sum_i^Z \sum_{j\neq i}^Z \mathcal{V}_{ij} \qquad (2.2)$$

where

$$\mathcal{V}_i = -\frac{Ze^2}{4\pi\varepsilon_0 r_i} \qquad (2.3)$$

is the electrostatic attraction between the nucleus and the ith electron and

$$\mathcal{V}_{ij} = \frac{e^2}{4\pi\varepsilon_0 r_{ij}} \qquad (2.4)$$

is the electrostatic repulsion between the ith and jth electrons. We note that each term $\mathcal{V}_{ij}$ occurs twice in the sums over i and j in eq. (2.2), instead of once. This 'double counting' is compensated by inserting a factor $\frac{1}{2}$ in that equation. In these potentials

$$r_i = \sqrt{x_i^2 + y_i^2 + z_i^2}$$
$$r_{ij} = \sqrt{(x_i - x_j)^2 + (y_i - y_j)^2 + (z_i - z_j)^2}$$

Finally m_e is the mass of the electron and $-e$ its charge while $h = 2\pi\hbar$ is Planck's constant. We use rationalized Giorgi (SI) units. Then

$$4\pi\varepsilon_0 = c^{-2} \cdot 10^7$$
$$c = 2.998 \cdot 10^8 \text{ m s}^{-1} = \text{the velocity of light}$$
$$e = 1.602 \cdot 10^{-19} \text{ C}$$
$$m_e = 9.110 \cdot 10^{-31} \text{ kg}$$
$$\hbar = 1.055 \cdot 10^{-34} \text{ J s}$$

THE CENTRAL FORCE MODEL

Owing to the mutual electrostatic repulsion between the electrons, the calculation of the eigenstates and energy levels of an atom is a many particle problem that cannot be

2.1 THE CHEMICAL MODEL

solved exactly. Hence, one has to resort to approximative techniques. A remarkably good insight into the electronic structure of atoms is obtained using the central force model [41]. In this model the mutual electrostatic repulsion between the electrons is simply assumed to screen the attraction of the nucleus, i.e. to reduce this attraction by a factor α, which is the same for all electrons. As a result the Hamiltonian reduces to a sum of independent single electron Hamiltonians $\mathcal{H}'_i$:

$$\mathcal{H} \approx \mathcal{H}' = \sum_i^Z \mathcal{H}'_i \qquad (2.5)$$

where

$$\mathcal{H}'_i = -\frac{\hbar^2}{2m_e}\Delta_i + \alpha \mathcal{V}_i \qquad (2.6)$$

and $\mathcal{V}_i$ is given by eq. (2.3).

Thus the Hamiltonian splits into independent parts, each equal to the Hamiltonian of the hydrogen atom given in eqs (A.1) and (A.2), except for the factor e^2 in the potential being replaced by αZe^2. Hence, the energy levels and the wave functions are the same as given in eqs (A.3)–(A.7) and in Tables A.1–A.3, except that the factor e^2 is replaced by αZe^2. Thus, the energy levels of the individual electrons are given by

$$E'_n = -\frac{Ry'}{n^2} \qquad (2.7)$$

where $-Ry'$ is the energy E'_1 of the lowest level and equal to the potential $\mathcal{V}_i$ at a distance

$$r_i = 2a' = 2\frac{4\pi\varepsilon_0 \hbar^2}{m_e \alpha Ze^2} = 2\frac{a}{\alpha Z} \quad \text{where} \quad a = \frac{4\pi\varepsilon_0 \hbar^2}{m_e e^2} = 0.5292 \cdot 10^{-10} \text{ m} \qquad (2.8)$$

is the Bohr radius. So

$$Ry' = \frac{\alpha Ze^2}{8\pi\varepsilon_0 a'} = \alpha^2 Z^2 Ry \quad \text{where} \quad \frac{Ry}{e} = \frac{e}{8\pi\varepsilon_0 a} = 13.606 \text{ eV} \qquad (2.9)$$

defines the Rydberg Ry.

Also the eigenfunctions of the electrons are same as given in Tables A.1–A.3 except for replacing the Bohr radii a by a' as defined by eq. (2.8). Thus, the hydrogen wave functions are not only the eigenfunctions of an electron in a hydrogen atom, but in the central force model they also represent the wave functions of the electrons in an atom. Therefore they are called atomic orbitals. For our present purposes it is sufficient to recall eq. (A.7) showing that the atomic orbitals can be written as

$$\Psi_{nl\mu} = \rho_{nl}(r_i) F_{l\mu}(\theta_i, \phi_i) \qquad (2.10)$$

where we use spherical coordinates

$$x_i = r_i \sin\theta_i \cos\phi_i,$$
$$y_i = r_i \sin\theta_i \sin\phi_i$$
$$z_i = r_i \cos\theta_i$$

Table 2.1 The angular dependent part $F_{l\mu}(\theta,\phi)$ of the nine atomic orbitals with lowest energy.

$nl\mu$		$\Psi_{nl\mu}$	$F_{l\mu}(\theta,\phi)$
$n00$	ns	Ψ_{ns}	$\dfrac{1}{\sqrt{4\pi}}$
$n1z$	np_z	Ψ_{np_z}	$\sqrt{\dfrac{3}{4\pi}}\, z$
$n1x$	np_x	Ψ_{np_x}	$\sqrt{\dfrac{3}{4\pi}}\, x$
$n1y$	np_y	Ψ_{np_y}	$\sqrt{\dfrac{3}{4\pi}}\, y$

The radial parts $\rho_{nl}(r)$ of the atomic orbitals are given in Table A.1 and the angular dependent part $F_{l\mu}(\theta,\phi)$ in Table A.3. Here we only need the latter. Therefore it is repeated for convenience in Table 2.1. As in appendix A we present both current notations $l = 0, 1, 2, \ldots$ and $l = s, p, d, \ldots$ in the first and second columns, respectively. We note that in this simple central force model all single electron states with the same principal quantum number n are degenerate. Hence, for such a given value of n not only the states $\Psi_{nl\mu}$ are eigenstates of the single electron Hamiltonian $\mathcal{H}'_i$ but also linear combinations

$$\Psi_{n\nu} = \sum_{l,\mu} c_{\nu l\mu} \Psi_{nl\mu} \tag{2.11}$$

THE PAULI EXCLUSION PRINCIPLE

After having determined the eigenstates of the individual electrons in the atom, we obtain the ground state Ψ of that atom as a whole by using the Pauli exclusion principle. At this point it is important to realize, that all single electron Hamiltonians $\mathcal{H}'_i$ have the same set of orthogonal eigenstates $\Psi_{n\nu}$. Moreover, electrons have spin $s = \frac{1}{2}$, so each of these eigenstates is two-fold degenerate. Hence each electron may occupy one of the orthogonal states $\alpha_s \Psi_{n\nu}$ where $\alpha_s = \alpha_+$ or α_- denotes the spin state. Now, according to the Pauli principle, different electrons may occupy different, orthogonal states only. Then, to obtain the ground state of the atom, the single electron states $\alpha_s \Psi_{n\nu}$ are filled one by one, beginning at the lowest energy. Subsequently one calculates the antisymmetrized product of the filled states,

$$\Psi = \sum_p (-1)^p P_p \prod_{s,n,\nu} \alpha_s \Psi_{n\nu} \tag{2.12}$$

Here we sum over all possible permutations, represented by the permutation operator P_p that performs p elementary transpositions, where just two electrons are interchanged.

2.1 THE CHEMICAL MODEL

We now consider a silicon atom. Silicon has a nuclear charge $Z = 14$ and hence it has 14 electrons to accommodate. Using the Pauli exclusion principle and the twofold degeneracy due to the electron spin, we obtain the atomic ground state by filling the seven lowest atomic orbitals with two electrons each. First we fill the $1s$, $2s$ and $2p_x$, $2p_y$ and $2p_z$ atomic orbitals with 10 electrons. The thus formed Si^{4+} ion is precisely the silicon core. The next step is more difficult, because we cannot decide which orbitals will be occupied by the four remaining valence electrons. The reason is that the next four orbitals, the $3s$ and the $3p_x$, $3p_y$ and $3p_z$ orbitals are degenerate and provide eight states which the valence electrons may occupy. Worse, the so-called $3d$ orbitals—not given in Table 2.1 and corresponding to quantum numbers $n = 3$ and $l = 2$—are also degenerate with the $3s$ and $3p$ orbitals and provide 10 more degenerate states for the valence electrons to occupy.

HYBRID ORBITALS

In order to decide how the four valence electrons fill these states, we invoke the mutual repulsion $\frac{1}{2}\sum_{i \neq j} V_{ij}$ between the valence electrons. For a specific ground state Ψ of the atom the expectation value of this mutual repulsion given by

$$\Delta E = \cdots \int d\mathbf{r}_i \cdots \int d\mathbf{r}_j \cdots \Psi^* \left[\frac{1}{2} \sum_{j \neq i} V_{ij} \right] \Psi \quad (2.13)$$

In the ground state of the atom ΔE should be minimal. We now note that according to eqs (2.11) and (2.12) the ground state is composed of linear combinations $\Psi_{n\nu}$ of atomic orbitals. Hence, we should choose the constants $c_{\nu l \mu}$ in these linear combinations in such a way that ΔE is minimal. We will not pursue the calculation, but give the result only. The four valence electrons are found to occupy four so-called hybrid sp_3 orbitals:

$$\Psi_{sp_3(1,1,1)} = \frac{1}{2}[\Psi_{3s} + \Psi_{3p_x} + \Psi_{3p_y} + \Psi_{3p_z}]$$

$$= \frac{1}{4\sqrt{\pi}}[\rho_{30}(r) + \sqrt{3} \cdot \rho_{31}(r) \cdot \{+x+y+z\}]$$

$$\Psi_{sp_3(-1,-1,1)} = \frac{1}{2}[\Psi_{3s} - \Psi_{3p_x} - \Psi_{3p_y} + \Psi_{3p_z}]$$

$$= \frac{1}{4\sqrt{\pi}}[\rho_{30}(r) + \sqrt{3} \cdot \rho_{31}(r) \cdot \{-x-y+z\}]$$

$$\Psi_{sp_3(-1,1,-1)} = \frac{1}{2}[\Psi_{3s} - \Psi_{3p_x} + \Psi_{3p_y} - \Psi_{3p_z}]$$

$$= \frac{1}{4\sqrt{\pi}}[\rho_{30}(r) + \sqrt{3} \cdot \rho_{31}(r) \cdot \{-x+y-z\}]$$

$$\Psi_{sp_3(1,-1,-1)} = \frac{1}{2}[\Psi_{3s} + \Psi_{3p_x} - \Psi_{3p_y} - \Psi_{3p_z}]$$

$$= \frac{1}{4\sqrt{\pi}}[\rho_{30}(r) + \sqrt{3} \cdot \rho_{31}(r) \cdot \{+x-y-z\}] \quad (2.14)$$

where we use eq. (2.10) and Table 2.1 to write these hybrid orbitals explicitly. Moreover, we introduce the notation (a, b, c) to indicate the direction—but *not* the amplitude of the vector $\mathbf{r} = a\mathbf{x} + b\mathbf{y} + c\mathbf{z}$.

In order to understand why these orbitals minimize the mutual repulsion between the four valence electrons, we use an artist's impression of the hybrid sp_3 orbitals given in Fig. 2.1. We see that each of the four sp_3 orbitals, and hence the electron density, protrudes in a particular direction:

$$\Psi_{sp_3(1,1,1)} \quad \text{in the} \quad (1, 1, 1)\text{-direction}$$
$$\Psi_{sp_3(-1,-1,1)} \quad \text{in the} \quad (-1, -1, 1)\text{-direction}$$
$$\Psi_{sp_3(-1,1,-1)} \quad \text{in the} \quad (-1, 1, -1)\text{-direction}$$
$$\Psi_{sp_3(1,-1,-1)} \quad \text{in the} \quad (1, -1, -1)\text{-direction}$$

As seen in Fig. 2.1 these four directions span a regular tetrahedron. It is easily verified that the electrostatic repulsion between four electrons is minimized when each of them is located at one of the vertices of a regular tetrahedron. Thus, when each valence electron occupies a different sp_3 orbital, their mutual repulsion is also minimized.

2.1.2 Homopolar Binding in Silicon

LINEAR COMBINATION OF ATOMIC ORBITALS

The electrons with lowest energy, in particular the core electrons, occupy the orbitals with the smallest spatial extension. Therefore these electrons are not sensitive to the surroundings of the atom in the crystal. However, the electrons with higher energy, i.e. the valence electrons, occupy more extended orbitals and do indeed feel these surroundings. Thus, the latter may participate in the chemical binding between the atoms.

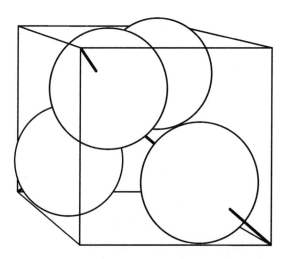

Figure 2.1 Artist's impression of the four sp_3 orbitals.

2.1 THE CHEMICAL MODEL

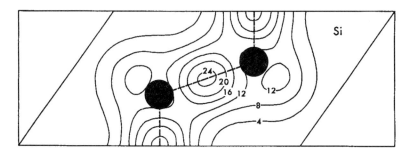

Figure 2.2 The electron density in silicon (adapted from ref. [10], p. 89, Fig. 8.8b, © Springer Verlag).

In semiconductors this chemical binding is of a homopolar nature. Then the valence electrons occupy wave functions yielding a high electronic charge density between the cores of neighbouring atoms. This is clearly illustrated in Fig. 2.2 showing a theoretically calculated electron density in silicon [10]. Here, the black discs represent two neighbouring cores while the valence electron density is depicted by lines of constant density. The latter shows a maximum at a point positioned exactly midway between these two cores. The cores have a charge $+4e$ and the valence electron clouds $-2e$. The crystal is held together like an ionic crystal: the total attraction between cores and valence electron clouds is larger than the mutual repulsion between cores.

Homopolar binding constitutes a many-particle problem that cannot be solved exactly. Hence, we have to resort to approximative methods. Here we consider the method of Linear Combination of Atomic Orbitals (LCAO). Then one first determines the atomic orbitals of the *individual* atoms and subsequently one constructs linear combinations of atomic orbitals belonging to *different* atoms. In this method the contribution of the various atomic orbitals to the linear combinations and the distance between the nuclei are treated as parameters. The ground state is obtained by specifying that the energy is a minimum as a function of the magnitude of these parameters.

We consider two neighbouring silicon atoms with their nuclei at positions $\mathbf{R}_1$ and $\mathbf{R}_2$. We first construct sp_3 orbitals for both atoms. Following the procedure indicated above, the next step in the LCAO method is to construct a linear combination of two such sp_3 orbitals yielding a maximum electron density exactly between the two atoms. This is achieved as follows. We choose

(a) the direction of the vector $\mathbf{R}_2 - \mathbf{R}_1$ along the $(1, 1, 1)$ direction
(b) the $\Psi_{sp_3(1,1,1)}(\mathbf{r} - \mathbf{R}_1)$ orbital for the atom at position $\mathbf{R}_1$
(c) the $\Psi_{sp_3(-1,-1,-1)}(\mathbf{r} - \mathbf{R}_2)$ orbital for the other atom

Here $\mathbf{r}$ is the position of the electron. Then, as shown in Fig. 2.3a, these orbitals have their maximum amplitude located exactly between both atoms. Next, we construct a linear combination

$$\Psi = c_1 \Psi_{sp_3(1,1,1)}(\mathbf{r} - \mathbf{R}_1) + c_2 \Psi_{sp_3(-1,-1,-1)}(\mathbf{r} - \mathbf{R}_2) \tag{2.15}$$

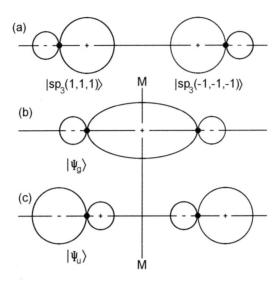

Figure 2.3 (a) The two sp_3 orbitals of the silicon atoms that participate in binding. (b) The bonding state Ψ_g. (c) The anti-bonding state Ψ_u.

The coefficients c_1 and c_2 are subsequently obtained by minimizing the energy. Without pursuing the calculation we give the result that this minimum is found for

$$c_1 = c_2 = \frac{1}{\sqrt{2(1+S)}} \tag{2.16}$$

where

$$S = \int d\mathbf{r} \Psi^*_{sp_3(1,1,1)}(\mathbf{r}-\mathbf{R}_1)\Psi_{sp_3(-1,-1,-1)}(\mathbf{r}-\mathbf{R}_2) \neq 0 \tag{2.17}$$

because the states $\Psi_{sp_3(1,1,1)}(\mathbf{r}-\mathbf{R}_1)$ and $\Psi_{sp_3(-1,-1,-1)}(\mathbf{r}-\mathbf{R}_2)$ are not orthonormal. As a result, the state

$$\Psi_g = \frac{1}{\sqrt{2(1+S)}}[\Psi_{sp_3(1,1,1)}(\mathbf{r}-\mathbf{R}_1) + \Psi_{sp_3(-1,-1,-1)}(\mathbf{r}-\mathbf{R}_2)] \tag{2.18}$$

corresponds to the lowest electron energy.

Finally, we determine the distance between the two nuclei in this state. For this purpose we calculate the energy in the state Ψ_g as a function of $\mathbf{R}_2 - \mathbf{R}_1$. The result has the general shape shown in Fig. 2.4. As the ground state corresponds to the minimum of E_g, the distance $\mathbf{R}_2 - \mathbf{R}_1$ between the two nuclei corresponds to the value where that minimum occurs.

BONDING AND ANTIBONDING STATES

Though they are not orthogonal, the two states $\Psi_{sp_3(1,1,1)}(\mathbf{r}-\mathbf{R}_1)$ and $\Psi_{sp_3(-1,-1,-1)}(\mathbf{r}-\mathbf{R}_2)$ still span a two-dimensional sub-space of Hilbert space.

2.1 THE CHEMICAL MODEL

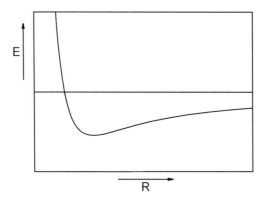

Figure 2.4 Sketch of the energy of two silicon cores bound by a valence electron as a function of the distance between the nuclei.

Hence, a second linear combination of these two states exists which is orthogonal to the ground state Ψ_g. It is found to be

$$\Psi_u = \frac{1}{\sqrt{2(1-S)}} [\Psi_{sp_3(1,1,1)}(\mathbf{r}-\mathbf{R}_1) - \Psi_{sp_3(-1,-1,-1)}(\mathbf{r}-\mathbf{R}_2)] \qquad (2.19)$$

Artists impressions of Ψ_g and Ψ_u are given in Fig. 2.3b and 2.3c. The two functions are cylindrically symmetric about the axis connecting the two nuclei. The indices g and u in the ground and first excited states denote the symmetry of the wave functions. They originate in the german words 'gerade' and 'ungerade', or even and uneven. For mirroring in the plane M, Ψ_g and Ψ_u are even and uneven functions, respectively, i.e. Ψ_g remains the same, while Ψ_u changes sign.

Fig. 2.3b clearly shows that the state Ψ_g is successful in retaining the electron between the two nuclei. The electrostatic attraction between the electron and the two nuclei is larger than the mutual repulsion between the cores, so the two silicon atoms are bound together. On the other hand, in the state Ψ_u the electron resides outside both nuclei. Now, there are no bonding forces and the atoms separate. For this reason Ψ_g is also called the bonding state and Ψ_u the anti-bonding state.

We now consider the number of electrons occupying such bonding states in a silicon crystal. As each silicon atom forms four sp_3 orbitals, it also forms four bonding and four anti-bonding states with four neighbouring silicon atoms. Because electrons have a spin $s = \frac{1}{2}$, each bonding state Ψ_g may be occupied by two electrons. Now each silicon atom provides four valence electrons for this purpose. Using these four electrons and four more from its neighbours, all bonding states are filled. Moreover, all electrons have been used, so the anti-bonding states are all empty.

A very important point to realize is that the bonding and anti-bonding states are intimately related to the valence band and the conduction band of the semiconductor. It is exactly the bonding states that together constitute the valence band, while all anti-bonding states together yield the conduction band.

The homopolar binding between silicon atoms via sp_3 orbitals leads immediately to the diamond structure of crystalline silicon. Because each silicon atom forms four sp_3 orbitals, it binds four other silicon atoms. As the maximum amplitudes of the bonding states are located at the vertices of a tetrahedron, the nearest neighbours of a silicon atom are also located at the vertices of a tetrahedron. When constructing a crystal, while requiring that the nearest neighbours of an atom form a tetrahedron, one naturally arrives at two possible crystal structures: the cubic diamond structure shown in Fig. 1.2 and the hexagonal wurtzite structure. The great majority of semiconductor devices are based on cubic silicon.

2.1.3 Other Semiconductors and Impurities

The other elements in the fourth column of the periodic system shown in Table 1.1, carbon, germanium and tin, may bind in the same way as silicon, so bonding states are also constructed from sp_3 orbitals. When carbon crystallizes with the diamond structure sp_3 orbitals are constructed from $2s$, $2p_x$, $2p_y$ and $2p_z$ orbitals. Germanium and tin crystallizing with the grey tin (α-Sn) structure use $4s$, $4p_x$, $4p_y$, $4p_z$ orbitals and $5s$, $5p_x$, $5p_y$, $5p_z$ orbitals, respectively, again leading to the diamond structure.

Several binary compounds have a very similar structure. As an example, we consider the compound AlP, consisting of aluminium and phosphorus atoms. The former element can be found in the third column of the periodic system and the latter in the fifth column. Therefore, it is called a III–V compound. As discussed above aluminium differs from silicon by having just one nuclear charge and one electron less, while phosphorus has just one nuclear charge and one electron more. The nucleus and 10 electrons form cores with a positive charge of $+3e$ and $+5e$, respectively. The remaining three valence electrons of the aluminium atom and the remaining five of the phosphorus atom fill bonding states that are constructed in precisely the same way as in the case of silicon. As there are exactly four such bonding states per phosphorus–aluminium pair and each of these states may contain two electrons, these states will just be filled. As a result of the great analogy between the electronic structures of Si and AlP, the crystal structure of these two substances is very similar. While Si crystallizes with the diamond structure shown in Fig. 1.2, AlP crystallizes with the zinc blende structure shown in Fig. 1.3. This latter structure can be easily derived from the diamond structure by replacing one half of the silicon atoms by aluminium atoms and the other half by phosphorus atoms such that the former are surrounded by the latter and vice versa. Other III–V compounds, crystallizing with the zinc blende structure, are GaAs, GaP and InP.

A number of compounds of elements of the second and sixth columns of the periodic table, the II–VI compounds, also crystallize with the zinc blende structure. Their electronic structure can be understood using similar arguments as given above for the case of AlP. Among others, ZnS belongs to this category and gave its name, zinc blende, to the crystal structure. Table 2.2 summarizes a number of elements and compounds crystallizing with the diamond or zinc blende structure, together with the length c of the edges of the cubic unit cell shown in Figs 1.2 and 1.3.

Using the description of the previous section one also understands why an aluminium atom in a silicon crystal acts as an acceptor or a phosphorus atom as a donor. Aluminium differs from silicon by having just one nuclear charge and one electron less, while phosphorus has just one nuclear charge and one electron more. Both have cores consisting of a nucleus and 10 electrons filling the $1s$, $2s$, $2p_x$, $2p_y$ and $2p_z$

2.2 THE SINGLE ELECTRON HAMILTONIAN

Table 2.2 Some semiconductors crystallizing in the diamond or zinc blende structure together with the size of their unit cell.

IV	c (nm)	III–V	c (nm)	II–VI	c (nm)
C	0.3567	AlP	0.5467	ZnS	0.5406
Si	0.5431	GaP	0.5447	CdS	0.5835
Ge	0.5657	InP	0.5869	ZnSe	0.5669
Sn	0.6489	AlAs	0.5639	CdSe	0.605
		GaAs	0.5654	ZnTe	0.6103
		InAs	0.6058	CdTe	0.6478

orbitals. Thus, the aluminium cores have a positive charge of $+3e$ and the phosphorus cores a charge $+5e$.

The remaining three electrons of the aluminium atom and the remaining five of the phosphorus atom are the valence electrons. They fill bonding states that are constructed in precisely the same way as in the case of silicon, i.e. from sp_3 orbitals that originate in $3s$, $3p_x$, $3p_y$ and $3p_z$ orbitals. As they need to fill four such bonding states per atom, the aluminium atom leaves one bonding state vacant, while the phosphorus atom has one electron to spare which must fill an anti-bonding state. Now the bonding states form the valence band and the anti-bonding states the conduction band. Thus the aluminium atom provides a hole in the valence band, while the phosphorus atom yields an electron in the conduction band.

2.2 The Single Electron Hamiltonian

The chemical treatment given above has the major flaw, that all electrons are described by localized wave functions. This may be reasonable for core electrons that are indeed localized near the atomic nuclei. However, this is not correct for valence electrons in bonding orbitals extending over larger distances in the crystal. This is illustrated in Fig. 2.2 showing the electron density in silicon [10]. All core electrons reside within the black discs, while outside these discs the electron density is completely due to valence electrons. As discussed above, the valence electron density shows a maximum between two neighbouring silicon cores, corresponding to the maximum amplitude of their bonding state. However, Fig. 2.2 also shows the large extension and overlap of different bonding states. This overlap leads to mixing of such states, and, as a result, the correct eigenfunctions of the valence electrons are linear combinations of many bonding states and extend over many unit cells. In order to obtain these correct eigenfunctions we need to consider the Hamiltonian of a complete crystal.

2.2.1 The Electron Hamiltonian of a Crystal

We construct the Hamiltonian for all electrons in a crystal making similar approximations to those made in the previous section treating the chemical approach. Thus we consider the nuclei to occupy fixed positions. Then their kinetic energy is zero and their mutual electrostatic repulsion is constant and we skip the corresponding terms in the Hamiltonian. Neglecting spin related terms as well, the Hamiltonian is given by

$$\mathcal{H} = -\frac{\hbar^2}{2m_e}\sum_i \Delta_i + \mathcal{V} \tag{2.20}$$

where

$$\Delta_i = \nabla_i \cdot \nabla_i \quad \text{where} \quad \nabla_i = \left(\frac{\partial}{\partial x_i}, \frac{\partial}{\partial y_i}, \frac{\partial}{\partial z_i}\right)$$

is the Laplace operator for the ith electron, which has coordinates $\mathbf{r}_i = (x_i, y_i, z_i)$. $\mathcal{V}$ contains all electrostatic potentials experienced by the electrons:

$$\mathcal{V} = \sum_i \mathcal{V}_i + \frac{1}{2}\sum_i \sum_{j \neq i} \mathcal{V}_{ij}. \tag{2.21}$$

Here, $\mathcal{V}_i$ is the electrostatic attraction exerted by all nuclei on the ith electron:

$$\mathcal{V}_i = \sum_l \mathcal{V}_{il} = -\sum_l \frac{Z_l e^2}{4\pi\varepsilon_0 r_{il}} \tag{2.22}$$

where $\mathbf{R}_l = (X_l, Y_l, Z_l)$ is the position of the lth nucleus and Z_l its charge, so

$$r_{il} = \sqrt{(x_i - X_l)^2 + (y_i - Y_l)^2 + (z_i - Z_l)^2}$$

is the distance between the ith electron and this nucleus. Furthermore, $\mathcal{V}_{ij}$ is the electrostatic repulsion between the ith and jth electrons:

$$\mathcal{V}_{ij} = \frac{e^2}{4\pi\varepsilon_0 r_{ij}} \tag{2.23}$$

where

$$r_{ij} = \sqrt{(x_i - x_j)^2 + (y_i - y_j)^2 + (z_i - z_j)^2}$$

is the distance between these electrons. We note that each term $\mathcal{V}_{ij}$ occurs twice in the sums over i and j in eq. (2.21), instead of once. In order to compensate for this 'double counting' a factor $\frac{1}{2}$ is inserted into the equation.

2.2.2 Self-consistent Methods

The Hamiltonian (2.20) represents a many particle problem, which is even more complicated than that of a single atom as given by eq. (2.1). Hence, we need approximative methods to handle it. As in the case of the single atom, we try to approximate the Hamiltonian by a sum of independent single electron Hamiltonians:

$$\mathcal{H} \approx \sum_i \mathcal{H}'_i(\mathbf{r}_i) \tag{2.24}$$

where

$$\mathcal{H}'_i(\mathbf{r}_i) = -\frac{\hbar^2}{2m_e}\Delta_i + \mathcal{V}'_i(\mathbf{r}_i) \tag{2.25}$$

2.2 THE SINGLE ELECTRON HAMILTONIAN

In these single electron Hamiltonians, the potentials (2.22) and (2.23) are replaced by effective potentials $V'_i(\mathbf{r}_i)$, that only depend on the position $\mathbf{r}_i$ of the ith electron and *not* the positions of the other electrons.

It is highly advantageous to restrict the approximation further by choosing the same effective potential for all electrons, i.e. $V'_i(\mathbf{r}_i)$ has the same functional dependence on $\mathbf{r}_i$ for all electrons. Then the index i can be omitted in $\mathcal{H}'_i$ and V'_i but *not* of course in $\mathbf{r}_i$ and

$$\mathcal{H}'_i(\mathbf{r}_i) = \mathcal{H}'(\mathbf{r}_i) = -\frac{\hbar^2}{2m_e}\Delta_i + V'(\mathbf{r}_i) \qquad (2.26)$$

This further restriction has the enormous advantage that one obtains the same set of orthonormal eigenfunctions for all electrons. This allows us to invoke the Pauli exclusion principle to construct the electronic ground state of the crystal. Let us denote the eigenstates of $\mathcal{H}'(\mathbf{r}_i)$ by Ψ_n, where n denotes the quantum numbers that are involved. Then the eigenfunctions of an individual electron are given by $\alpha_s \Psi_n(\mathbf{r}_i)$ where $\alpha_s = \alpha_+$ or α_- denotes the spin state. As a result the ground state of the crystal is obtained by filling these single electron states one by one, beginning at the lowest energy and ending when all electrons are exhausted. In the Hartree approximation, one ignores that the electrons are indistinguishable and we write for the the resulting ground state

$$\Psi = \prod_{s,n} \alpha_s \Psi_n.$$

In the Hartree–Fock approximation one takes into account that the electrons cannot be distinguished and one calculates the Slater sum of these products, i.e. one calculates the antisymmetrized product of the filled states

$$\Psi = \sum_p (-1)^p P_p \prod_{s,n} \alpha_s \Psi_n \qquad (2.27)$$

So we sum over all possible permutations, represented by the permutation operator P_p that performs p elementary permutations where just two electrons are interchanged. Clearly this procedure would have been impossible if we had chosen different potentials for different electrons, because the individual electron states would then not have been orthonormal.

In order to obtain an effective potential $V'(\mathbf{r}_i)$, one uses a self-consistent method. First, one chooses a trial potential for $V'(\mathbf{r}_i)$ and inserts it into eq. (2.25). Then the eigenfunctions Ψ_n of $\mathcal{H}'(\mathbf{r}_i)$ are calculated. Subsequently, in the next step, a new and hopefully better trial potential $V'(\mathbf{r}_i)$ is determined, using

$$V'(\mathbf{r}_i) = V_i + \int d\mathbf{r}_1 \cdots \int d\mathbf{r}_j \cdots \Psi^* \left[\frac{1}{2}\sum_{j\neq i} V_{ij}\right]\Psi \qquad (2.28)$$

This is the Hartree–Fock potential. It incorporates the full electrostatic attraction of the nuclei. However, the electrostatic repulsion of the other electrons is replaced by the repulsion of an electron cloud with a density determined by the ground state Ψ, that was found in the previous step. To improve $V'(\mathbf{r}_i)$ the procedure is repeated again and again. In each cycle one inserts $V'(\mathbf{r}_i)$ from eq. (2.28) in eq. (2.25) and determines new eigenfunctions of $\mathcal{H}'(\mathbf{r}_i)$. Eventually, one hopes that the procedure converges and that one obtains eigenfunctions that are a good approximation of the exact eigenstates of the crystal.

Before embarking on this procedure, one could question whether eq. (2.28) yields a *different* potential for each electron. This would be in contradiction with our intention and would inhibit the use of the procedure described above. We need to ask this question because of the exclusion of a term with $j = i$ in the sum in eq. (2.28). This exclusion is necessary because $\mathcal{V}_{ii}$ diverges. Using the Hartree–Fock approximation our problem is fictitious however, because the wave function Ψ is constructed in such a way that the electrons are indistinguishable. Hence, the result of the integration in eq. (2.28) is independent of i and as a result the functional shape of $\mathcal{V}'(\mathbf{r}_i)$ is *independent* of the choice of the electron. Hence, the functional shape of the single electron Hamiltonian $\mathcal{H}'(\mathbf{r}_i)$ is also the same for all electrons and as a result the same set of orthogonal wave functions applies for them.

2.2.3 Core and Valence Electrons

We may take advantage of the fact that the wave functions of the core electrons are very localized. This allows their wave functions to be calculated by treating the cores independently, i.e., neglecting their interaction with other cores. Then the wave functions of these core electrons are just the atomic orbitals discussed in the previous section.

Thus only the wave functions of the valence electrons have to be calculated using the Hartree–Fock potential for the crystal as a whole. It contains two types of electron–electron repulsion: first, those between the valence electrons and the core electrons; second, those between the valence electrons themselves. Thus, the Hartree–Fock potential splits into two parts. The part that the valence electrons experience from the cores is fixed once the atomic orbitals describing the core electrons are determined. The part representing the electrostatic repulsion between the valence electrons is still undetermined however, and must be obtained using a self consistent method.

Moreover, as the extension of the core wave functions is very small, their contribution to the Hartree–Fock potential may well be approximated by a smooth function. In its simplest form this so-called pseudo potential method uses the central force model described above. Then the interaction between the valence electrons and the silicon cores simply reduces to a shielded electrostatic potential of the nuclei. Introducing a shielding factor α_l for the lth core, the Hartree–Fock potential reduces to

$$\mathcal{V}'(\mathbf{r}_i) = -\sum_l \frac{\alpha_l Z_l e^2}{4\pi\varepsilon_0 r_{il}} + \int d\mathbf{r}_1 \cdots \int d\mathbf{r}_j \cdots \Psi^* \frac{1}{2} \sum_{j \neq i} \frac{e^2}{4\pi\varepsilon_0 r_{ij}} \Psi \qquad (2.29)$$

where Ψ is the total state of the valence electrons.

A simple estimate of α_l is obtained by assuming the wave functions of the core electrons to be extremely close to the nuclei. Then the cores of the atoms are reduced to central potentials of point charges $(Z_l - Z_c)e$, where Z_c is the number of core electrons per atom. In the case of silicon, the screening factor reduces to $\alpha_l = (Z_l - Z_c)/Z_l = 4/14$ for all cores.

2.3 The Crystal Lattice

In the above treatment we have assumed the positions of the nuclei in the crystal to be fixed. In lowest order such an approximation is very suitable to describe the electronic structure of semiconductors. However, many phenomena in semiconductors can only be understood if we include nuclear motion or so-called lattice vibrations. Therefore in the present section we consider these vibrations.

2.3.1 Nuclear Motion

THE FULL HAMILTONIAN OF A CRYSTAL

As soon as nuclear motion is assumed to take place, one needs to account for the nuclear kinetic energy and for the fact that the electrostatic repulsion between the nuclei is not constant anymore. Thus we have to add the corresponding terms to the Hamiltonian (2.20) of the crystal. As a result this Hamiltonian is extended to

$$\mathcal{H} = -\frac{\hbar^2}{2m_e} \sum_i \Delta_i - \sum_l \frac{\hbar^2}{2M_l} \Delta_l + \mathcal{V} \tag{2.30}$$

where

$$\Delta_i = \nabla_i \cdot \nabla_i \quad \text{and} \quad \nabla_i = \left(\frac{\partial}{\partial x_i}, \frac{\partial}{\partial y_i}, \frac{\partial}{\partial z_i}\right)$$

is the Laplace operator for the ith electron, which has coordinates $\mathbf{r}_i = (x_i, y_i, z_i)$. Furthermore

$$\Delta_l = \nabla_l \cdot \nabla_l \quad \text{where} \quad \nabla_l = (\nabla_{lx}, \nabla_{ly}, \nabla_{lz}) \equiv \left(\frac{\partial}{\partial X_l}, \frac{\partial}{\partial Y_l}, \frac{\partial}{\partial Z_l}\right)$$

is the Laplace operator for the lth nucleus at the position $\mathbf{R}_l = (X_l, Y_l, Z_l)$ which has a mass M_l. Note the short notation $\nabla_{lx} \equiv (\partial/\partial X_l)$ which is introduced to avoid lengthy formulae below. Now the potential $\mathcal{V}$ is given by

$$\mathcal{V} = \sum_i \mathcal{V}_i + \frac{1}{2} \sum_i \sum_{j \neq i} \mathcal{V}_{ij} + \frac{1}{2} \sum_l \sum_{l' \neq l} \mathcal{V}_{ll'} \tag{2.31}$$

Here, $\mathcal{V}_i$ is the electrostatic attraction (2.22) exerted by all nuclei on the ith electron, $\mathcal{V}_{ij}$ is the electrostatic repulsion (2.23) between the ith and jth electrons and

$$\mathcal{V}_{ll'} = \frac{ZZ'e^2}{4\pi\varepsilon_0 R_{ll'}} \tag{2.32}$$

represents the electrostatic repulsion between the nuclei, where

$$R_{ll'} = \sqrt{(X_l - X_{l'})^2 + (Y_l - Y_{l'})^2 + (Z_l - Z_{l'})^2}$$

So the lowest order term in the matrix element $\mathcal{H}_L^{00}$ is quadratic in the nuclear displacements $u_{l\mu}$. As this lowest order term is generally much larger than the following higher order terms, we may limit the series expansion to this quadratic term. Then

$$\mathcal{H}_L^{00} = -\sum_l \frac{\hbar^2}{2M_l} \Delta_l + \frac{1}{2} \sum_{l\mu,l'\mu'} \Phi_{l\mu,l'\mu'}^{00} u_{l\mu} u_{l'\mu'} \tag{2.44}$$

It should be stressed that the first order term can be made equal to zero in $\mathcal{H}_L^{00}$ only. The same is not simultaneously possible for the other diagonal matrix elements, because the equilibrium positions of the nuclei are generally different in different electronic states. On the other hand, the total number of valence electrons in a semiconductor crystal is very large—e.g. $2.00 \cdot 10^{23}$ in a silicon crystal of 1 cm^3—while in practical situations only a very small fraction—typically 10^{-5} or less—are excited to higher single electron states. Therefore the lattice vibrations are hardly affected by electronic excitations. This can be formalized by stating that the diagonal matrix elements are approximately the same for all relevant electronic states, i.e. that one may approximate

$$\mathcal{H}_L^{nn} \approx \mathcal{H}_L^{00} = \mathcal{H}_L = -\sum_l \frac{\hbar^2}{2M_l} \Delta_l + \frac{1}{2} \sum_{l\mu,l'\mu'} \Phi_{l\mu,l'\mu'} u_{l\mu} u_{l'\mu'} \tag{2.45}$$

where we skipped the indices nn in $\mathcal{H}_L^{nn}$ and $\Phi_{l\mu,l'\mu'}^{nn}$ because these quantities are approximately independent of the electronic state Ψ_n.

As mentioned above, in first-order perturbation theory $\mathcal{H}_L$ yields the correction on the energy of the crystal due to nuclear motion. Also $\mathcal{H}_L$ is still an operator with respect to the nuclear displacements $u_{l\mu}$. To calculate the correction on the energy due to $\mathcal{H}_L$, we need to determine its eigenvalues and eigenstates. Now, $\mathcal{H}_L$ has the shape of a multi-dimensional harmonic oscillator. In section 2.3.2 we will see how it can always be split into independent one-dimensional harmonic oscillators. As a result, the eigenvalues and eigenstates $\mathcal{H}_L$ can always be formally obtained.

As stated before the non-diagonal matrix elements $\mathcal{H}_L^{nm}$ cause transitions between electronic states. In this case the linear term in the displacements cannot be eliminated. However, this linear term is generally much larger than the following higher order terms and it is sufficient to limit the series expansion to this first order term. Then, for $n \neq m$, and using an index EL to stress the fact that this term represents interaction between electrons and lattice,

$$\mathcal{H}_L^{nm} \approx \mathcal{H}_{EL}^{nm} = \sum_{l\mu} \Phi_{l\mu}^{nm} u_{l\mu} \tag{2.46}$$

We note that this matrix element originates in the first order term of the expansion of the total potential $\mathcal{V}$. This first order term represents the change of this potential upon a displacement of the nuclei, or in other words, a deformation of the crystal. Therefore, this first-order term in the expansion of $\mathcal{V}$ is called the deformation potential and $\Phi_{l\mu}^{nm}$ a matrix element of this deformation potential. We also note that $\mathcal{H}_{EL}^{nm}$ is still an operator acting on the nuclear displacements $u_{l\mu}$ just as $\mathcal{H}_L$. To determine the nature of the transitions that are induced by $\mathcal{H}_{EL}^{nm}$ we need to calculate its matrix elements between the eigenstates of $\mathcal{H}_L$. This will be done in section 2.3.3.

2.3.2 Phonon Modes

REDUCTION TO ONE-DIMENSIONAL HARMONIC OSCILLATORS

Though the lattice Hamiltonian (2.45) is complicated, its eigenvalues and eigenstates can be determined formally in a straightforward manner. This is due to the fact that $\mathcal{H}_L$ is an n-dimensional harmonic oscillator which can always be reduced to a set of independent one-dimensional harmonic oscillators [8]. Furthermore, the one-dimensional harmonic oscillator is a textbook problem in Quantum Mechanics, which is summarized in appendix B. In this section we present the reduction of $\mathcal{H}_L$ to a set of independent one-dimensional harmonic oscillators and subsequently we calculate its eigenstates and energy levels.

We start by introducing a simplified notation where the nuclear displacements are denoted by a single index λ. It takes the values

$$3l+1, \quad 3l+2, \quad 3l+3$$

for the displacements of the lth nucleus in the x, y and z directions, respectively. Then, in this new notation all index pairs (l,μ) and (l',μ') are replaced by single indices λ and λ', respectively. As a result the lattice Hamiltonian (2.45) is written as

$$\mathcal{H}_L = -\sum_\lambda \frac{\hbar^2}{2M_\lambda}\left(\frac{\partial}{\partial u_\lambda}\right)^2 + \frac{1}{2}\sum_{\lambda,\lambda'} \Phi_{\lambda\lambda'} u_\lambda u_{\lambda'} \quad (2.47)$$

Here M_λ is equal to the mass M_l of the nucleus for which $\lambda = 3l+1, 3l+2$ or $3l+3$. It is instructive to compare this Hamiltonian with the Hamiltonian for the one-dimensional harmonic oscillator given by eq. (B.1). In the present case the displacements u_λ, the masses M_λ and the force constants $\Phi_{\lambda\lambda'}$ replace the coordinate x, the mass m and the force constant C_0 in eq. (B.1). As in eqs. (B.4) and (B.2) we introduce reduced displacements

$$w_\lambda = \sqrt{\frac{M_\lambda}{\hbar}}\, u_\lambda \quad (2.48)$$

and reduced interaction constants

$$D_{\lambda\lambda'} = \frac{1}{\sqrt{M_\lambda M_{\lambda'}}}\Phi_{\lambda\lambda'} \quad (2.49)$$

So we obtain the multidimensional equivalent of eq. (B.6),

$$\mathcal{H}_L = \frac{\hbar}{2}\left[-\sum_\lambda \left(\frac{\partial}{\partial w_\lambda}\right)^2 + \sum_{\lambda,\lambda'} D_{\lambda\lambda'} w_\lambda w_{\lambda'}\right] \quad (2.50)$$

Note that the reduced displacements w_λ have dimension $[t]^{1/2}$ and that the reduced interaction constants $D_{\lambda\lambda'}$ have dimension $[t]^{-2}$ or the square of frequency.

As the electrostatic potentials in the Hamiltonian of the crystal are real functions of real variables, its derivatives, the interaction constants $D_{\lambda\lambda'}$, are also real. Moreover, it follows from eqs (2.40) that $\Phi^{nm}_{l\mu,l'\mu'} = \Phi^{nm}_{l'\mu',l\mu}$, so after introducing single indices λ and λ', $\Phi^{nm}_{\lambda,\lambda'} = \Phi^{nm}_{\lambda',\lambda}$, or using eq. (2.49), $D_{\lambda\lambda'} = D_{\lambda'\lambda}$. Hence these interaction constants constitute a real symmetric matrix. Such a real symmetric matrix has real eigenvalues, which we denote by ω_j^2. Consequently a unitary matrix can be found which transforms the matrix with elements $D_{\lambda\lambda'}$ to a real diagonal matrix with elements $\omega_j^2 \delta_{j,j'}$. Let $U_{\lambda j}$ be the elements of this unitary matrix, then

$$D_{\lambda\lambda'} = \sum_{j,j'} U_{\lambda j} \omega_j^2 \delta_{j,j'} U_{j'\lambda'}^{-1} = \sum_j U_{\lambda j} \omega_j^2 U_{\lambda' j}^* \qquad (2.51)$$

where, owing to unitarity

$$U_{j\lambda}^{-1} = U_{\lambda j}^* \qquad (2.52)$$

We now show that the Hamiltonian (2.50) splits into independent one-dimensional harmonic oscillators when transformed by the unitary matrix with elements $U_{\lambda j}$. We first define generalized displacements

$$q_j = \sum_\lambda U_{j\lambda}^{-1} w_\lambda = \sum_\lambda U_{\lambda j}^* w_\lambda \qquad (2.53)$$

so

$$w_\lambda = \sum_j U_{\lambda j} q_j, \quad \text{while} \quad \frac{\partial}{\partial w_\lambda} = \sum_j \frac{\partial q_j}{\partial w_\lambda} \frac{\partial}{\partial q_j} = \sum_j U_{\lambda j}^* \frac{\partial}{\partial q_j} \qquad (2.54)$$

Note that the generalized displacements q_j have the same dimension as the reduced displacements w_λ, namely $[t]^{1/2}$. Next we transform the kinetic energy term,

$$-\frac{\hbar}{2} \sum_\lambda \left(\frac{\partial}{\partial w_\lambda}\right)^2 = -\frac{\hbar}{2} \sum_\lambda \left(\sum_j U_{\lambda j}^* \frac{\partial}{\partial q_j}\right)^* \left(\sum_{j'} U_{\lambda j'}^* \frac{\partial}{\partial q_{j'}}\right)$$

$$= -\frac{\hbar}{2} \sum_{j,j'} \left(\sum_\lambda U_{\lambda j} U_{\lambda j'}^*\right) \left(\frac{\partial}{\partial q_j}\right)^* \frac{\partial}{\partial q_{j'}}$$

$$= -\frac{\hbar}{2} \sum_{j,j'} \delta_{j,j'} \left(\frac{\partial}{\partial q_j}\right)^* \frac{\partial}{\partial q_{j'}}$$

$$= -\frac{\hbar}{2} \sum_j \left|\frac{\partial}{\partial q_j}\right|^2 \qquad (2.55)$$

Note that the first step uses w_λ as real. Finally we transform the potential energy term,

$$\frac{\hbar}{2} \sum_{\lambda,\lambda'} D_{\lambda\lambda'} w_\lambda w_{\lambda'} = \frac{\hbar}{2} \sum_{\lambda,\lambda'} \left(\sum_j U_{\lambda j} \omega_j^2 U_{j\lambda'}^*\right) w_\lambda w_{\lambda'}$$

$$= \frac{\hbar}{2} \sum_j \omega_j^2 \left(\sum_\lambda U_{\lambda j} w_\lambda\right) \left(\sum_{\lambda'} U_{j\lambda'}^* w_{\lambda'}\right)$$

$$= \frac{\hbar}{2} \sum_j \omega_j^2 q_j^* q_j = \frac{\hbar}{2} \sum_j \omega_j^2 |q_j|^2 \qquad (2.56)$$

2.3 THE CRYSTAL LATTICE

As a result the lattice Hamiltonian splits into independent terms, one for each value of j,

$$\mathcal{H}_L = \frac{\hbar}{2} \sum_j \left[-\left|\frac{\partial}{\partial q_j}\right|^2 + \omega_j^2 |q_j|^2 \right] \quad (2.57)$$

For a stable crystal the eigenvalues ω_j^2 of the matrix with elements $D_{\lambda\lambda'}$ are not only real but also positive. To show this we suppose that one them, e.g. ω_j^2, is negative. Then the crystal gains energy by increasing the corresponding generalized displacement q_j to infinity. However, using eqs (2.48) and (2.53), then at least one of the nuclear displacements u_λ also increases to infinity. Physically this means that the crystal falls apart, contradicting the assumption that it is stable. Hence, for a stable crystal $\omega_j^2 > 0$ and consequently ω_j is real.

Also the generalized displacements can be chosen to be real. This follows from the fact that the matrix elements $D_{\lambda\lambda'}$ are real, so the elements $U_{\lambda j}$ of the transformation matrix can also be chosen to be real. As a result the lattice Hamiltonian $\mathcal{H}_L$ reduces to a set of independent one-dimensional harmonic oscillators described by oscillator frequencies ω_j,

$$\mathcal{H}_L = \frac{\hbar}{2} \sum_j \left[-\left(\frac{\partial}{\partial q_j}\right)^2 + \omega_j^2 q_j^2 \right] \quad (2.58)$$

LATTICE MODES

Each of these one-dimensional harmonic oscillators is called a lattice mode. Their eigenstates and eigenvalues are treated in appendix B. Using the results of that section we find the energy of the lattice to be given by

$$E_L = \sum_j \hbar\omega_j \left(n_j + \frac{1}{2}\right) \quad \text{where} \quad n_j = 0, 1, 2, 3, \ldots \quad (2.59)$$

corresponding to an eigenstate

$$\chi = \prod_j |n_j\rangle = \prod_j \frac{1}{\sqrt{n_j!}} (a_j^\dagger)^{n_j} |0\rangle \quad (2.60)$$

where

$$a_j = \frac{1}{\sqrt{2}} \left(\sqrt{\omega_j} q_j + \frac{1}{\sqrt{\omega_j}} \frac{\partial}{\partial q_j} \right) \quad (2.61)$$

$$a_j^\dagger = \frac{1}{\sqrt{2}} \left(\sqrt{\omega_j} q_j - \frac{1}{\sqrt{\omega_j}} \frac{\partial}{\partial q_j} \right) \quad (2.62)$$

are, respectively, the annihilation and creation operators corresponding to the jth harmonic oscillator. Clearly they are dimensionless because q_j has the dimension $[t]^{1/2}$. It is seen from eq. (2.60) that the latter operator increases the quantum number n_j by +1 and hence the lattice energy by $+\hbar\omega_j$. Similarly, the former operator changes these quantities by -1 and $-\hbar\omega_j$.

SECOND QUANTIZATION AND PHONONS

We see that the energy of the crystal lattice is quantized and can only be increased or decreased with energy quanta equal to $\hbar\omega_j$. We may compare this outcome with the description of light which is also quantized into energy quanta equal to $\hbar\omega$. In the case of light we are accustomed to interpret these quanta as particles that are called photons. This induces us to interpret the quanta $\hbar\omega_j$ of lattice energy as particles as well, now to be called phonons. In this language, a crystal in the state described by eq. (2.60) is said to contain n_j phonons of energy $\hbar\omega_j$. Furthermore, $a_j^\dagger$ is called a phonon creation operator because it increases the lattice energy by one quantum $\hbar\omega_j$, i.e. it creates a phonon with energy $\hbar\omega_j$, while a_j is called a phonon annihilation operator because it annihilates such a phonon.

Moreover, when the jth lattice mode is excited into state $|n_j\rangle$, one could say that n_j phonons with energy $\hbar\omega_j$ *occupy* this lattice mode. Hence, in a language where quanta of lattice energy are interpreted as particles, one may interpret lattice modes as eigenstates of these particles. In particular, the jth lattice mode is interpreted as the jth eigenstate of a phonon, while $\hbar\omega_j$ is the energy corresponding to this eigenstate. Thus, the role of particles and eigenstates has been reversed. While originally one-dimensional harmonic oscillators played the role of particles and the eigenstates of these harmonic oscillators represented their eigenstates, now phonons are the particles, while the harmonic oscillators form their eigenstates. This reversal of the role of particles and eigenstates is often referred to as second quantization.

One may wonder whether phonons are bosons or fermions. This question can be answered in several ways. First, we observe that many phonons may occupy the same lattice mode, now also called phonon mode. This excludes phonons to be fermions for which multiple occupation is not allowed because of the Pauli exclusion principle. Next we consider eq. (2.60) and it is seen that interchanging two phonon creation operators, i.e. interchanging two phonons, never changes the sign of χ. Both arguments show that phonons must necessarily be bosons. For a more elaborate discussion the reader is referred to the treatment of photons by Dirac [15].

2.3.3 Electron–Phonon Coupling

THE NON-DIAGONAL LATTICE TERMS

As already discussed above, in first order perturbation theory the non-diagonal matrix elements $\mathcal{H}_{EL}^{nm}$ cause transitions between different electronic states. We remember that $\mathcal{H}_{EL}^{nm}$ is still an operator acting on the nuclear displacements $u_{l\mu}$. Therefore, to understand the nature of the transitions caused by $\mathcal{H}_{EL}^{nm}$ we need to calculate its matrix elements between the eigenstates of the lattice Hamiltonian $\mathcal{H}_L$. For this purpose we perform the same transformations as in the previous section. First, we rewrite $\mathcal{H}_{EL}^{nm}$ using the simplified notation where the indices $l\mu$ are replaced by λ. So

$$\mathcal{H}_{EL}^{nm} = \sum_\lambda \Phi_\lambda^{nm} u_\lambda \qquad (2.63)$$

2.3 THE CRYSTAL LATTICE

Next, we introduce reduced displacements

$$w_\lambda = \sqrt{\frac{M_\lambda}{\hbar}} u_\lambda \quad \text{so} \quad u_\lambda = \sqrt{\frac{\hbar}{M_\lambda}} w_\lambda \qquad (2.64)$$

and define

$$D_\lambda^{nm} = \sqrt{N} \Phi_\lambda^{nm} \quad \text{so} \quad \Phi_\lambda^{nm} = \frac{1}{\sqrt{N}} D_\lambda^{nm} \qquad (2.65)$$

Note that while w_λ has dimension $[t]^{1/2}$, D_λ^{nm} has the same dimension as Φ_λ^{nm}. The extra factor $\sqrt{N}$, where N is the total number of atoms in the crystal is inserted in order to render the final result independent of the size of the crystal. How this works out will be shown below. Using definitions (2.64) and (2.65), we find

$$\mathcal{H}_{EL}^{nm} = \sum_\lambda \Phi_\lambda^{nm} u_\lambda = \sum_\lambda \sqrt{\frac{\hbar}{M_\lambda}} \Phi_\lambda^{nm} w_\lambda = \sum_\lambda \sqrt{\frac{\hbar}{NM_\lambda}} D_\lambda^{nm} w_\lambda \qquad (2.66)$$

Subsequently, we transform to generalized nuclear displacements, using the unitary matrix $U_{\lambda j}$ introduced in the previous section. As a result

$$\mathcal{H}_{EL}^{nm} = \sum_\lambda \sqrt{\frac{\hbar}{NM_\lambda}} D_\lambda^{nm} w_\lambda$$

$$= \sum_\lambda \sqrt{\frac{\hbar}{NM_\lambda}} D_\lambda^{nm} \sum_j U_{\lambda j} q_j$$

$$= \frac{\hbar}{\sqrt{N}} \sum_j \left(\sum_\lambda \frac{1}{\sqrt{\hbar M_\lambda}} D_\lambda^{nm} U_{\lambda j} \right) q_j$$

$$= \frac{\hbar}{\sqrt{N}} \sum_j \Xi_j^{nm} \sqrt{\omega_j} q_j \qquad (2.67)$$

where we define

$$\Xi_j^{nm} = \sum_\lambda \frac{1}{\sqrt{\hbar \omega_j M_\lambda}} D_\lambda^{nm} U_{\lambda j} \qquad (2.68)$$

Here, the normalization factor $\sqrt{\hbar \omega_j M_\lambda}$ ensures that Ξ_j^{nm} has dimension $[t]^{-1}$ or frequency. Finally, we rewrite q_j in phonon creation and annihilation operators using eqs. (2.61) and (2.62). Then

$$\mathcal{H}_{EL}^{nm} = \frac{\hbar}{\sqrt{2N}} \sum_j \Xi_j^{nm} (a_j + a_j^\dagger) \qquad (2.69)$$

EMISSION AND ABSORPTION OF PHONONS BY ELECTRONS

The form of $\mathcal{H}_{EL}^{nm}$ obtained allows us to identify which type of transitions it induces. For this purpose we calculate the matrix elements of $\mathcal{H}_{EL}^{nm}$ between two phonon states.

According to appendix B the matrix elements of a_j and $a_j^\dagger$ between the states $|n_j\rangle$ and $|m_j\rangle$ are given by

$$\langle n_j | a_j | m_j \rangle = \sqrt{n_j + 1}\, \delta_{m_j, n_j + 1}$$
$$\langle n_j | a_j^\dagger | m_j \rangle = \sqrt{n_j}\, \delta_{m_j, n_j - 1} \qquad (2.70)$$

Hence, the only non-zero matrix elements of $\mathcal{H}_{EL}^{nm}$ are of the type

$$\langle n_j | \mathcal{H}_{EL}^{nm} | n_j - 1 \rangle = \langle n_j - 1 | \mathcal{H}_{EL}^{nm} | n_j \rangle = \hbar \sqrt{\frac{n_j}{2N}} \Xi_j^{nm} \qquad (2.71)$$

mixing adjacent phonon states $|n_j\rangle$ and $|n_j - 1\rangle$ on the one hand and electron states Ψ_n and Ψ_m on the other hand. Hence $\mathcal{H}_{EL}^{nm}$ causes transitions where simultaneously electrons switch states between Ψ_n and Ψ_m while phonons switch states between $|n_j\rangle$ and $|n_j - 1\rangle$. The situation is schematically illustrated in Fig. 2.5 representing the total energy as a function of a single lattice coordinate X_l for two different electron states Ψ_n and Ψ_m. For clarity the two cases are depicted adjacent to each other. The minima of the parabolas represent the eigenvalues E_n and E_m of the electron Hamiltonian in these two states. The parabolas themselves represent the potential term of the lattice Hamiltonian $\mathcal{H}_L$. The horizontal lines represent the energy levels (2.59) of the lattice Hamiltonian. Their spacing is the same for each of the two electron states. According to the discussion above, $\mathcal{H}_{EL}^{nm}$ causes transitions where simultaneously the electrons are excited from state Ψ_n to state Ψ_m while a phonon with energy $\hbar\omega_j$ is annihilated. The reverse may occur where the electron falls back from state Ψ_m to Ψ_n while a phonon with energy $\hbar\omega_j$ is created. Energy conservation will require that such transitions only take place if

$$E_m - E_n = \hbar\omega_j \qquad (2.72)$$

The rates for such phonon absorption and emission processes follow from first-order time-dependent perturbation theory. Taking the initial phonon state to be $|n_j\rangle$ and the electrons to be initially in the low energy state Ψ_n, we find for the rate for

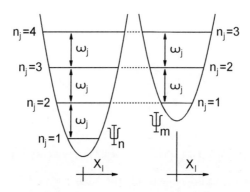

Figure 2.5 Electron–phonon transitions.

2.3 THE CRYSTAL LATTICE

phonon absorption,

$$W_{n \to m} = \frac{2\pi}{\hbar} |\langle n_j | \mathcal{H}_{EL}^{nm} | n_j - 1 \rangle|^2 \delta(E_m - E_n - \hbar\omega_j)$$

$$= \frac{2\pi}{\hbar} \frac{n_j}{2N} |\hbar \Xi_j^{nm}|^2 \, \delta(E_n - E_m - \hbar\omega_j) \quad (2.73)$$

Here, we inserted the matrix element (2.71). Furthermore, the Dirac δ-function $\delta(E_m - E_n - \hbar\omega_j)$ represents energy conservation as required by eq. (2.72). Inversely, assuming the initial phonon state to be the same, but the electrons to be initially in the high energy state Ψ_m, the rate for phonon emission is given by

$$W_{m \to n} = \frac{2\pi}{\hbar} |\langle n_j | \mathcal{H}_{EL}^{nm} | n_j + 1 \rangle|^2 \delta(E_m - E_n - \hbar\omega_j)$$

$$= \frac{2\pi}{\hbar} \frac{n_j + 1}{2N} |\hbar \Xi_j^{nm}|^2 \, \delta(E_m - E_n - \hbar\omega_j) \quad (2.74)$$

Now, the need to insert the factor $\sqrt{N}$ in eq. (2.65) becomes clear. On the one hand one expects the resulting transition probability to be independent of crystal size. On the other hand, the number of phonon modes obeying the δ-functions in eqs (2.73) and (2.74) is expected to be proportional to this size. By inserting the factor $\sqrt{N}$ in eq. (2.65), we have added a factor $1/N$ in eqs (2.73) and (2.74), so the growth of the number of phonon modes with size is neutralized. Thus, the interaction constants Ξ_j^{nm} do not only have dimension $[t^{-1}]$, but they are truly material parameters, that are independent of crystal size.

It is seen that the rate for phonon emission is larger than the rate for phonon absorption by a factor

$$\frac{W_{m \to n}}{W_{n \to m}} = \frac{n_j + 1}{n_j} \quad (2.75)$$

It is precisely this difference between the phonon emission and absorption rates which ensures thermal equilibrium to be established between the lattice degrees of freedom and the electron degrees of freedom. For the crystal to be in complete thermal equilibrium, the total number of phonon emissions should be equal to the total number of phonon absorptions. This will only occur when the higher probability of phonon emission is compensated by a lower probability of finding the electrons in the high energy state. This compensation is achieved with a thermal distribution among the electron states while the corresponding 'electron temperature' is equal to the temperature of the crystal lattice.

POLAR AND NON-POLAR SEMICONDUCTORS

Chapter 7 will be devoted to a detailed discussion of the interaction constants Ξ_j^{nm}. However, already at this point an important remark should be made with respect to the nature of $\mathcal{H}_{EL}^{nm}$. Its origin, the first order term in the series expansion (2.34) of the electrostatic interactions in the crystal, is qualitatively different for elemental semiconductors like Si and Ge on the one hand and composite semiconductors like GaAs on the other hand. In elemental semiconductors all cores carry the same charge $+4e$.

Hence, deformation of the crystal does *not* create an electrical polarization nor a long range electrical field. Therefore these materials are called non-polar semiconductors. As a result, the first order term involves short range fields only, which average out when integrated over a larger volume. On the other hand, in composite semiconductors like GaAs we have two types of cores with different charges $+3e$ and $+5e$ respectively. Deformation of the crystal yields a macroscopic electrical polarization and a long range electric field, which acts directly on the electrons. As a consequence electron–phonon interactions are much stronger for these latter materials that are called polar semiconductors.

3 Band Structure

3.1 Translational Symmetry

3.1.1 Primitive Cells and Space Lattices

CRYSTAL STRUCTURE

Even using the approximative methods described in the previous chapter, the calculation of the eigenstates and energy levels of a crystal seems a hopeless task, because even the smallest crystal contains an enormous number of atoms, e.g. $0.50 \cdot 10^{23}$ cm^{-3} for silicon. Fortunately, crystalline solids are characterized by an important property, translational symmetry, helping us to reduce the complexity of this problem greatly.

Roughly, translational symmetry corresponds to the fact that a crystal consists of many equivalent unit cells, like those shown in Figs 1.2 and 1.3. To make optimum use of translational symmetry it is advantageous to find the smallest possible unit cell, the so-called primitive cell. For the diamond and zinc blende structures these primitive cells are four times smaller than the cubic unit cells. Fig. 3.1a shows how a primitive cell is constructed from the cubic unit cell. It has the shape of an elongated parallelepiped defined by three basis vectors $\mathbf{a}_1$, $\mathbf{a}_2$ and $\mathbf{a}_3$ with a length $a = c/\sqrt{2}$, where c is the length of the edges of the cubic unit cell. Fig. 3.1a also shows how the four vertices 1 to 4 span a regular tetrahedron. The same holds for the four vertices 5 to 8, while the six vertices 2 to 7 span a regular octahedron.

Thus a crystal consists of a large number of primitive cells put together in the way shown in Fig. 3.2. We describe it in two steps. First, we consider the vertices of all primitive cells. This set of discrete points is called the space lattice of the crystal. Then we define the atomic positions with respect to these space lattice points. These relative positions are of course the same in each primitive cell and are called the basis. We note that the position of the origin of the space lattice may be freely chosen at any position in the crystal. As a result the basis depends on this choice. Normally one chooses the point of highest symmetry as the origin of the primitive cell.

For both the diamond and zinc blende structure the basis consists of two sites for atoms. In the zinc blende structure these sites are occupied by different types of atoms, e.g. Ga and As, while in the diamond structure both sites are occupied by the same type of atom, e.g. Si. In the zinc blende structure such an atomic site is also the point of highest, i.e. tetrahedral, symmetry in the primitive cell. Therefore, for this structure

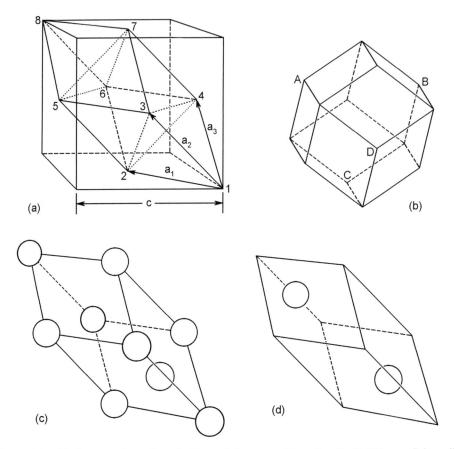

Figure 3.1 (a) Construction of a primitive cell from a cubic unit cell. (b) Wigner–Seitz cell. (c) Primitive cell for the zinc blende structure. (d) Primitive cell for the diamond structure.

it is customary to choose the origin of the primitive cell at such an atomic site. Then, as shown in Fig. 3.1c, the other atomic site is positioned in the centre of one of the regular tetrahedra. In the diamond structure atomic sites have the same high, tetrahedral, symmetry as in the zinc blende structure. However, a point of even higher, i.e. cubic, symmetry is found midway between two neighbouring atoms. Therefore, for the diamond structure the origin of the primitive cell is often chosen at this point of cubic symmetry. The atomic sites are in the positions shown in Fig. 3.1d. In much of what follows the choice of the origin is irrelevant. In the few cases where we need to make a choice, we will always choose an atomic site—with tetrahedral symmetry—as the origin. Thus, we obtain the same treatment for the zinc blende as for the diamond lattice. Still, in one single case—electron-phonon coupling, to be treated in chapter 7—we need to take into account that the symmetry of the cubic diamond structure is higher than the symmetry of the tetrahedral zinc blende structure.

Also other crystal structures are described by a primitive cell with the shape shown in Fig. 3.1a. If the cell contains one atom only we obtain the simple face centred cubic (fcc) lattice which is typical for many metals like Al. The rock salt structure, typical

3.1 TRANSLATIONAL SYMMETRY

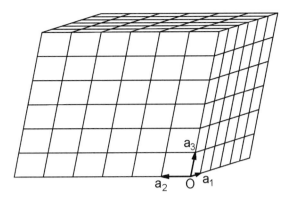

Figure 3.2 The space lattice of a crystal.

for a large number of ionically bound solids like NaCl, is obtained by putting one atomic site at the origin of the primitive cell and the second at the centre of the octahedron instead of one of the two tetrahedrons.

Not only is the primitive cell shown in Fig. 3.1a used to describe the fcc, diamond, zinc blende and rock salt structures, but also another cell, the dodecahedron shown in Fig. 3.1b. To construct this so-called Wigner–Seitz cell, we consider the space lattice shown in Fig. 3.2. We choose a single space lattice point and connect it to neighbouring space lattice points. Next, we construct the perpendicular bisecting planes of these connecting lines. Our new primitive cell is now precisely the space enclosed by these planes. If we choose one of the atomic sites at the centre of this new primitive cell, the other site is located at four vertices forming a tetrahedron, e.g. A, B, C, and D in Fig. 3.1b.

SPACE LATTICES

We now proceed with a more rigorous description of translational symmetry. We first consider an infinitely large crystal. We describe it using the primitive cells and space lattice shown in Figs 3.1a and 3.2 and defined by three basis vectors $\mathbf{a}_1$, $\mathbf{a}_2$ and $\mathbf{a}_3$. We note that these vectors are not necessarily orthogonal. Then the volume of the primitive cell is equal to

$$V_C = \mathbf{a}_1 \cdot (\mathbf{a}_2 \times \mathbf{a}_3) \tag{3.1}$$

We choose the origin of our frame of reference at one of the space lattice points denoted by O in Fig. 3.2. An arbitrary position $\mathbf{r}$ within the crystal can be decomposed in its components along the $\mathbf{a}_1$, $\mathbf{a}_2$ and $\mathbf{a}_3$ axes

$$\mathbf{r} = \sum_\alpha x_\alpha \mathbf{a}_\alpha \tag{3.2}$$

This decomposition corresponds to a transformation from an orthogonal to a non-orthogonal frame of reference. In the first $\mathbf{r}$ has components ξ_1, ξ_2 and ξ_3 along orthogonal axes and the unit of length is 1 m. In the second $\mathbf{r}$ has components x_1, x_2 and x_3 along the $\mathbf{a}_1$, $\mathbf{a}_2$ and $\mathbf{a}_3$ vectors and the unit of length along each vector is equal to the length of that vector. Thus in the new frame of reference the unit of volume is equal to the volume V_C of the primitive cell. The Jacobian of this transformation is equal to

$$\begin{vmatrix} \left(\dfrac{\partial \xi_1}{\partial x_1}\right) & \left(\dfrac{\partial \xi_1}{\partial x_2}\right) & \left(\dfrac{\partial \xi_1}{\partial x_3}\right) \\ \left(\dfrac{\partial \xi_2}{\partial x_1}\right) & \left(\dfrac{\partial \xi_2}{\partial x_2}\right) & \left(\dfrac{\partial \xi_2}{\partial x_3}\right) \\ \left(\dfrac{\partial \xi_3}{\partial x_1}\right) & \left(\dfrac{\partial \xi_3}{\partial x_2}\right) & \left(\dfrac{\partial \xi_3}{\partial x_3}\right) \end{vmatrix} = \mathbf{a}_1 \cdot (\mathbf{a}_2 \times \mathbf{a}_3) = V_C \qquad (3.3)$$

Hence any space integral can be written as

$$\frac{1}{V_C}\int d\mathbf{r} = \int dx_1 \int dx_2 \int dx_3 \qquad (3.4)$$

The space lattice is defined by the special positions for which $x_1 = l_1$, $x_2 = l_2$ and $x_3 = l_3$ are integers and hence number the primitive cells. Space lattice vectors

$$\mathbf{R}_l = \sum_\alpha l_\alpha \mathbf{a}_\alpha \qquad (3.5)$$

connect the origin of our frame of reference with these space lattice points.

We say that a property of the crystal is translationally invariant if it is invariant for a translation of the frame of reference along a lattice vector. Obviously, the positions of the nuclei are translationally invariant: if one encounters a nucleus at a position $\mathbf{r}$, one is sure to find the same type of nucleus at the position $\mathbf{r} + \mathbf{R}_l$. As a direct result, the total electron density and the effective potential in the single electron Hamiltonian (2.28) are translationally invariant:

$$\mathcal{V}'(\mathbf{r}_i) = \mathcal{V}'(\mathbf{r}_i + \mathbf{R}_l) \qquad (3.6)$$

One should realize however, that not all properties of the crystal show such a translational symmetry. In particular wave functions of valence electrons extend over many unit cells and change when translating along a lattice vector.

3.1.2 Reciprocal Space and Brillouin Zones

RECIPROCAL SPACE LATTICES

As is well known, when dealing with periodic functions, it is often advantageous to consider its Fourier transform. In the following subsections we discuss how such Fourier transforms can be used to describe translational symmetry in crystals. We start by introducing the reciprocal lattice which is defined in reciprocal space, which has coordinates $\mathbf{k}$ with dimension 'inverse length' contrary to coordinates $\mathbf{r}$ in normal space having the dimension of 'length'. Thus, coordinates $\mathbf{k}$ in reciprocal lattice have the physical meaning of 'wave vector', while coordinates $\mathbf{r}$ in normal space have the physical meaning of 'position'.

Like normal space, also reciprocal space is constructed from primitive cells, now called Brillouin zones, spanned by three basis vectors $\mathbf{b}_1$, $\mathbf{b}_2$ and $\mathbf{b}_3$. Again these

3.1 TRANSLATIONAL SYMMETRY

vectors are not necessarily orthogonal. The volume of a Brillouin zone is then given by

$$V_B = \mathbf{b}_1 \cdot (\mathbf{b}_2 \times \mathbf{b}_3) \quad (3.7)$$

As in normal space an arbitrary position $\mathbf{k}$ within reciprocal space can be decomposed in its components along the $\mathbf{b}_1$, $\mathbf{b}_2$ and $\mathbf{b}_3$ axes

$$\mathbf{k} = \sum_\beta y_\beta \mathbf{b}_\beta \quad (3.8)$$

And again this decomposition corresponds to a transformation from an orthogonal to a non-orthogonal frame of reference. In the first $\mathbf{k}$ has components η_1, η_2 and η_3 along orthogonal axes and the unit of length is 1 m^{-1}. In the second $\mathbf{k}$ has components y_1, y_2 and y_3 along the $\mathbf{b}_1$, $\mathbf{b}_2$ and $\mathbf{b}_3$ vectors and the unit of length along each vector is equal to the length of that vector. Thus in the new frame of reference the unit of volume is equal to the volume V_B of the Brillouin zone. The Jacobian of this transformation is equal to

$$\begin{vmatrix} \left(\frac{\partial \eta_1}{\partial y_1}\right) & \left(\frac{\partial \eta_1}{\partial y_2}\right) & \left(\frac{\partial \eta_1}{\partial y_3}\right) \\ \left(\frac{\partial \eta_2}{\partial y_1}\right) & \left(\frac{\partial \eta_2}{\partial y_2}\right) & \left(\frac{\partial \eta_2}{\partial y_3}\right) \\ \left(\frac{\partial \eta_3}{\partial y_1}\right) & \left(\frac{\partial \eta_3}{\partial y_2}\right) & \left(\frac{\partial \eta_3}{\partial y_3}\right) \end{vmatrix} = \mathbf{b}_1 \cdot (\mathbf{b}_2 \times \mathbf{b}_3) = V_B \quad (3.9)$$

Hence any integral in reciprocal space can be written as

$$\frac{1}{V_B} \int d\mathbf{k} = \int dy_1 \int dy_2 \int dy_3 \quad (3.10)$$

Also in reciprocal space we define a space lattice. This reciprocal space lattice is defined by the special positions for which $y_1 = h_1$, $y_2 = h_2$ and $y_3 = h_3$ are integers numbering the Brillouin zones. The Brillouin zone at the origin where $h_1 = h_2 = h_3 = 0$ is called the first Brillouin zone. Reciprocal space lattice vectors

$$\mathbf{K}_h = \sum_\beta h_\beta \mathbf{b}_\beta \quad (3.11)$$

connect the origin with these reciprocal space lattice points.

THE RELATION BETWEEN PRIMITIVE CELLS AND BRILLOUIN ZONES

We need to relate the reciprocal lattice to the normal lattice. For this purpose we define the unit of 'length' in the reciprocal lattice and the shape of the first Brillouin zone by relating $\mathbf{b}_1$, $\mathbf{b}_2$ and $\mathbf{b}_3$ to $\mathbf{a}_1$, $\mathbf{a}_2$ and $\mathbf{a}_3$:

$$\mathbf{a}_\alpha \cdot \mathbf{b}_\beta = 2\pi \delta_{\alpha\beta} \quad \text{where} \quad \alpha, \beta = 1, 2, 3 \quad (3.12)$$

Using this relation we can also express the volume V_B of a Brillouin zone in the volume V_C of a primitive cell:

$$V_B = \frac{(2\pi)^3}{V_C} \quad (3.13)$$

As an example we construct the first Brillouin zones for the diamond and zinc blende structures. We start with the cubic unit cell shown in Figs 1.2 and 1.3. We choose our frame of reference parallel to the edges of this cell, so it is spanned by the vectors

$$\mathbf{c}_1 = (c, 0, 0)$$
$$\mathbf{c}_2 = (0, c, 0) \quad (3.14)$$
$$\mathbf{c}_3 = (0, 0, c)$$

From this cubic cell we construct the primitive cell shown in Fig. 3.1a. Using the same frame of reference, it is spanned by the vectors

$$\mathbf{a}_1 = \left(\frac{c}{2}, \frac{c}{2}, 0\right)$$
$$\mathbf{a}_2 = \left(\frac{c}{2}, 0, \frac{c}{2}\right) \quad (3.15)$$
$$\mathbf{a}_3 = \left(0, \frac{c}{2}, \frac{c}{2}\right)$$

Subsequently, we construct the first Brillouin zone by requiring eq. (3.12) to hold. Then

$$\mathbf{b}_1 = \pi\left(\frac{2}{c}, \frac{2}{c}, -\frac{2}{c}\right)$$
$$\mathbf{b}_2 = \pi\left(\frac{2}{c}, -\frac{2}{c}, \frac{2}{c}\right) \quad (3.16)$$
$$\mathbf{b}_3 = \pi\left(-\frac{2}{c}, \frac{2}{c}, \frac{2}{c}\right)$$

The resulting Brillouin zone is shown in Fig. 3.3a. In normal space it would have been the primitive cell of a so-called body centred cubic (bcc) lattice.

The Brillouin zone shown in Fig. 3.3a has the shape of a parallelepiped. One may construct a different Brillouin zone by using the recipe to construct a Wigner–Seitz cell. The result is shown in Fig. 3.3b. This representation of the Brillouin zone is commonly encountered in the literature. Some high symmetry points in the zone are usually designated by a special notation given in Table 3.1. These special points prove to be useful to represent functions of the coordinate $\mathbf{k}$ in reciprocal space. An example is shown in Fig. 3.4.

3.1.3 Transforming from Normal to Reciprocal Space

FOURIER TRANSFORMS

Now we have defined reciprocal space, we proceed by relating functions $\Psi(\mathbf{r})$ of position in normal space to functions $\Phi(\mathbf{k})$ of position in reciprocal space. This is

3.1 TRANSLATIONAL SYMMETRY

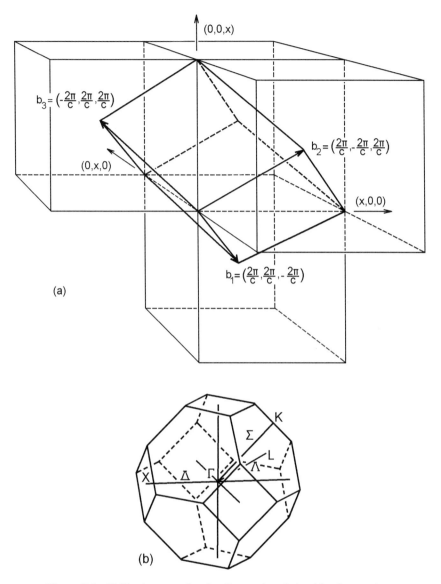

Figure 3.3 Brillouin zones for the diamond and zinc blende structure.

usually done using a Fourier transform pair,

$$\Psi(\mathbf{r}) = \frac{1}{V_B}\int_\infty d\mathbf{k}\, \Phi(\mathbf{k})\, e^{i\mathbf{k}\cdot\mathbf{r}} = \frac{V_C}{(2\pi)^3}\int_\infty d\mathbf{k}\, \Phi(\mathbf{k})\, e^{i\mathbf{k}\cdot\mathbf{r}}$$

$$\Phi(\mathbf{k}) = \frac{1}{V_C}\int_\infty d\mathbf{r}\, \Psi(\mathbf{r})\, e^{-i\mathbf{k}\cdot\mathbf{r}}$$

where we use eq. (3.13) to relate the volume V_B of a Brillouin zone to the volume V_C of a primitive cell. Furthermore $\int_\infty$ denotes integration over full normal or reciprocal space.

Table 3.1 Coordinates of special points in the first Brillouin zone.

Γ	$(0, 0, 0)$
Λ	$\dfrac{\pi}{c}(x, x, x)$
L	$\dfrac{\pi}{c}(1, 1, 1)$
Δ	$\dfrac{2\pi}{c}(x, 0, 0)$
X	$\dfrac{2\pi}{c}(1, 0, 0)$
Σ	$\dfrac{3\pi}{2c}(0, x, x)$
K	$\dfrac{3\pi}{2c}(0, 1, 1)$

However, this Fourier transform pair is not practical for quantum mechanical problems. The reason is that quantum mechanics requires transformations to be unitary, in order for the normalization of wave functions to be conserved under such transformations. Now we intend to use reciprocal space to obtain the eigenstates of the the single electron Hamiltonian (2.26) and the crystal lattice Hamiltonian (2.45). Therefore, we replace the standard Fourier transform pair given above by the transform pair,

$$\Psi(\mathbf{r}) = \frac{1}{\sqrt{V_C V_B}} \int_\infty d\mathbf{k}\, \Phi(\mathbf{k})\, e^{i\mathbf{k}\cdot\mathbf{r}} = \frac{1}{(2\pi)^{3/2}} \int_\infty d\mathbf{k}\, \Phi(\mathbf{k})\, e^{i\mathbf{k}\cdot\mathbf{r}} \qquad (3.17)$$

$$\Phi(\mathbf{k}) = \frac{1}{\sqrt{V_C V_B}} \int_\infty d\mathbf{r}\, \Psi(\mathbf{r})\, e^{-i\mathbf{k}\cdot\mathbf{r}} = \frac{1}{(2\pi)^{3/2}} \int_\infty d\mathbf{r}\, \Psi(\mathbf{r})\, e^{-i\mathbf{k}\cdot\mathbf{r}} \qquad (3.18)$$

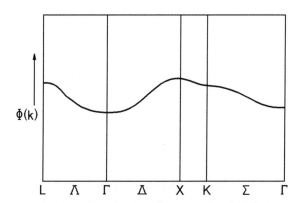

Figure 3.4 Representation of a function in reciprocal space.

3.1 TRANSLATIONAL SYMMETRY

Clearly, this transformation is unitary, because the inverse of the transformation operator is equal to its Hermitian adjoint, i.e. it obeys

$$U^{-1} = \frac{1}{(2\pi)^{3/2}} e^{-i\mathbf{k}\cdot\mathbf{r}} = \left[\frac{1}{(2\pi)^{3/2}} e^{i\mathbf{r}\cdot\mathbf{k}}\right]^* = U^\dagger \qquad (3.19)$$

DIRAC δ-FUNCTIONS

The Fourier transform pair (3.17) and (3.18) gives rise to two Dirac δ-functions. We start by inserting the latter equation in the former to obtain

$$\Psi(\mathbf{r}) = \frac{1}{(2\pi)^{3/2}} \int_\infty d\mathbf{k}\, \Phi(\mathbf{k})\, e^{i\mathbf{k}\cdot\mathbf{r}}$$

$$= \frac{1}{(2\pi)^{3/2}} \int_\infty d\mathbf{k} \left(\frac{1}{(2\pi)^{3/2}} \int_\infty d\mathbf{r}'\, \Psi(\mathbf{r}')\, e^{-i\mathbf{k}\cdot\mathbf{r}'}\right) e^{i\mathbf{k}\cdot\mathbf{r}}$$

$$= \int_\infty d\mathbf{r}'\, \Psi(\mathbf{r}') \left(\frac{1}{(2\pi)^3} \int_\infty d\mathbf{k}\, e^{i\mathbf{k}\cdot(\mathbf{r}-\mathbf{r}')}\right) \qquad (3.20)$$

Thus,

$$\delta(\mathbf{r} - \mathbf{r}') = \frac{1}{(2\pi)^3} \int_\infty d\mathbf{k}\, e^{i\mathbf{k}\cdot(\mathbf{r}-\mathbf{r}')} \qquad (3.21)$$

is a three-dimensional Dirac δ-function. Similarly, inserting eq. (3.17) in eq. (3.18), we obtain another Dirac δ-function,

$$\delta(\mathbf{k} - \mathbf{k}') = \frac{1}{(2\pi)^3} \int_\infty d\mathbf{r}\, e^{-i(\mathbf{k}-\mathbf{k}')\cdot\mathbf{r}} \qquad (3.22)$$

NORMALIZATION OF WAVE FUNCTIONS

Once these Dirac δ-functions are obtained it is easy to prove that the Fourier transform pair (3.17) and (3.18) conserves the normalization of wave functions. We consider a normalized wave function $\Psi(\mathbf{r})$. Then,

$$\int_\infty d\mathbf{r}\, |\Psi(\mathbf{r})|^2 = 1 \qquad (3.23)$$

Now, inserting first eq. (3.17) and next the Dirac δ-function (3.22), we find

$$\int_\infty d\mathbf{r}\, |\Psi(\mathbf{r})|^2 = \int_\infty d\mathbf{r} \left[\frac{1}{(2\pi)^{3/2}} \int_\infty d\mathbf{k}\, \Phi(\mathbf{k})\, e^{i\mathbf{k}\cdot\mathbf{r}}\right] \left[\frac{1}{(2\pi)^{3/2}} \int_\infty d\mathbf{k}'\, \Phi(\mathbf{k}')\, e^{i\mathbf{k}'\cdot\mathbf{r}}\right]^*$$

$$= \int_\infty d\mathbf{r} \left[\frac{1}{(2\pi)^{3/2}} \int_\infty d\mathbf{k}\, \Phi(\mathbf{k})\, e^{i\mathbf{k}\cdot\mathbf{r}}\right] \left[\frac{1}{(2\pi)^{3/2}} \int_\infty d\mathbf{k}'\, \Phi^*(\mathbf{k}')\, e^{-i\mathbf{k}'\cdot\mathbf{r}}\right]$$

$$= \int_\infty d\mathbf{k} \int_\infty d\mathbf{k}'\, \Phi(\mathbf{k})\, \Phi^*(\mathbf{k}') \frac{1}{(2\pi)^3} \int_\infty d\mathbf{r}\, e^{i(\mathbf{k}-\mathbf{k}')\cdot\mathbf{r}}$$

$$= \int_\infty d\mathbf{k} \int_\infty d\mathbf{k}'\, \Phi(\mathbf{k})\, \Phi^*(\mathbf{k}')\, \delta(\mathbf{k}-\mathbf{k}') = \int_\infty d\mathbf{k}\, |\Phi(\mathbf{k})|^2 \qquad (3.24)$$

So,

$$\int_\infty d\mathbf{k} |\Phi(\mathbf{k})|^2 = 1 \tag{3.25}$$

showing that the normalization of the wave function is retained after transformation to reciprocal space.

3.1.4 Functions obeying Translational Symmetry

DISCRETE FUNCTIONS IN RECIPROCAL SPACE

The above derived formalism is especially powerful when handling functions that obey the translational symmetry of the crystal lattice. An important example of such a function is the potential term in the single electron Hamiltonian. Formally, translational symmetry implies that such functions obey

$$\Psi(\mathbf{r}) = \Psi(\mathbf{r} + \mathbf{R}_l) \tag{3.26}$$

Periodic symmetry implies that Fourier transforms to reciprocal space have a special shape. This shape is found by using eq. (3.17) to transform both $\Psi(\mathbf{r})$ and $\Psi(\mathbf{r} + \mathbf{R}_l)$. Then

$$\Psi(\mathbf{r}) = \frac{1}{(2\pi)^{3/2}} \int_\infty d\mathbf{k}\, \Phi(\mathbf{k})\, e^{i\mathbf{k}\cdot\mathbf{r}} \tag{3.27}$$

and

$$\Psi(\mathbf{r} + \mathbf{R}_l) = \frac{1}{(2\pi)^{3/2}} \int_\infty d\mathbf{k}\, \Phi(\mathbf{k})\, e^{i\mathbf{k}\cdot(\mathbf{r}+\mathbf{R}_l)}$$

$$= \frac{1}{(2\pi)^{3/2}} \int_\infty d\mathbf{k}\, \Phi(\mathbf{k})\, e^{i\mathbf{k}\cdot\mathbf{r}} e^{i\mathbf{k}\cdot\mathbf{R}_l} \tag{3.28}$$

Translational symmetry in normal space requires that these results are equal for any space lattice vector $\mathbf{R}_l$. Or that

$$e^{i\mathbf{k}\cdot\mathbf{R}_l} = 1 \tag{3.29}$$

so $\mathbf{k}\cdot\mathbf{R}_l$ should be equal to 2π times an integer. Now,

$$\mathbf{k}\cdot\mathbf{r} = \left(\sum_\beta y_\beta \mathbf{b}_\beta\right) \cdot \left(\sum_\alpha x_\alpha \mathbf{a}_\alpha\right) = \sum_\beta \sum_\alpha y_\beta x_\alpha \mathbf{b}_\beta \cdot \mathbf{a}_\alpha$$

$$= \sum_\beta \sum_\alpha y_\beta x_\alpha 2\pi \delta_{\beta\alpha} = 2\pi \sum_\alpha y_\alpha x_\alpha \tag{3.30}$$

Hence,

$$\mathbf{k}\cdot\mathbf{R}_l = 2\pi \sum_\alpha y_\alpha l_\alpha \tag{3.31}$$

where, as discussed in section 3.1.1, l_α are integers. Clearly, eq. (3.29) can only be fulfilled when all three components y_α are also integers h_α or when $\mathbf{k}$ is a reciprocal space lattice vector $\mathbf{K}_h$. Hence translational symmetry of $\Psi(\mathbf{r})$ implies that its Fourier transform $\Phi(\mathbf{k}) = 0$ unless $\mathbf{k} = \mathbf{K}_h$. These conditions can be mathematically expressed

3.1 TRANSLATIONAL SYMMETRY

using the Dirac δ-functions derived above,

$$\Phi(\mathbf{k}) = (2\pi)^{3/2} \sum_h \phi_h \delta(\mathbf{k} - \mathbf{K}_h) \tag{3.32}$$

where we add a normalization constant $(2\pi)^{3/2}$. As we will see below this choice is not arbitrary, but such that, (as above) the normalization of wave functions is retained upon Fourier transformation.

FOURIER SUMS

A general expression for $\Psi(\mathbf{r})$ is now directly obtained by inserting equation (3.32) for $\Phi(\mathbf{k})$ in the Fourier transform (3.17). We find

$$\Psi(\mathbf{r}) = \frac{1}{(2\pi)^{3/2}} \int_\infty d\mathbf{k}\, \Phi(\mathbf{k})\, e^{i\mathbf{k}\cdot\mathbf{r}}$$

$$= \frac{1}{(2\pi)^{3/2}} \int_\infty d\mathbf{k} \left((2\pi)^{3/2} \sum_h \phi_h \delta(\mathbf{k} - \mathbf{K}_h) \right) e^{i\mathbf{k}\cdot\mathbf{r}}$$

$$= \sum_h \phi_h \int_\infty d\mathbf{k}\, \delta(\mathbf{k} - \mathbf{K}_h)\, e^{i\mathbf{k}\cdot\mathbf{r}} = \sum_h \phi_h e^{i\mathbf{K}_h\cdot\mathbf{r}} \tag{3.33}$$

Thus, $\Psi(\mathbf{r})$ can be written as a Fourier sum and transforms to a sum of δ-functions in reciprocal space, each term positioned at a reciprocal space lattice point. If $\Psi(\mathbf{r})$ represents a wave function, then the Fourier coefficients ϕ_h represent the components of the corresponding state vector in Hilbert space. Thus each reciprocal space lattice point represents a dimension in Hilbert space.

It is now easily shown that the relative intensities ϕ_h of these δ-functions are given by

$$\phi_h = \frac{1}{V_C} \int_{V_C} d\mathbf{r}\, e^{-i\mathbf{K}_h\cdot\mathbf{r}} \Psi(\mathbf{r}) \tag{3.34}$$

where $\int_{V_C}$ denotes integration over a single primitive cell. Note that $\Psi(\mathbf{r})$ shows lattice periodicity, so we may take any arbitrary primitive cell. For convenience we choose the cell at the origin. We check eq. (3.34) by inserting eq. (3.33) in it and subsequently using eq. (3.30) to change to coordinates x_1, x_2 and x_3 and $y_1 = h_1$, $y_2 = h_2$ and $y_3 = h_3$. Then

$$\frac{1}{V_C} \int_{V_C} d\mathbf{r}\, \Psi(\mathbf{r})\, e^{-i\mathbf{K}_h\cdot\mathbf{r}} = \frac{1}{V_C} \int_{V_C} d\mathbf{r} \left(\sum_{h'} \phi_{h'} e^{i\mathbf{K}_{h'}\cdot\mathbf{r}} \right) e^{-i\mathbf{K}_h\cdot\mathbf{r}}$$

$$= \sum_{h'} \phi_{h'} \frac{1}{V_C} \int_{V_C} d\mathbf{r}\, e^{i(\mathbf{K}_{h'} - \mathbf{K}_h)\cdot\mathbf{r}}$$

$$= \sum_{h'} \phi_{h'} \int_0^1 dx_1 \int_0^1 dx_2 \int_0^1 dx_3\, e^{2\pi i \sum_\alpha (h'_\alpha - h_\alpha) x_\alpha}$$

$$= \sum_{h'} \phi_{h'} \prod_\alpha \left[\int_0^1 dx_\alpha e^{2\pi i (h'_\alpha - h_\alpha) x_\alpha} \right]$$

$$= \sum_{h'} \phi_{h'} \prod_\alpha \delta_{h'_\alpha, h_\alpha} = \sum_{h'} \phi_{h'} \delta_{h',h} = \phi_h$$

where we use the Kronecker δ-function

$$\delta_{h',h} \equiv \prod_\alpha \delta_{h'_\alpha,h_\alpha} = \prod_\alpha \int_0^1 dx_\alpha e^{2\pi i(h'_\alpha - h_\alpha)x_\alpha} \tag{3.35}$$

which is proven as follows. If $h_\alpha \neq h'_\alpha$, the integral over x_α yields zero because the exponential function oscillates with a wavelength which is exactly an integral fraction of the integration interval. If $h_\alpha = h'_\alpha$ the exponential function is simply equal to 1 and hence the integral also yields 1.

The Fourier transform pair (3.33) and (3.34) allows the derivation of an alternative expression for the Dirac δ-function which will be useful in future derivations. We insert eq. (3.34) in eq. (3.33) to find

$$\begin{aligned}
\Psi(\mathbf{r}) &= \sum_h \phi_h e^{i\mathbf{K}_h \cdot \mathbf{r}} \\
&= \sum_h \left(\frac{1}{V_C} \int_{V_C} d\mathbf{r}' e^{-i\mathbf{K}_h \cdot \mathbf{r}'} \Psi(\mathbf{r}') \right) e^{i\mathbf{K}_h \cdot \mathbf{r}} \\
&= \frac{1}{V_C} \int_{V_C} d\mathbf{r}' \left(\sum_h e^{i\mathbf{K}_h \cdot \mathbf{r}} e^{-i\mathbf{K}_h \cdot \mathbf{r}'} \right) \Psi(\mathbf{r}') \\
&= \int_{V_C} d\mathbf{r}' \left(\frac{1}{V_C} \sum_h e^{i\mathbf{K}_h \cdot (\mathbf{r} - \mathbf{r}')} \right) \Psi(\mathbf{r}')
\end{aligned} \tag{3.36}$$

proving that

$$\delta(\mathbf{r} - \mathbf{r}') = \frac{1}{V_C} \sum_h e^{i\mathbf{K}_h \cdot (\mathbf{r} - \mathbf{r}')} . \tag{3.37}$$

NORMALIZATION OF THE FOURIER COEFFICIENTS

The Kronecker δ-function (3.35) allows us to determine how normalization of wave functions is conserved by the transformation pair (3.33) and (3.34). First, we note that a wave function $\Psi(\mathbf{r})$ is now a periodic function, so it extends over infinite space. Therefore, we cannot normalize it using eq. (3.23), as that equation would necessarily lead to $\Psi(\mathbf{r}) = 0$ for all $\mathbf{r}$. Here, we will normalize $\Psi(\mathbf{r})$ in a unit volume,

$$\int_{\text{unit volume}} d\mathbf{r} |\Psi(\mathbf{r})|^2 = \frac{1}{V_C} \int_{V_C} d\mathbf{r} |\Psi(\mathbf{r})|^2 = 1 \tag{3.38}$$

Now we expand this normalization convention using eq. (3.33) and the Kronecker δ-function (3.35). Then,

$$\begin{aligned}
\frac{1}{V_C} \int_{V_C} d\mathbf{r} |\Psi(\mathbf{r})|^2 &= \frac{1}{V_C} \int_{V_C} d\mathbf{r} \left(\sum_h \phi_h e^{i\mathbf{K}_h \cdot \mathbf{r}} \right) \left(\sum_{h'} \phi_{h'}^* e^{-i\mathbf{K}_{h'} \cdot \mathbf{r}} \right) \\
&= \sum_h \sum_{h'} \phi_h \phi_{h'}^* \frac{1}{V_C} \int_{V_C} d\mathbf{r} \, e^{i(\mathbf{K}_h - \mathbf{K}_{h'}) \cdot \mathbf{r}} \\
&= \sum_h \sum_{h'} \phi_h \phi_{h'}^* \delta_{h,h'} = \sum_h |\phi_h|^2
\end{aligned} \tag{3.39}$$

3.1 TRANSLATIONAL SYMMETRY

Thus, the choice of the normalization constant $(2\pi)^{3/2}$ in eq. (3.32), combined with the normalization convention (3.38), yields that the quantum mechanical state vector with components ϕ_h is normalized,

$$\sum_h |\phi_h|^2 = 1 \qquad (3.40)$$

3.1.5 Discrete Functions in Finite Crystals

DISCRETE FUNCTIONS IN NORMAL SPACE

Another case showing the strength of the introduction of reciprocal space, concerns functions that are defined on discrete positions in the crystal. An important example are functions of the nuclear displacements. This situation is completely analogous to the one described in the previous section. We start by defining such a discrete function using Dirac δ-functions located at space lattice points $\mathbf{R}_l$. Then

$$\Psi(\mathbf{r}) = (2\pi)^{3/2} \sum_l \psi_l \, \delta(\mathbf{r} - \mathbf{R}_l) \qquad (3.41)$$

where, as in eq. (3.32), a normalization factor $(2\pi)^{3/2}$ is required to obtain a correct normalization of transformed wave functions. The Fourier transform of $\Psi(\mathbf{r})$ to reciprocal space is now easily calculated using eq. (3.18). Equivalently to eq. (3.33) we find

$$\Phi(\mathbf{k}) = \sum_l \psi_l e^{-i\mathbf{k}\cdot\mathbf{R}_l} \qquad (3.42)$$

which displays translational symmetry, but now in reciprocal space, contrary to the case treated in the previous subsection. So,

$$\Phi(\mathbf{k} + \mathbf{K}_h) = \Phi(\mathbf{k}) \qquad (3.43)$$

Next we consider the inverse transformation. Similarly to eq. (3.34) one finds

$$\psi_l = \frac{1}{V_B} \int_{V_B} d\mathbf{k} \, \Phi(\mathbf{k}) \, e^{i\mathbf{k}\cdot\mathbf{R}_l} \qquad (3.44)$$

where $\int_{V_B}$ denotes integration over the first Brillouin zone. Finally, we note that the resulting Fourier transform pair (3.42) and (3.44) gives rise to a Kronecker δ-function

$$\delta_{l',l} = \prod_\alpha \delta_{l'_\alpha, l_\alpha} = \prod_\alpha \int_0^1 dy_\alpha \, e^{-2\pi i y_\alpha (l'_\alpha - l_\alpha)} \qquad (3.45)$$

and a Dirac δ-function

$$\delta(\mathbf{k} - \mathbf{k}') = \frac{1}{V_B} \sum_l e^{-i(\mathbf{k}-\mathbf{k}')\cdot\mathbf{R}_l} \qquad (3.46)$$

Moreover, equivalently to eqs (3.38) and (3.40), normalization is conserved by this Fourier transformation. Hence, if the function $\Phi(\mathbf{k})$ is normalized in a unit volume of reciprocal space,

$$\frac{1}{V_B} \int_{V_B} d\mathbf{k} |\Phi(\mathbf{k})|^2 = 1 \qquad (3.47)$$

then the quantum mechanical state vector with components ψ_l is also normalized,

$$\sum_l |\psi_l|^2 = 1 \tag{3.48}$$

FINITE CRYSTALS AND PERIODIC BOUNDARY CONDITIONS

In infinitely large crystals wave functions may extend over infinitely large space. The normalization conventions (3.38) and (3.39) solve this problem for wave functions obeying translational symmetry, by normalizing them in a unit volume. In the case of discrete functions the normalization conventions (3.47) and (3.48) pose problems, however. It is directly seen from eq. (3.48) that $\psi_l \to 0$, if the wave function is extended to infinite space. Therefore, to keep ψ_l finite we need to restrict the crystal to a finite size. To treat this restriction we consider the hypothetical crystal shown in Fig. 3.5. It consists of L_1 primitive cells along the $\mathbf{a}_1$ direction and L_2 and L_3 along the other two directions. Thus it contains

$$N = \prod_\alpha L_\alpha \tag{3.49}$$

primitive cells. We choose the origin of our frame of reference at a vertex of the crystal indicated by O in Fig. 3.5. Positions and space lattice points are again defined by eqs (3.2) and (3.5). For positions and space lattice points to lie within the crystal, we clearly require

$$0 \leq x_\alpha, \quad l_\alpha < L_\alpha \tag{3.50}$$

In many cases the exact boundary conditions at the surfaces of the crystal are not critical and we may choose them to our liking. Mathematically most convenient are so-called periodic boundary conditions. These imply that one fills all space with crystals as shown in Fig. 3.5, all identical and neatly packed together. Then any property of the crystal at position $\mathbf{r}$ reproduces upon a translation over a whole number of crystals, i.e. a distance $\sum_\alpha m_\alpha L_\alpha \mathbf{a}_\alpha$. In particular a quantum mechanical wave function $\Psi(\mathbf{r})$ obeys

$$\Psi(\mathbf{r}) = \Psi\left(\mathbf{r} + \sum_\alpha m_\alpha L_\alpha \mathbf{a}_\alpha\right) \tag{3.51}$$

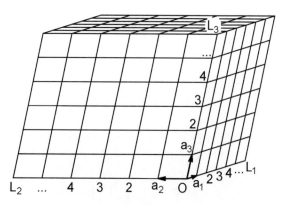

Figure 3.5 A crystal containing $N = L_1 L_2 L_3$ primitive cells.

3.1 TRANSLATIONAL SYMMETRY

Thus, the 'primitive cell' applying for arbitrary wave functions consists of the complete crystal. So it is N times larger than the one applying for functions obeying translational symmetry as treated in section 3.1.4. Following section 3.1.2, these huge 'primitive cells' give rise to extremely small 'Brillouin zones' that are N times smaller than the real Brillouin zones. So normal lattice vectors $\mathbf{R}_l = (l_1\mathbf{a}_1, l_2\mathbf{a}_2, l_3\mathbf{a}_3)$ are replaced by giant lattice vectors $(m_1 L_1\mathbf{a}_1, m_2 L_2\mathbf{a}_2, m_3 L_3\mathbf{a}_3)$, while normal reciprocal lattice vectors $\mathbf{K}_h$ are substituted by dwarf reciprocal lattice vectors

$$\mathbf{k}_q = (q_1/L_1)\mathbf{b}_1 + (q_2/L_2)\mathbf{b}_2 + (q_3/L_3)\mathbf{b}_3 \tag{3.52}$$

where the quantities q_α are integers numbering them. Clearly a Brillouin zone contains exactly N dwarf reciprocal lattice vectors.

UNITARY TRANSFORMATION MATRICES

This observation allows us to repeat the treatment of section 3.1.4. Then, equivalently to eq. (3.32), the transform $\Phi(\mathbf{k})$ of a discrete quantum mechanical wave function to reciprocal space can be written as a set of Dirac δ-functions located at discrete points,

$$\Phi(\mathbf{k}) = \frac{V_B}{\sqrt{N}} \sum_q^N \phi_q \delta(\mathbf{k} - \mathbf{k}_q) \tag{3.53}$$

Because the wave function is discrete in normal space, its transform $\Phi(\mathbf{k})$ is periodic in reciprocal space and it is sufficient to define this transform in the first Brillouin zone only. Therefore the sum over q is restricted to the N discrete points in this zone. Furthermore, the factor $V_B/\sqrt{N}$ scales the amplitudes of the discrete components ϕ_q. As will become apparent shortly, its present particular form has been chosen in order to render the transformation to reciprocal space unitary.

Using eq. (3.53), eq. (3.44) reduces to

$$\begin{aligned}
\psi_l &= \frac{1}{V_B} \int_{V_B} d\mathbf{k} \left[\frac{V_B}{\sqrt{N}} \sum_q^N \phi_q \delta(\mathbf{k} - \mathbf{k}_q) \right] e^{i\mathbf{k}\cdot\mathbf{R}_l} \\
&= \frac{1}{\sqrt{N}} \sum_q^N \phi_q \int_{V_B} d\mathbf{k}\, \delta(\mathbf{k} - \mathbf{k}_q)\, e^{i\mathbf{k}\cdot\mathbf{R}_l} \\
&= \frac{1}{\sqrt{N}} \sum_q^N \phi_q\, e^{i\mathbf{k}_q\cdot\mathbf{R}_l} = \sum_q^N U_{lq} \phi_q
\end{aligned} \tag{3.54}$$

so it corresponds to transformation by a $N \times N$ matrix with elements $U_{lq} = (1/\sqrt{N}) \exp(i\mathbf{k}_q \cdot \mathbf{R}_l)$. It is easily proven that this matrix is unitary. We calculate

$$\begin{aligned}
\sum_l^N U_{lq}^* U_{lq'} &= \frac{1}{N} \sum_l^N e^{-i\mathbf{k}_q \cdot \mathbf{R}_l} e^{i\mathbf{k}_{q'} \cdot \mathbf{R}_l} = \frac{1}{N} \sum_l^N e^{i(\mathbf{k}_{q'} - \mathbf{k}_q)\cdot \mathbf{R}_l} \\
&= \frac{1}{N} \sum_l^N e^{2\pi i \sum_\alpha (q'_\alpha - q_\alpha) l_\alpha / L_\alpha} = \frac{1}{N} \prod_\alpha \sum_{l_\alpha}^{L_\alpha} e^{2\pi i (q'_\alpha - q_\alpha) l_\alpha / L_\alpha} \\
&= \frac{1}{N} \prod_\alpha \frac{1 - e^{2\pi i (q'_\alpha - q_\alpha) L_\alpha / L_\alpha}}{1 - e^{2\pi i (q'_\alpha - q_\alpha)/L_\alpha}} = \frac{1}{N} \prod_\alpha \frac{1 - e^{2\pi i (q'_\alpha - q_\alpha)}}{1 - e^{2\pi i (q'_\alpha - q_\alpha)/L_\alpha}}
\end{aligned} \tag{3.55}$$

Now q_α and q'_α are integers, so the exponent in the numerator is equal to one and the result vanishes. There is one exception however: when $q_\alpha = q'_\alpha$, also the denominator vanishes and a more careful evaluation should be made. Then we find

$$\lim_{q_\alpha \to q'_\alpha} \frac{1 - e^{2\pi i(q'_\alpha - q_\alpha)}}{1 - e^{2\pi i(q'_\alpha - q_\alpha)/L_\alpha}} = \lim_{q_\alpha \to q'_\alpha} \frac{1 - 1 - 2\pi i(q'_\alpha - q_\alpha)}{1 - 1 - 2\pi i(q'_\alpha - q_\alpha)/L_\alpha}$$

$$= \lim_{q_\alpha \to q'_\alpha} \frac{(q'_\alpha - q_\alpha)}{(q'_\alpha - q_\alpha)/L_\alpha} = L_\alpha \quad (3.56)$$

Hence,

$$\sum_l^N U_{lq}^* U_{lq'} = \frac{1}{N} \prod_\alpha L_\alpha \delta_{q_\alpha, q'_\alpha} = \prod_\alpha \delta_{q_\alpha, q'_\alpha} \equiv \delta_{q,q'} \quad (3.57)$$

proving that the $N \times N$ matrix with elements U_{ql} is unitary. Clearly, this is due to the specific choice of the factor $V_B/\sqrt{N}$ in eq. (3.53). As a result the inverse transformation may now be written as

$$\phi_q = \sum_l^N U_{lq}^* \psi_l = \frac{1}{\sqrt{N}} \sum_l^N \psi_l \, e^{-i\mathbf{k}_q \cdot \mathbf{R}_l} \quad (3.58)$$

while the transformation also conserves the normalization of wave vectors, as is seen by calculating

$$\sum_l^N |\psi_l|^2 = \sum_l^N \left(\sum_q^N U_{lq} \phi_q \right) \left(\sum_{q'}^N U_{lq'}^* \phi_{q'}^* \right)$$

$$= \sum_q^N \sum_{q'}^N \phi_q \phi_{q'}^* \sum_l^N U_{lq} U_{lq'}^* = \sum_q^N \sum_{q'}^N \phi_q \phi_{q'}^* \delta_{q,q'} = \sum_q^N |\phi_q|^2 \quad (3.59)$$

3.2 Bloch Functions and Band Structure

3.2.1 The Single Electron Hamiltonian in Reciprocal Space

The introduction of the reciprocal lattice allows us to derive some general properties of the eigenfunctions of the single electron Hamiltonian (2.26),

$$\mathcal{H}'(\mathbf{r}) = -\frac{\hbar^2}{2m_e} \Delta + \mathcal{V}'(\mathbf{r}) \quad (3.60)$$

where for simplicity of notation we omit the index i identifying a particular electron. In the present section we derive such a general property: the eigenfunctions of $\mathcal{H}'(\mathbf{r})$ are so-called Bloch functions. We start with the Schrödinger equation

$$\mathcal{H}'(\mathbf{r})\Psi(\mathbf{r}) = \left[-\frac{\hbar^2}{2m_e} \Delta + \mathcal{V}'(\mathbf{r}) \right] \Psi(\mathbf{r}) = E \, \Psi(\mathbf{r}) \quad (3.61)$$

3.2 BLOCH FUNCTIONS AND BAND STRUCTURE

Here, E is an energy level of the single electron Hamiltonian, $\Psi(\mathbf{r})$ the corresponding eigenstate, while we do not indicate explicitly the quantum numbers associated with this state. Next we transform this equation to reciprocal space. Then

$$\frac{1}{(2\pi)^{3/2}} \int_\infty d\mathbf{r}\, e^{-i\mathbf{k}\cdot\mathbf{r}} \left[-\frac{\hbar^2}{2m_e}\Delta + \mathcal{V}'(\mathbf{r})\right] \Psi(\mathbf{r}) = \frac{1}{(2\pi)^{3/2}} \int_\infty d\mathbf{r}\, e^{-i\mathbf{k}\cdot\mathbf{r}} E\Psi(\mathbf{r}) \quad (3.62)$$

We rewrite this result following eqs (3.17) and (3.18) and introducing the transform of $\Psi(\mathbf{r})$ to reciprocal space,

$$\Phi(\mathbf{k}) = \frac{1}{(2\pi)^{3/2}} \int_\infty d\mathbf{r}\, e^{-i\mathbf{k}\cdot\mathbf{r}} \Psi(\mathbf{r}) \quad (3.63)$$

while

$$\Psi(\mathbf{r}) = \frac{1}{(2\pi)^{3/2}} \int_\infty d\mathbf{k}\, e^{i\mathbf{k}\cdot\mathbf{r}} \Phi(\mathbf{k}) \quad (3.64)$$

Then the right-hand side of the transformed Schrödinger equation (3.61) reduces to

$$E\,\Phi(\mathbf{k}) \quad (3.65)$$

Transforming the left hand side is more complicated. To illustrate the procedure we start with the extremely simple case that the potential term in $\mathcal{H}'(\mathbf{r})$ is equal to zero, i.e. the case of a 'free' electron in the crystal.

3.2.2 Free Electrons

THE KINETIC ENERGY IN RECIPROCAL SPACE

For free electrons the left hand side of eq. (3.62) contains one term only, representing the kinetic energy. It is rewritten using eq. (3.64). We find

$$\frac{1}{(2\pi)^{3/2}} \int_\infty d\mathbf{r}\, e^{-i\mathbf{k}\cdot\mathbf{r}} \left[-\frac{\hbar^2}{2m_e}\Delta \Psi(\mathbf{r})\right]$$

$$= \frac{1}{(2\pi)^{3/2}} \int_\infty d\mathbf{r}\, e^{-i\mathbf{k}\cdot\mathbf{r}} \left[-\frac{\hbar^2}{2m_e}\Delta \frac{1}{(2\pi)^{3/2}} \int_\infty d\mathbf{k}'\, e^{i\mathbf{k}'\cdot\mathbf{r}} \Phi(\mathbf{k}')\right]$$

$$= \frac{\hbar^2}{2m_e} \int_\infty d\mathbf{k}'\, \Phi(\mathbf{k}') \frac{1}{(2\pi)^3} \int_\infty d\mathbf{r}\, e^{-i\mathbf{k}\cdot\mathbf{r}}(\Delta e^{i\mathbf{k}'\cdot\mathbf{r}})$$

$$= -\frac{\hbar^2}{2m_e} \int_\infty d\mathbf{k}'\, \Phi(\mathbf{k}') \frac{1}{(2\pi)^3} \int_\infty d\mathbf{r}\, e^{-i\mathbf{k}\cdot\mathbf{r}}(i k')^2 e^{i\mathbf{k}'\cdot\mathbf{r}}$$

$$= \frac{\hbar^2}{2m_e} \int_\infty d\mathbf{k}'\, (k')^2 \Phi(\mathbf{k}') \frac{1}{(2\pi)^3} \int_\infty d\mathbf{r}\, e^{i(\mathbf{k}'-\mathbf{k})\cdot\mathbf{r}}$$

$$= \frac{\hbar^2}{2m_e} \int_\infty d\mathbf{k}'\, (k')^2 \Phi(\mathbf{k}')\delta(\mathbf{k}'-\mathbf{k})$$

$$= \frac{\hbar^2 k^2}{2m_e} \Phi(\mathbf{k}) \quad (3.66)$$

Here we used the Dirac δ-function (3.22) and introduced the notation $k^2 = \mathbf{k} \cdot \mathbf{k}$ and $(k')^2 = \mathbf{k}' \cdot \mathbf{k}'$. As a result the transformed Schrödinger equation (3.62) is reduced to a set of completely independent linear equations

$$\frac{\hbar^2 k^2}{2m_e} \Phi(\mathbf{k}) = E \, \Phi(\mathbf{k}) \tag{3.67}$$

We note that $\mathbf{k}$ is generally not in the first Brillouin zone. Let $\mathbf{h} = (h_1, h_2, h_3)$ denote the Brillouin zone where $\mathbf{k}$ is positioned. Then $\mathbf{k}$ can be split into a reciprocal lattice vector $\mathbf{K}_h = \sum_\alpha h_\alpha b_\alpha$ and a part $\mathbf{k}_0$ denoting its equivalent position in the first Brillouin zone,

$$\mathbf{k} = \mathbf{K}_h + \mathbf{k}_0 = \sum_\alpha h_\alpha b_\alpha + \mathbf{k}_0 \tag{3.68}$$

Then eq. (3.67) becomes

$$\frac{\hbar^2 (\mathbf{k}_0 + \mathbf{K}_h)^2}{2m_e} \Phi(\mathbf{k}_0 + \mathbf{K}_h) = E \, \Phi(\mathbf{k}_0 + \mathbf{K}_h) \tag{3.69}$$

Again these are all independent equations, one for each value of $\mathbf{k}_0$ and $\mathbf{h}$. Thus $\mathbf{k}_0$ *and* $\mathbf{h}$ *are good* quantum numbers for the eigenstates of the free electron.

BAND STRUCTURE OF FREE ELECTRONS

Eq. (3.69) is easily solved. For the energy we find

$$E = E_{\mathbf{k}_0}^h = \frac{\hbar^2 |\mathbf{k}_0 + \mathbf{K}_h|^2}{2m_e} \tag{3.70}$$

Because $\mathbf{k}_0$ and $\mathbf{h}$ are good quantum numbers, the corresponding state obeys

$$\Phi(\mathbf{k}_0 + \mathbf{K}_h) = (2\pi)^{3/2} \delta[\mathbf{k} - (\mathbf{k}_0 + \mathbf{K}_h)]$$

where we add a normalization factor $(2\pi)^{3/2}$ as in eq. (3.32). Then, using eq. (3.64), the wave function is given by

$$\Psi(\mathbf{r}) = \Psi_{\mathbf{k}_0}^h(\mathbf{r}) = e^{i(\mathbf{k}_0 + \mathbf{K}_h) \cdot \mathbf{r}} \tag{3.71}$$

Note that our normalization convention operates in such a way that the electron density

$$\rho_{\mathbf{k}_0}^h(\mathbf{r}) = |\Psi_{\mathbf{k}_0}^h(\mathbf{r})|^2 = 1 \tag{3.72}$$

corresponding to one electron per unit volume for each electron state. Note furthermore that the classical value of the energy of a free electron is equal to $p^2/2m_e$. Hence $\hbar(\mathbf{k}_0 + \mathbf{K}_h)$ can be interpreted as the momentum of the electron. In other words, reciprocal space corresponds to momentum space for free electrons. It is customary to denote the part $\hbar \mathbf{k}_0$ as the crystal momentum of the electron.

Fig. 3.6 shows the energy $E_{\mathbf{k}_0}^h$ as a function of the value of $\mathbf{k}_0$ in the first Brillouin zone for the diamond or zinc blende structure. The Brillouin zone has the shape of the Wigner–Seitz cell of Fig. 3.3 and the energy is represented for the special points Γ, Λ, etc., in that cell using the method of Fig. 3.4. It is interesting to note how the curves fold back at the borders of the Brillouin zone. Fig. 3.6 represents a so-called band

3.2 BLOCH FUNCTIONS AND BAND STRUCTURE

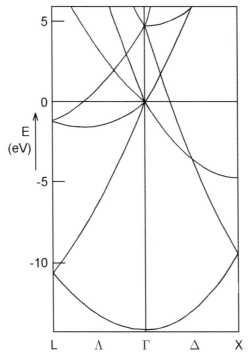

Figure 3.6 The energy levels of a free electron in a diamond or zinc blende crystal.

structure, here of the hypothetical case of electrons in a diamond or zinc blende structure where all potentials vanish. This specific band structure is called an empty lattice structure.

One easily determines the horizontal and vertical scales in Fig. 3.6. For GaAs the length of the cubic unit cell $c = 0.5654$ nm. Then, according to Table 3.1, $|\mathbf{k}| = 2\pi/c = 2\pi \cdot 1.769 \cdot 10^9$ m^{-1} in the X-point at the boundary of the first Brillouin zone. Using eq. (3.70) and

$$e = 1.602 \cdot 10^{-19} \text{C}$$

$$m_e = 9.110 \cdot 10^{-31} \text{ kg}$$

$$\hbar = 1.055 \cdot 10^{-34} \text{J s}$$

the lowest energy of an electron at the X-point is found to be

$$E_{k_0}^h/e = 4.711 \text{ eV}$$

3.2.3 Electrons in a Periodic Potential

THE POTENTIAL IN RECIPROCAL SPACE

Next we consider the full Schrödinger equation (3.61) including the potential term. We now also need to transform this latter term to reciprocal space. The potential in the single electron Hamiltonian shows translational symmetry in normal space because the nuclear positions and hence the electron density show this symmetry. Such functions obeying the translational symmetry of the crystal lattice were dealt with in section 3.1.4 and we will use the results of that section here. According to eqs (3.33) and (3.34) the potential in the single electron Hamiltonian may be written as a Fourier sum

$$\mathcal{V}'(\mathbf{r}) = \sum_h \mathcal{W}_h \exp[i\mathbf{K}_h \cdot \mathbf{r}] \tag{3.73}$$

where the relative intensities $\mathcal{W}_h$ of the Fourier components are given by

$$\mathcal{W}_h = \frac{1}{V_C} \int_{V_C} d\mathbf{r}\, \mathcal{V}'(\mathbf{r})\, e^{-i\mathbf{K}_h \cdot \mathbf{r}} \tag{3.74}$$

Eq. (3.74) together with the results of the previous sections allow us to obtain the transform of the Schrödinger equation to reciprocal space. We start with eq. (3.62) and insert expression (3.65) for the right-hand side. Then

$$\frac{1}{(2\pi)^{3/2}} \int_\infty d\mathbf{r}\, e^{-i\mathbf{k}\cdot\mathbf{r}} \left[-\frac{\hbar^2}{2m_e} \Delta + \mathcal{V}'(\mathbf{r}) \right] \Psi(\mathbf{r}) = E\, \Phi(\mathbf{k}) \tag{3.75}$$

The first term on the left-hand side represents the kinetic energy. It remains the same upon switching on the potential, so we may still insert eq. (3.66) to rewrite it. Hence,

$$\frac{\hbar^2 k^2}{2m_e} \Phi(\mathbf{k}) + \frac{1}{(2\pi)^{3/2}} \int_\infty d\mathbf{r}\, e^{-i\mathbf{k}\cdot\mathbf{r}} \mathcal{V}'(\mathbf{r}) \Psi(\mathbf{r}) = E\, \Phi(\mathbf{k}) \tag{3.76}$$

The remaining integral on the left hand side of eq. (3.76) contains the potential energy. We rewrite it using the transforms to reciprocal space of $\mathcal{V}'(\mathbf{r})$ and $\Psi(\mathbf{r})$ as given by eqs (3.73) and (3.63). We find

$$\frac{1}{(2\pi)^{3/2}} \int_\infty d\mathbf{r}\, e^{-i\mathbf{k}\cdot\mathbf{r}} \mathcal{V}'(\mathbf{r}) \Psi(\mathbf{r}) = \frac{1}{(2\pi)^{3/2}} \int_\infty d\mathbf{r}\, e^{-i\mathbf{k}\cdot\mathbf{r}} \sum_h \mathcal{W}_h e^{i\mathbf{K}_h \cdot \mathbf{r}} \Psi(\mathbf{r})$$

$$= \sum_h \mathcal{W}_h \frac{1}{(2\pi)^{3/2}} \int_\infty d\mathbf{r}\, e^{-i\mathbf{k}\cdot\mathbf{r}} e^{i\mathbf{K}_h \cdot \mathbf{r}} \Psi(\mathbf{r})$$

$$= \sum_h \mathcal{W}_h \frac{1}{(2\pi)^{3/2}} \int_\infty d\mathbf{r}\, e^{-i(\mathbf{k}-\mathbf{K}_h)\cdot\mathbf{r}} \Psi(\mathbf{r})$$

$$= \sum_h \mathcal{W}_h\, \Phi(\mathbf{k} - \mathbf{K}_h) \tag{3.77}$$

As a result the Schrödinger equation transforms to a set of coupled linear equations

$$\frac{\hbar^2 k^2}{2m_e} \Phi(\mathbf{k}) + \sum_h \mathcal{W}_h\, \Phi(\mathbf{k} - \mathbf{K}_h) = E\, \Phi(\mathbf{k}) \tag{3.78}$$

BLOCH FUNCTIONS

We obtain more insight in these equations if we split $\mathbf{k}$ into a reciprocal lattice vector $\mathbf{K}_{h'} = \sum_\alpha h'_\alpha \mathbf{b}_\alpha$ and a part $\mathbf{k}_0$ denoting its equivalent position in the first Brillouin zone,

$$\mathbf{k} = \mathbf{K}_{h'} + \mathbf{k}_0 = \sum_\alpha h'_\alpha \mathbf{b}_\alpha + \mathbf{k}_0 \qquad (3.79)$$

Then eq. (3.78) becomes

$$\frac{\hbar^2(\mathbf{k}_0 + \mathbf{K}_{h'})^2}{2m_e} \Phi(\mathbf{k}_0 + \mathbf{K}_{h'}) + \sum_h \mathcal{W}_h \, \Phi(\mathbf{k}_0 + \mathbf{K}_{h'} - \mathbf{K}_h) = E \, \Phi(\mathbf{k}_0 + \mathbf{K}_{h'}) \qquad (3.80)$$

Next, we write

$$\mathbf{K}_{h''} = \mathbf{K}_{h'} - \mathbf{K}_h$$

and renumber the summation on the left hand side. This poses no problems as the summations extend from $-\infty$ to $+\infty$. As a result the transformed Schrödinger equation is rewritten as,

$$\sum_{h''} \left[\frac{\hbar^2(\mathbf{k}_0 + \mathbf{K}_{h''})^2}{2m_e} \delta_{h',h''} + \mathcal{W}_{h'-h''} \right] \Phi(\mathbf{k}_0 + \mathbf{K}_{h''}) = E \, \Phi(\mathbf{k}_0 + \mathbf{K}_{h'}) \qquad (3.81)$$

where $\delta_{h',h''}$ is a Kronecker δ-function.

Upon inspection of eq. (3.81) it is seen that the Schrödinger equation is transformed to a set of independent matrix equations, one for each value of $\mathbf{k}_0$. This implies that $\mathbf{k}_0$ is a *good* quantum number. Each of these matrix equations constitutes an independent eigenvalue problem where the eigenvalues and eigenvectors of the matrix with elements

$$\frac{\hbar^2(\mathbf{k}_0 + \mathbf{K}_{h''})^2}{2m_e} \delta_{h',h''} + \mathcal{W}_{h'-h''} \qquad (3.82)$$

are calculated. The non-diagonal matrix elements $\mathcal{W}_{h'-h''}$ mix states with different values of $\mathbf{K}_h$, so this quantity is *not* a good quantum number contrary to the case of free electrons treated in section 3.2.2. Upon solving such a matrix equation, we obtain a set of eigenvectors with components

$$\phi_{\mathbf{k}_0}^{nh'} = \Phi_n(\mathbf{k}_0 + \mathbf{K}_{h'}) \qquad (3.83)$$

and corresponding eigenvalues $E_{\mathbf{k}_0}^n$. Here the subscript $\mathbf{k}_0$ indicates the matrix equation which is solved, while the so-called band index n numbers the eigenvectors resulting from this matrix equation and finally h' identifies the components of these eigenvectors. The eigenvalues $E_{\mathbf{k}_0}^n$ are of course the energy levels of the single electron Hamiltonian. The eigenvectors represent a discrete function in reciprocal space which can be written in a similar way as eq. (3.32),

$$\Phi_n(\mathbf{k}) = (2\pi)^{3/2} \sum_{h'} \phi_{\mathbf{k}_0}^{nh'} \delta(\mathbf{k} - \mathbf{k}_0 - \mathbf{K}_{h'}) \qquad (3.84)$$

These are the eigenfunctions of the single electron Hamiltonian in reciprocal space. The eigenfunctions in normal space are found by transforming them using eq. (3.17). One finds

$$\frac{1}{(2\pi)^{3/2}} \int_\infty d\mathbf{k}\, e^{i\mathbf{k}\cdot\mathbf{r}} \Phi_n(\mathbf{k}) = \frac{1}{(2\pi)^{3/2}} \int_\infty d\mathbf{k}\, e^{i\mathbf{k}\cdot\mathbf{r}} (2\pi)^{3/2} \sum_{h'} \phi_{k_0}^{nh'} \delta(\mathbf{k} - \mathbf{k}_0 - \mathbf{K}_{h'})$$

$$= \sum_{h'} \phi_{k_0}^{nh'} e^{i(\mathbf{k}_0+\mathbf{K}_{h'})\cdot\mathbf{r}} = \left(\sum_{h'} \phi_{k_0}^{nh'} e^{i\mathbf{K}_{h'}\cdot\mathbf{r}} \right) e^{i\mathbf{k}_0\cdot\mathbf{r}}$$

$$= u_{k_0}^n(\mathbf{r}) e^{i\mathbf{k}_0\cdot\mathbf{r}} = b_{k_0}^n(\mathbf{r}) \qquad (3.85)$$

The resulting functions $b_{k_0}^n(\mathbf{r})$ are called Bloch functions. They are characterized by the property that the part

$$u_{k_0}^n(\mathbf{r}) = \sum_{h'} \phi_{k_0}^{nh'} e^{i\mathbf{K}_{h'}\cdot\mathbf{r}} \qquad (3.86)$$

obeys translational symmetry,

$$u_{k_0}^n(\mathbf{r}) = u_k^n(\mathbf{r} + \mathbf{R}_l) = u_{k_0}^n\left(\mathbf{r} + \sum_\alpha l_\alpha \mathbf{a}_\alpha\right) \qquad (3.87)$$

because it has the shape given by eq. (3.33). Consequently the electron density

$$\rho_{k_0}^n(\mathbf{r}) = |b_{k_0}^n(\mathbf{r})|^2 = |u_{k_0}^n(\mathbf{r})|^2 \, e^{i\mathbf{k}_0\cdot\mathbf{r}} \, e^{-i\mathbf{k}_0\cdot\mathbf{r}} = |u_{k_0}^n(\mathbf{r})|^2 \qquad (3.88)$$

is always translationally symmetric. This is fortunate, as we used translational symmetry of the electron density to argue that the single electron potential should be translationally symmetric.

Note that, according to eq. (3.85), Bloch functions are linear combinations of free electron wave functions, where the constants $\phi_{k_0}^{nh'}$ are the components of the eigenvectors of a matrix equation (3.81). By normalizing these eigenvectors according to eq. (3.40), the normalization convention for Bloch functions is rendered the same as the convention for free electron wave functions. Then, as in section 3.2.2 we normalize in such a way that $\rho_{k_0}^n(\mathbf{r})$ corresponds to one electron per Bloch state per unit volume.

FINITE CRYSTALS

Until now we have considered infinitely large crystals only. For some properties of Bloch functions we need to consider the finite crystal size, however. We treat finite size in the way described in section 3.1.5 and introduce periodic boundary conditions. Then a Bloch function $b_{k_0}^n(\mathbf{r})$ obeys

$$b_{k_0}^n(\mathbf{r}) = b_{k_0}^n\left(\mathbf{r} + \sum_\alpha m_\alpha L_\alpha \mathbf{a}_\alpha\right) \qquad (3.89)$$

where L_1, L_2 and L_3 are the number of primitive cells in the crystal along the $\mathbf{a}_1$, $\mathbf{a}_2$ and $\mathbf{a}_3$ axes, respectively. As a result reciprocal lattice vectors take discrete values

$$\mathbf{k}_q = (q_1/L_1)\mathbf{b}_1 + (q_2/L_2)\mathbf{b}_2 + (q_3/L_3)\mathbf{b}_3 \qquad (3.90)$$

only, where the quantities q_α are integers.

3.2 BLOCH FUNCTIONS AND BAND STRUCTURE

Then, equivalently to eq. (3.53), the transform of a Bloch function to reciprocal space can be written as a set of Dirac δ-functions located at discrete points,

$$\Phi_n(\mathbf{k}) = \frac{V_B}{\sqrt{N}} \sum_q \phi_q \delta(\mathbf{k} - \mathbf{k}_q) \qquad (3.91)$$

where, $N = \prod_\alpha L_\alpha$ is the number of primitive cells in the crystal. However, according to eq. (3.84)

$$\Phi_n(\mathbf{k}) = (2\pi)^{3/2} \sum_h \phi_{k_0}^{nh} \delta(\mathbf{k} - \mathbf{k}_0 - \mathbf{K}_h) \qquad (3.92)$$

where $\mathbf{k}_0$ lies in the first Brillouin zone. Clearly, only a finite number of $\mathbf{k}_0$ values and corresponding Bloch functions is allowed, namely those for which $\mathbf{k}_0 = \mathbf{k}_q$, where $\mathbf{k}_q$ lies in the first Brillouin zone. Consequently the number of allowed Bloch functions is equal to the number N of primitive cells in the crystal. Hence, *the number of eigenstates $b_{k_0}^n(\mathbf{r})$ for a given band index n is equal to the number of primitive cells in the crystal*. This result is important in several ways. First, it implies that *the density of Bloch states is constant in reciprocal space*. Once the energy is solved as a function of $\mathbf{k}_0$ this allows a direct calculation of the density of Bloch states as a function of energy. Second, it allows us to calculate which states are full and which are empty, once we know the number of electrons. Finally, it shows us that the number of eigenstates becomes very small when the crystal is diminished to an extremely small size. This is important for mesoscopic structures, e.g. quantum wells that are essentially crystals that are only a few primitive cells thick in one direction, or quantum wires that are equally small in two dimensions, or quantum dots that are equally small in all three directions.

At this point we should note that finite crystals may require an adaptation of the normalization convention. In a finite crystal we expect a Bloch state to extend over the total volume NV_C of the crystal. Hence, the electron density corresponding to this Bloch state should yield one electron per Bloch state in a volume NV_C. Now the Bloch functions (3.86) yield an electron density of one electron per Bloch function per unit volume. Therefore we need to add a normalization factor $1/\sqrt{NV_C}$ in order achieve a correct normalization in a finite crystal.

3.2.4 Band Structure of Typical Semiconductors

BAND STRUCTURE CALCULATIONS

The solution of the transformed Schrödinger equation (3.81) is feasible for a real semiconductor crystal because the problem is separated into a large number of independent smaller problems for each value of $\mathbf{k}_0$. Still such so-called band structure calculations require extensive digital computations before the energy $E_{k_0}^n$ of an electron and its eigenstate $b_{k_0}^n(\mathbf{r})$ is obtained.

Ideally, band structure calculations are performed from first principles. Then using a self-consistent method as described in section 2.2.2, one tries to obtain the periodic potential and the eigenstates of the electrons in the crystal. Depending on the type of crystal, different approaches are used in such *ab initio* band structure calculations. If the periodic potential is relatively weak compared to the kinetic energy of the electrons, it is advantageous to start the calculations using free electron states as basis states. Then the periodic potential is switched on as a perturbation. This method, the nearly free electron model (NFEM) cannot be used if the periodic potential is very strong so the electrons are strongly bound to the atomic nuclei. Then, one needs to use the tight binding model (TBM). Here one starts the calculations employing linear combinations of atomic orbitals (LCAO). In both cases the valence electrons may rarely get close to the atomic nuclei. Hence the precise shape of the periodic potential near these nuclei is unimportant. In the pseudopotential method the periodic potential is smoothed at the nuclear positions, facilitating the calculations.

In semiconductors often a more pragmatic method is used, where the periodic potential is fitted to obtain a band structure in accordance with experimental results. For a detailed discussion of this latter method the reader is referred to ref. [10] which also gives extensive results for a large number of semiconductors. Here we only show some results from that reference. Figs 3.7a–3.10a show the calculated band structures

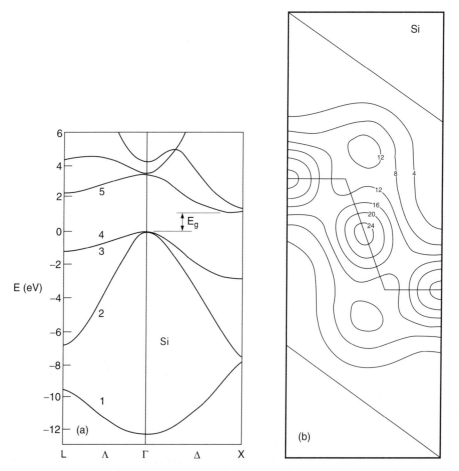

Figure 3.7 (a) The band structure of Si (adapted from Fig. 8.2, p. 81, ref. [10], © Springer Verlag). (b) The valence electron density in Si (Fig. 8.8b, p. 89, ref. [10], © Springer Verlag).

3.2 BLOCH FUNCTIONS AND BAND STRUCTURE

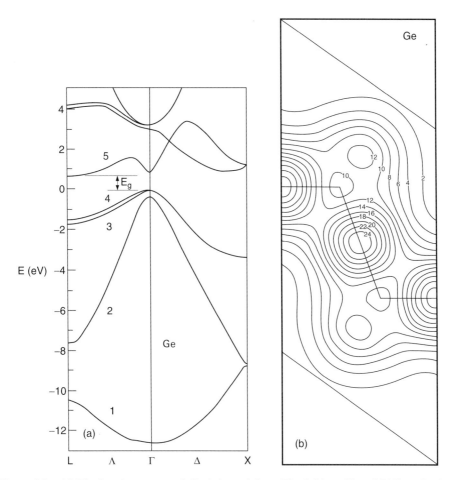

Figure 3.8 (a) The band structure of Ge (adapted from Fig. 8.11, p. 92, ref. [10], © Springer Verlag). (b) The valence electron density in Ge (Fig. 8.16, p. 97, ref. [10], © Springer Verlag).

of Si, Ge, GaAs and ZnSe, respectively, as a function of $\mathbf{k}_0$ using the method of Fig. 3.4. Furthermore, the index n is chosen to increase with increasing energy. For each value of n the energy $E_{\mathbf{k}_0}^n$ displays the eye striking feature that it is a continuous function of $\mathbf{k}_0$. Each of these curves is called an energy band.

It is interesting to see how the results of ref. [10] evolve from the empty lattice band structure, i.e. the band structure in the absence of a periodic potential. Fig. 3.11a repeats the empty lattice structure. In Fig. 3.11b, eight components of the periodic potential $\mathcal{W}_h$ have been switched on. They correspond to

$$\mathbf{K}_h = \pm\mathbf{b}_1 = \pm(1, 1, -1)(2\pi/c)$$
$$\mathbf{K}_h = \pm\mathbf{b}_2 = \pm(1, -1, 1)(2\pi/c)$$
$$\mathbf{K}_h = \pm\mathbf{b}_3 = \pm(-1, 1, 1)(2\pi/c)$$
$$\mathbf{K}_h = \pm(1, 1, 1)(2\pi/c)$$

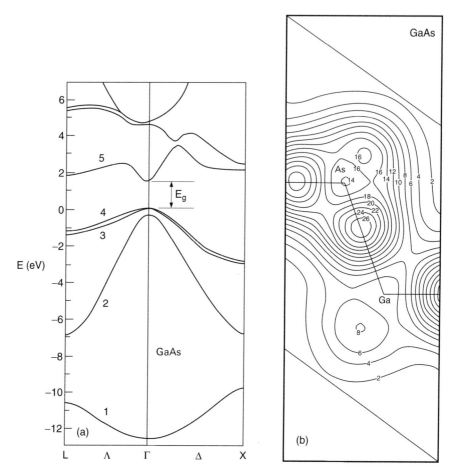

Figure 3.9 (a) The band structure of GaAs (adapted from Fig. 8.21, p. 103, ref. [10], © Springer Verlag). (b) The valence electron density in GaAs (Fig. 8.29, p. 112, ref. [10], © Springer Verlag).

and are fully equivalent under the cubic and tetrahedral symmetry of the diamond and zinc blende structures. Therefore they have all been given the same value of 2.8 eV. Though all other components of $\mathcal{W}_h$ are still zero, a strong resemblance with the more precise band structures in Figs 3.7a–3.10a can already be observed. An even more striking resemblance is obtained in Fig. 3.11c, where we switched on a second set of 12 equivalent components of the periodic potential $\mathcal{W}_h$. They correspond to $\mathbf{K}_h = (0, 2, 2)(2\pi/c)$ and all reciprocal lattice vectors that are equivalent to it under the cubic and tetrahedral symmetry of the diamond and zinc blende structures. In Fig. 3.11c these components of the periodic potential have been given a value of 0.35 eV. Thus, it is not surprising that the results of ref. [10] could be obtained with a very small number of independent components of the periodic potential $\mathcal{W}_h$.

3.2 BLOCH FUNCTIONS AND BAND STRUCTURE

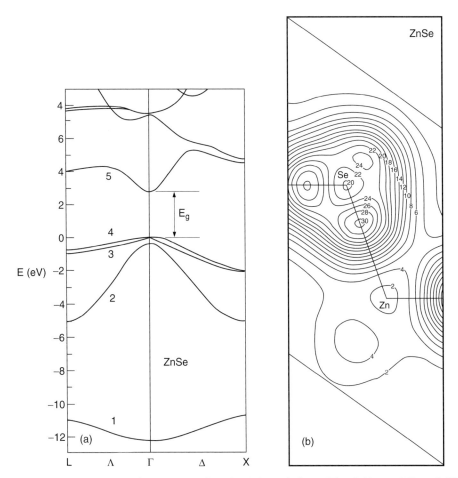

Figure 3.10 (a) The band structure of ZnSe (adapted from Fig. 8.30, p. 113, ref. [10], © Springer Verlag). (b) The valence electron density in ZnSe (Fig. 8.33, p. 116, ref. [10], © Springer Verlag).

VALENCE AND CONDUCTION BANDS

It is important to know which bands are full and which are empty in the ground state. First, we note that each primitive cell contains two atoms and hence eight valence electrons, irrespective of whether it concerns elemental semiconductors like Si or composite semiconductors like GaAs. Next, we note that each Bloch state may contain two electrons because they have spin $\frac{1}{2}$. Hence, we need to fill four Bloch states per unit cell in order to obtain the ground state. If a crystal consists of N primitive cells, we hence need to fill $4N$ Bloch states. However, we found in section 3.2.3 that the number of independent k_0 values is equal to the number of primitive cells. Hence, one full band contains N Bloch states and filling four bands corresponds to filling $4N$ Bloch states. Thus, for the ground state we find that the four lowest bands are full, while the others are empty. Upon inspection of Figs 3.7a–3.10a, we see immediately

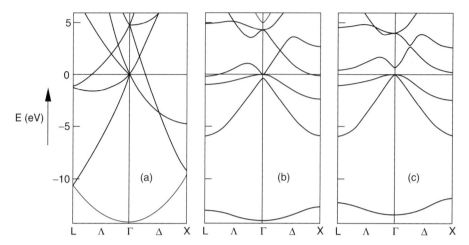

Figure 3.11 (a) Empty lattice structure. (b) Band structure after switching on one set of equivalent components of the periodic potential. (c) Band structure after switching on a second set of equivalent components of the periodic potential.

that all states corresponding to the energy bands denoted by $n = 1, 2, 3$ and 4 are full and all higher bands are empty.

Once we have determined which bands are full and which bands are empty, we may invoke eq. (3.88) to determine the electron density in the crystal. Figs 3.7b–3.10b give electron densities obtained in this way in ref. [10]. When regarding these results it is interesting to note that for Si and Ge the electron density corresponds strikingly well with the bonding states shown in Fig. 2.3b. For GaAs the resemblance is still visible, but the electron density is shifted towards the As core which has a higher charge than the Ga core. In ZnSe the electron density is shifted so far towards the highly charged Se core, that one could interpret this core together with the surrounding valence electrons as a Se^{-2} ion!

Of paramount importance for semiconductors is the feature which is visible in Figs 3.7a and 3.10a between the energy bands denoted by $n = 4$ and 5. There we see an energy interval E_g where no energy levels are found. This is the so-called energy gap. In the ground state all bands below this gap ($n = 1, 2, 3,$ and 4) are full, while all higher bands ($n \geq 5$) are empty. Thus the former are the valence bands, while the latter are the conduction bands. The smallest electronic excitation of a semiconductor involves the excitation of an electron from the top of the valence band to the bottom of the conduction band. For this purpose the energy of that electron has to be increased by E_g. The result of this excitation is an electron at the bottom of the conduction band and a hole at the top of the valence band. Clearly low energy electrons and holes reside near the band extremes. Therefore these band extremes are particularly important for the calculation of the behaviour of these low energy charge carriers. In chapter 4 we will discuss the $\mathbf{k}\cdot\mathbf{p}$ method which is specifically designed to treat the band structure at such band extremes.

In this respect we note an essential difference between the band structures of Si and Ge and those of GaAs and ZnSe. While in the latter two cases both the top of the

3.2 BLOCH FUNCTIONS AND BAND STRUCTURE

valence band and the bottom of the conduction band occur at the Γ-point, in the former case the top of the valence band and the bottom of the conduction band occur at different positions in the Brillouin zone. Thus, in GaAs and ZnSe an electron may be promoted from the top of the valence band to the bottom of the conduction band while conserving its value of k_0 while in Ge and Si the same transition involves a change of k_0. Because of this property, GaAs and ZnSe are called direct semiconductors and their band gap is called a direct band gap, while Ge and Si are called indirect semiconductors having an indirect band gap.

3.2.5 Electrons and Holes

As discussed above the lowest excitation in a pure semiconductor consists of promoting an electron from the valence band to the conduction band. This results in an electron near the bottom of the conduction band and a hole near the top of the valence band. However, a large energy gap E_g between the valence and conduction bands must be bridged to create such an excite state. Therefore, in the description of excited states the concentration of these electrons and holes can always be assumed to be extremely small. This allows us to employ an extreme simplification where we assume a single electron in the conduction band or a single hole in the valence band, while the crystal is otherwise in the ground state. As discussed in section 2.1.3, donor and acceptor impurities also provide electrons and holes. However, in practical cases the concentration of donor and acceptor impurities is very low, so then it is also acceptable to use a description of a single electron or a single hole in a crystal which is otherwise in the ground state.

Electrons in the conduction band and holes in the valence band are responsible for some of the most crucial properties of semiconductors. As an example we consider the conductivity. In chapter 8 we will see that the sum of the velocities of all electrons in a full band is equal to zero, so there is no net motion of the electrons in such a full band. Hence, a full band cannot yield conductivity and neither can an empty band. Thus, in its ground state, when its valence band is completely filled and its conduction band is empty, a pure semiconductor behaves as an insulator. However, when electrons are excited to the conduction band, both the valence band and the conduction band become partly filled and electrical transport may occur. The resulting conduction can be described as the motion of the electrons in the conduction band and of the holes in the valence band that were created by the excitation. Similarly, because donor or acceptor impurities provide electrons in the conduction band or holes in the valence band, these impurities are also responsible for electrical conductivity.

To describe electrical transport we need to consider the various interactions of the electrons and holes. Note that it is not only the applied electric and magnetic fields that influence the motion of the electrons in the valence band and the conduction band. It is also influenced by the electron–lattice interaction treated in section 2.3.3. For example, when electrons gain too much kinetic energy they may emit a phonon, i.e. they may excite a phonon mode into a higher state. As a result electron–phonon interaction tends to slow down the motion of electrons and holes. Donor and acceptor impurities complicate the picture further. For although they provide extra charge carriers, the remaining cores of these impurities have a different charge than the surrounding cores of the host crystal. The cores of donor impurities represent a net charge $+e$ while acceptor impurities represent a net charge $-e$. As a result, electrons moving through the crystal experience Rutherford scattering at these charged impurities, resulting in extra slowing down.

ELECTRON WAVE FUNCTIONS

As a first step in treating such effects we need to consider the proper wave functions describing electrons and holes. First, we consider a single electron in one of the Bloch states $b_{k_0}^n(\mathbf{r})$ in the nth conduction band. In principle electrons are indistinguishable, which does not allow us to determine which one of all electrons in the crystal occupies this state $b_{k_0}^n(\mathbf{r})$. However, as there is only one electron in the conduction band and it occupies a Bloch state with a large extension in space, we may assume that it barely influences the wave functions of the remaining electrons that are all in the valence band. Therefore we venture to identify the electron in the conduction band with the index 0, and to approximate the total state of the crystal by

$$\Psi_{k_0}^n = \Psi \, b_{k_0}^n(\mathbf{r}_0) \tag{3.93}$$

where Ψ represents the state of all electrons in the valence band. We note that this approximation corresponds to replacing the Hartree–Fock treatment described in section 2.2.2 by the simpler Hartree treatment.

Clearly, the result of any process involving this electron, can be described as a transition to another Bloch state $b_{k'}^m(\mathbf{r}_0)$ in the mth conduction band. Here, the important point is that in many cases the responsible interaction has no influence on a full band. As we will eventually see in section 8.1.2, an electric field can cause no electron motion in a full band. Hence, the state Ψ of the electrons in the valence band remains unchanged. Thus, the only relevant transitions are those from the state $\Psi_{k_0}^n$ to a state

$$\Psi_{k'_0}^m = \Psi \, b_{k'_0}^m(\mathbf{r}_0) \tag{3.94}$$

To describe such a process quantum mechanically, we need to calculate the probability of this transition. Suppose that an interaction $\mathcal{V}(\mathbf{r}_0, \mathbf{r}_1, \ldots, \mathbf{r}_i, \ldots)$ causes the process. Then, according to first-order perturbation theory, the transition probability is determined by the matrix element of $\mathcal{V}(\mathbf{r}_0, \mathbf{r}_1, \ldots, \mathbf{r}_i, \ldots)$ between the two states,

$$V_{nk,mk'} = \int d\mathbf{r}_0 \int d\mathbf{r}_1, \ldots \int d\mathbf{r}_i, \ldots [\Psi^* b_{k_0}^{n*}(\mathbf{r}_0)] \mathcal{V}(\mathbf{r}_0, \mathbf{r}_1, \ldots, \mathbf{r}_i, \ldots)[\Psi b_{k'_0}^m(\mathbf{r}_0)] \tag{3.95}$$

Now because the wave function Ψ describing the valence band remains the same, the integrals over the positions $\mathbf{r}_1, \ldots, \mathbf{r}_i, \ldots$ will be independent of the transitions we are studying. As a result, we can describe the transition of the electron in the conduction band with a matrix element

$$V_{nk,mk'} = \int d\mathbf{r}_0 \, b_{k_0}^{n*}(\mathbf{r}_0) \overline{\mathcal{V}}(\mathbf{r}_0) b_{k'_0}^m(\mathbf{r}_0) \tag{3.96}$$

where

$$\overline{\mathcal{V}}(\mathbf{r}_0) = \int d\mathbf{r}_1, \ldots \int d\mathbf{r}_i, \ldots \Psi^* \mathcal{V}(\mathbf{r}_0, \mathbf{r}_1, \ldots, \mathbf{r}_i, \ldots) \Psi \tag{3.97}$$

3.2 BLOCH FUNCTIONS AND BAND STRUCTURE

is independent of the choice of the states $b_{k_0}^n(\mathbf{r}_0)$ and $b_{k'_0}^m(\mathbf{r}_0)$. Thus, only the Bloch states of the conduction band electron itself are needed to calculate the transition probability and hence to determine the process involving this electron. Of course, we do need to consider the precise shape of $\overline{\mathcal{V}}(\mathbf{r}_0)$ for each of the interactions involved. In chapter 5 we will consider interactions with externally applied electromagnetic fields, while chapter 7 will be devoted to electron–phonon interactions.

HOLE WAVE FUNCTIONS

Next, we consider a single hole in the valence band. Before we venture a description of the wave function of the valence band containing this hole, we first ask ourselves another question: 'How do we know that the hole exists?' The only way to find out, is to do an experiment and, as we know from quantum mechanics, any experiment yielding information on the hole is bound to change its state. So the only relevant questions concern transitions of the hole between states and not the states themselves. Let us now consider the case that some interaction causes a transition of the hole from the Bloch state $b_{k_0}^n(\mathbf{r})$ in the nth valence band to another Bloch state $b_{k'_0}^m(\mathbf{r})$ in the mth valence band. In reality, of course something else happens: an electron is transferred from the state $b_{k'_0}^m(\mathbf{r}_0)$ in the mth valence band to the state $b_{k_0}^n(\mathbf{r})$ in the nth valence band. This is shown schematically in Fig. 3.12. Again we venture to identify this electron with the index 0, and to approximate the total initial electronic state of the crystal by

$$\Psi_{k'_0}^m = \Psi\, b_{k'_0}^m(\mathbf{r}_0) \tag{3.98}$$

where Ψ now represents the total state of all electrons in the various valence bands excluding the states $b_{k_0}^n(\mathbf{r})$ and $b_{k'_0}^m(\mathbf{r})$. As above we invoke the argument that in many cases the responsible interaction has no influence on a full band. Now, the argument is more subtle however. In the present case the band is not full, so the interaction does change its state. However, the only possible change of state is exactly the transition described above. Therefore, in the total final electronic state the wave function Ψ is again the same and this final state may be approximated by

$$\Psi_{k_0}^n = \Psi\, b_{k_0}^n(\mathbf{r}_0) \tag{3.99}$$

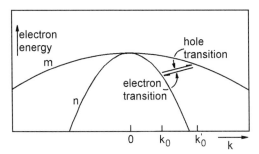

Figure 3.12 Electron *versus* hole transition.

Therefore, if the interaction $V(\mathbf{r}_0, \mathbf{r}_1, \ldots, \mathbf{r}_i, \ldots)$ is responsible for the transition, the matrix element determining the probability of this transition is given by

$$V_{mk',nk} = \int d\mathbf{r}_0 \int d\mathbf{r}_1, \ldots \int d\mathbf{r}_i, \ldots [\Psi^* b_{k'_0}^{m*}(\mathbf{r}_0)] V(\mathbf{r}_0, \mathbf{r}_1, \ldots, \mathbf{r}_i, \ldots)[\Psi b_{k_0}^n(\mathbf{r}_0)]$$

$$= \int d\mathbf{r}_0 \, b_{k'_0}^{m*}(\mathbf{r}_0) \overline{V}(\mathbf{r}_0) b_{k_0}^n(\mathbf{r}_0) \tag{3.100}$$

where again

$$\overline{V}(\mathbf{r}_0) = \int d\mathbf{r}_1, \ldots \int d\mathbf{r}_i, \ldots, \Psi^* V(\mathbf{r}_0, \mathbf{r}_1, \ldots, \mathbf{r}_i, \ldots) \Psi \tag{3.101}$$

This is the complex conjugate of the matrix element which we obtain for the transition of an isolated hole from the Bloch state $b_{k_0}^n(\mathbf{r}_0)$ to the Bloch state $b_{k'_0}^m(\mathbf{r}_0)$. In this latter case,

$$V_{nk,mk'} = \int d\mathbf{r}_0 \, b_{k_0}^{n*}(\mathbf{r}_0) \overline{V}(\mathbf{r}_0) b_{k'_0}^m(\mathbf{r}_0) = V_{mk',nk}^* \tag{3.102}$$

Now according to first order perturbation theory, the probability of the transition is proportional to $|V_{nk,mk'}|^2$, which is the same for the hole and the electron. Therefore, the picture of holes in Bloch states $b_{k_0}^n(\mathbf{r}_0)$ can be used profitably.

3.2.6 Superlattices

Using an epitaxial growth technique, like molecular beam epitaxy (MBE) and chemical vapour deposition (CVD), one grows semiconductor crystals by adding one atomic monolayer after another. Thus, one may not only grow single crystals of semiconductors, but also more complicated structures where different materials alternate on an atomic scale. An example of such a superlattice is shown in Fig. 3.13. It consists of alternating layers of GaAs and AlAs, each exactly two primitive cells thick. Fig. 3.13a shows the projection of the atomic structure on the $(1, -1, 0)$ plane. As is seen from this figure layers 1 and 2 consist of GaAs, while in layers 3 and 4 the gallium atoms are replaced by aluminium atoms. The corresponding primitive cells are also indicated as cells 1, 2, 3 and 4. Here cell 4 is partly behind cell 3, while the latter is partly behind cell 2, etc. Because the latter two cells contain different atoms from the former two, this artificial structure must be described by a new and *four times larger* primitive cell which consists of all these four primitive cells. This supercell is indicated by a thick line.

To understand the band structure of this superlattice we consider the description of this layered structure in reciprocal space. According to eq. 3.13 the size of the Brillouin zone is now *four times smaller* than that of the composing materials, i.e. the original Brillouin zone is now split into four smaller ones. To see what happens with the energy levels, we start with the oversimplified case that the periodic potential is the same in both materials. Then, the energy levels are the same as for the individual materials and the new smaller Brillouin zone merely represents a change in the representation of these energy levels. The procedure is the same as followed in section 3.2.2 for a free electron in a crystal lattice. There the energy levels extending into infinite reciprocal space were folded back into the first Brillouin zone to obtain Fig. 3.6. Now

3.2 BLOCH FUNCTIONS AND BAND STRUCTURE

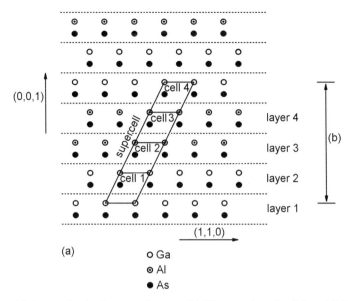

Figure 3.13 (a) A superlattice in normal space. (b) The wavelength of the additional periodic potential due to the difference of the composing materials.

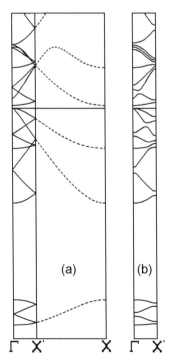

Figure 3.14 Band structure of a superlattice (a) ignoring the difference and (b) including the difference of the composing materials.

the energy levels in the old larger Brillouin zone corresponding to the composing materials are folded back into the new smaller Brillouin zone corresponding to the superlattice. This folding process is shown in Fig. 3.14a. Here the dotted lines represent the levels in the original larger Brillouin zone while the drawn lines represent the band structure in the new smaller Brillouin zone.

In a real superlattice the two composing materials are of course not identical and one expects the periodic potential to have components with a wavelength shown in Fig. 3.13b. These extra components change the band structure of Fig. 3.14a, so it obtains the general shape given in Fig. 3.14b. Note that the change from Fig. 3.14a to Fig. 3.14b is very similar to the change from Fig. 3.11a to Fig 3.11b, i.e. the effect of the potential is to reduce the number of crossings of energy levels. As a result the original bands are split into a number of new bands separated by newly arising gaps. These new bands are generally called minibands.

For Fig. 3.14 we chose a hypothetical indirect semiconductor with the conduction band minimum in the X-point. For the growth direction we chose the $(1,0,0)$-direction. These choices were made in order to show another feature of superlattices. We see from Fig. 3.14 that the conduction band minimum is folded back to the Γ-point. Thus, while the individual components of the superlattice are indirect semiconductors, the resulting superlattice may represent a direct semiconductor.

Semiconductor multilayers have given rise to a vast literature in physics as well as in electrical engineering. For a further study the reader is referred to e.g., the books of Bastard [2], Ridley [33] and Weisbuch and Vinter [40].

4 The k·p-Approximation

4.1 General Formalism

In a pure semiconductor the valence band is full while the conduction band is empty. Hence the lowest excitations of a semiconductor correspond to promoting electrons from the top of the valence band to the bottom of the conduction band. Thus two types of charge carriers are created, electrons in the conduction band and holes, i.e. empty electron states, in the valence band. Many phenomena in semiconductors involve low energy charge carriers only. Then the holes remain close to the top of the valence band and the electrons close to the bottom of the conduction band. Clearly, for these phenomena it is sufficient to know the band structure at these band extrema. In all semiconductors the maximum of the valence band is found at the Γ-point, which is the centre of the first Brillouin zone when represented by the Wigner–Seitz method. Thus, holes in the top of this band are in Bloch states $b_k^n(\mathbf{r})$ where $\mathbf{k}$ is close to this Γ-point. As discussed in section 3.2.4 the minimum of the conduction band of direct semiconductors is also located at this special point in reciprocal space. Hence, for these semiconductors the electrons at the bottom of this band are also described by Bloch states $b_k^n(\mathbf{r})$ where $\mathbf{k}$ is close to this Γ-point. Thus, for the description of low energy charge carriers in direct semiconductors we only need to know the band structure near the Γ point.

The present chapter is devoted to an approximate method to perform band structure calculations near the Γ-point. This method, the **k·p**-approximation, is a perturbation treatment where the eigenvalues and eigenfunctions of the single electron Hamiltonian are expanded as a function of **k** [19]. Note that for Bloch functions the position **k** is always restricted to the first Brillouin zone, so henceforward we skip any index 0 denoting that these vectors are located in this zone.

4.1.1 The k·p-Hamiltonian

We consider the time independent Schrödinger equation (3.61) yielding the band energies E_k^n and Bloch functions $b_k^n(\mathbf{r})$

$$\mathcal{H}'(\mathbf{r})b_k^n(\mathbf{r}) = E_k^n b_k^n(\mathbf{r}) \qquad (4.1)$$

In this equation

$$\mathcal{H}'(\mathbf{r}) = -\frac{\hbar^2}{2m_e}\Delta + \mathcal{V}'(\mathbf{r}) = \frac{1}{2m_e}\left(\frac{\hbar}{i}\nabla\right)\cdot\left(\frac{\hbar}{i}\nabla\right) + \mathcal{V}'(\mathbf{r}) \qquad (4.2)$$

is the single electron Hamiltonian (3.60), where the electrostatic potential $\mathcal{V}'$ obeys the translational symmetry of the crystal lattice. Furthermore

$$b_k^n(\mathbf{r}) = e^{i\mathbf{k}\cdot\mathbf{r}}u_k^n(\mathbf{r}) \qquad (4.3)$$

is a Bloch function, characterized by $u_k^n(\mathbf{r})$ obeying the same translational symmetry. Clearly, the proposed expansion as a function of $\mathbf{k}$ necessarily restricts these $\mathbf{k}$ values to the first Brillouin zone. Therefore, we have simplified our notation and we skipped the index 0 denoting that $\mathbf{k}$ is in this first Brillouin zone.

In the $\mathbf{k}\cdot\mathbf{p}$ method a new Hamiltonian is constructed which has the same eigenvalues E_k^n as $\mathcal{H}'(\mathbf{r})$ but eigenfunctions $u_k^n(\mathbf{r})$ instead of $b_k^n(\mathbf{r})$. To derive this so-called $\mathbf{k}\cdot\mathbf{p}$-Hamiltonian, we start by calculating

$$\begin{aligned}\left(\frac{\hbar}{i}\nabla\right)e^{i\mathbf{k}\cdot\mathbf{r}}u_k^n(\mathbf{r}) &= e^{i\mathbf{k}\cdot\mathbf{r}}\left(\frac{\hbar}{i}\nabla u_k^n(\mathbf{r})\right) + \left(\frac{\hbar}{i}\nabla e^{i\mathbf{k}\cdot\mathbf{r}}\right)u_k^n(\mathbf{r}) \\ &= e^{i\mathbf{k}\cdot\mathbf{r}}\left(\frac{\hbar}{i}\nabla u_k^n(\mathbf{r})\right) + e^{i\mathbf{k}\cdot\mathbf{r}}(\hbar\mathbf{k})u_k^n(\mathbf{r}) \\ &= e^{i\mathbf{k}\cdot\mathbf{r}}\left(\frac{\hbar}{i}\nabla + \hbar\mathbf{k}\right)u_k^n(\mathbf{r})\end{aligned} \qquad (4.4)$$

Then

$$\begin{aligned}\left(\frac{\hbar}{i}\nabla\right)\cdot\left(\frac{\hbar}{i}\nabla\right)e^{i\mathbf{k}\cdot\mathbf{r}}u_k^n(\mathbf{r}) &= e^{i\mathbf{k}\cdot\mathbf{r}}\left(\frac{\hbar}{i}\nabla + \hbar\mathbf{k}\right)\cdot\left(\frac{\hbar}{i}\nabla + \hbar\mathbf{k}\right)u_k^n(\mathbf{r}) \\ &= e^{i\mathbf{k}\cdot\mathbf{r}}\left[\left(\frac{\hbar}{i}\nabla\right)\cdot\left(\frac{\hbar}{i}\nabla\right) + 2\hbar\mathbf{k}\cdot\left(\frac{\hbar}{i}\nabla\right) + \hbar^2\mathbf{k}\cdot\mathbf{k}\right]u_k^n(\mathbf{r}) \\ &= e^{i\mathbf{k}\cdot\mathbf{r}}\left[-\hbar^2\Delta + 2\frac{\hbar^2}{i}(\mathbf{k}\cdot\nabla) + \hbar^2|\mathbf{k}|^2\right]u_k^n(\mathbf{r})\end{aligned} \qquad (4.5)$$

Insertion of this expression in the Schrödinger equation (4.1) yields

$$e^{i\mathbf{k}\cdot\mathbf{r}}\left[-\frac{\hbar^2}{2m_e}\Delta + \frac{\hbar^2}{im_e}(\mathbf{k}\cdot\nabla) + \frac{\hbar^2}{2m_e}|\mathbf{k}|^2 + \mathcal{V}'(\mathbf{r})\right]u_k^n(\mathbf{r}) = e^{i\mathbf{k}\cdot\mathbf{r}}E_k^n u_k^n(\mathbf{r})$$

Now the exponential function can be divided out, and we obtain a new Schrödinger equation,

$$[\mathcal{H}^{(0)}(\mathbf{r}) + \mathcal{H}^{(1)}(\mathbf{r})]u_k^n(\mathbf{r}) = E_k^n u_k^n(\mathbf{r}) \qquad (4.6)$$

Here $\mathcal{H}^{(0)}(\mathbf{r}) + \mathcal{H}^{(1)}(\mathbf{r})$ is the desired $\mathbf{k}\cdot\mathbf{p}$ Hamiltonian, where

$$\mathcal{H}^{(0)}(\mathbf{r}) = \mathcal{H}'(\mathbf{r}) = -\frac{\hbar^2}{2m_e}\Delta + \mathcal{V}'(\mathbf{r}) \qquad (4.7)$$

4.1 GENERAL FORMALISM

is the single electron Hamiltonian (4.2) and

$$\mathcal{H}^{(1)}(\mathbf{r}) = \frac{\hbar^2}{im_e}(\mathbf{k}\cdot\nabla) + \frac{\hbar^2}{2m_e}|\mathbf{k}|^2 \qquad (4.8)$$

is an additional term originating in the procedure presented in eq. (4.5).

Like the original Schrödinger equation (4.1), this new equation (4.6) allows the calculation of the eigenfunctions and the energies of the electrons. The advantage of eq. (4.6) above eq. (4.1) is twofold. First, in the new equation both the $\mathbf{k}\cdot\mathbf{p}$-Hamiltonian $\mathcal{H}^{(0)}(\mathbf{r}) + \mathcal{H}^{(1)}(\mathbf{r})$ and its eigenfunctions $u_k^n(\mathbf{r})$ obey the periodicity of the crystal lattice. Hence eq. (4.6) needs to be solved for a single primitive cell only, while eq. (4.1) must be solved for the whole crystal. For semiconductors like Si and GaAs this implies that the number of electrons involved in the problem reduces from $8N$ to only 8, i.e. by many orders of magnitude! Secondly, $\mathcal{H}^{(1)}(\mathbf{r})$ reduces to zero for $\mathbf{k} \to 0$, so for states near the Γ-point, i.e. the centre of the first Brillouin zone, $\mathcal{H}^{(1)}(\mathbf{r})$ is small. Then, perturbation methods may be used to determine its influence. In that case, the unperturbed problem

$$\mathcal{H}^{(0)} u_k^n(\mathbf{r}) = E_k^n u_k^n(\mathbf{r}) \qquad (4.9)$$

corresponds to solving the Schrödinger equation for the special situation that

$$\mathbf{k} = 0 \quad \text{so} \quad u_k^n(\mathbf{r}) = u_0^n(\mathbf{r}) \quad \text{and} \quad E_k^n = E_0^n \qquad (4.10)$$

or to solving

$$\mathcal{H}^{(0)} u_0^n(\mathbf{r}) = E_0^n u_0^n(\mathbf{r}) \qquad (4.11)$$

while the perturbation $\mathcal{H}^{(1)}$ yields the corrections on E_0^n and $u_0^n(\mathbf{r})$ when $\mathbf{k} \neq 0$.

4.1.2 Basis States and Matrix Elements

PROPERTIES OF THE UNPERTURBED EIGENSTATES

Before embarking on using eq. (4.6) to solve the eigenstates and energy levels of the valence electrons in semiconductors, we consider some properties of its eigenfunctions $u_k^n(\mathbf{r})$ in some more detail.

First, the unperturbed states $u_0^n(\mathbf{r})$ can always be chosen to be real functions of $\mathbf{r}$. To prove this, we start using basic quantum mechanics to show the following lemma: If $u_0^n(\mathbf{r})$ is a complex eigenstate of the real and Hermitian operator $\mathcal{H}^{(0)}$, then its complex conjugate $u_0^{n*}(\mathbf{r})$ is also an eigenstate of $\mathcal{H}^{(0)}$ and moreover $u_0^n(\mathbf{r})$ and $u_0^{n*}(\mathbf{r})$ are degenerate. To prove this lemma we take the complex conjugate of eq. (4.11):

$$[\mathcal{H}^{(0)} u_0^n(\mathbf{r})]^* = E_0^{n*} u_0^{n*}(\mathbf{r}) \qquad (4.12)$$

Now $\mathcal{H}^{(0)}$ is a real linear operator, so

$$[\mathcal{H}^{(0)} u_0^n(\mathbf{r})]^* = \mathcal{H}^{(0)} \Re\{u_0^n(\mathbf{r})\} - i\mathcal{H}^{(0)} \Im\{u_0^n(\mathbf{r})\}$$
$$= \mathcal{H}^{(0)}[\Re\{u_0^n(\mathbf{r})\} - i\Im\{u_0^n(\mathbf{r})\}] = \mathcal{H}^{(0)} u_0^{n*}(\mathbf{r}) \qquad (4.13)$$

Furthermore, $\mathcal{H}^{(0)}$ is a Hermitian operator, so its eigenvalues E_0^n are real,

$$E_0^{n*} = E_0^n \tag{4.14}$$

Inserting these two results, we find

$$\mathcal{H}^{(0)} u_0^{n*}(\mathbf{r}) = E_0^n u_0^{n*}(\mathbf{r}) \tag{4.15}$$

which, upon comparison with eq. (4.11), proves our lemma. We now continue by considering the two degenerate eigenstates $u_0^n(\mathbf{r})$ and $u_0^{n*}(\mathbf{r})$. In general these eigenstates may be complex. However, because of degeneracy also two orthogonal and *real* linear combinations

$$\frac{1}{\sqrt{2}}[u_0^n(\mathbf{r}) + u_0^{n*}(\mathbf{r})] \quad \text{and} \tag{4.16}$$

$$\frac{1}{i\sqrt{2}}[u_0^n(\mathbf{r}) - u_0^{n*}(\mathbf{r})] \tag{4.17}$$

represent correct eigenstates. Thus, the eigenstates of $\mathcal{H}^{(0)}$ can always be chosen to be real functions of $\mathbf{r}$. From now on we will assume that such real eigenfunctions have been chosen.

To understand the second property of the functions $u_k^n(\mathbf{r})$, we consider Hilbert space as spanned by all Bloch states $b_k^n(\mathbf{r})$. This Hilbert space consists of orthogonal subspaces characterized by the value of the vector $\mathbf{k}$ in the first Brillouin zone in reciprocal space. Each subspace is spanned by an orthonormal set of basis vectors. As discussed in sections 3.2.2 and 3.2.3 these basis vectors might be either the set of free electron states $\exp[i(\mathbf{K}_h + \mathbf{k}) \cdot \mathbf{r}]$ or the set of Bloch states $b_k^n(\mathbf{r})$. Let us now consider the subspace characterized by quantum number $\mathbf{k}$. In the former case its basis states are numbered by quantum numbers $\mathbf{h}$, while in the latter case they are numbered by quantum numbers n. We will now show the important property that the functions $u_k^n(\mathbf{r})$ also form a complete and orthonormal set of basis functions for this subspace of Hilbert space. As we will see below, this property allows the introduction of Dirac brackets to rewrite the Schrödinger equation (4.6) thus achieving a considerable simplification of the notation.

To prove the orthonormality and completeness of the set of functions $u_k^n(\mathbf{r})$, we expand them in the shape (3.86),

$$u_k^n(\mathbf{r}) = \sum_h \phi_k^{nh} e^{i\mathbf{K}_h \cdot \mathbf{r}} \tag{4.18}$$

In this expression the quantities ϕ_k^{nh} are the components of eigenvectors that are solutions of the matrix equations (3.81). As in chapter 3, we normalize these eigenvectors using eq. (3.40). Then, according to basic quantum mechanics these eigenvectors are orthonormal and complete [11],

$$\sum_h \phi_k^{nh*} \phi_k^{mh} = \delta_{n,m} \tag{4.19}$$

$$\sum_n \phi_k^{nh*} \phi_k^{nh'} = \delta_{h,h'} \tag{4.20}$$

4.1 GENERAL FORMALISM

where the Kronecker δ-functions $\delta_{n,m}$ and $\delta_{h,h'}$ correspond to, say, eq. (3.35). The orthonormality of the functions $u_k^n(\mathbf{r})$ follows from the first property (4.19)

$$\frac{1}{V_C}\int_{V_C} d\mathbf{r}\, u_k^{n*}(\mathbf{r})u_k^m(\mathbf{r}) = \frac{1}{V_C}\int_{V_C} d\mathbf{r} \sum_h \phi_k^{nh*} e^{-i\mathbf{K}_h\cdot\mathbf{r}} \sum_{h'} \phi_k^{mh'} e^{i\mathbf{K}_{h'}\cdot\mathbf{r}}$$

$$= \sum_h \sum_{h'} \phi_k^{nh*}\phi_k^{mh'} \frac{1}{V_C}\int_{V_C} d\mathbf{r}\, e^{i(\mathbf{K}_{h'}-\mathbf{K}_h)\cdot\mathbf{r}}$$

$$= \sum_h \sum_{h'} \phi_k^{nh*}\phi_k^{mh'} \delta_{h',h}$$

$$= \sum_h \phi_k^{nh*}\phi_k^{mh} = \delta_{n,m} \quad (4.21)$$

Here we inserted expression (3.35) for the Kronecker δ-function $\delta_{h',h}$. Note that, as the Bloch functions $b_k^n(\mathbf{r})$ in chapter 3, here the functions $u_k^n(\mathbf{r})$ are normalized in a unit volume. Similarly, the completeness of the functions $u_k^n(\mathbf{r})$ follows from the second property (4.20),

$$\sum_n u_k^{n*}(\mathbf{r})u_k^n(\mathbf{r}')$$

$$= \sum_n \sum_h \phi_k^{nh*} e^{-i\mathbf{K}_h\cdot\mathbf{r}} \sum_{h'} \phi_k^{nh'*} e^{i\mathbf{K}_{h'}\cdot\mathbf{r}'}$$

$$= \sum_h \sum_{h'} e^{-i\mathbf{K}_h\cdot\mathbf{r}+i\mathbf{K}_{h'}\cdot\mathbf{r}'} \sum_n \phi_k^{nh*}\phi_k^{nh'*}$$

$$= \sum_h \sum_{h'} e^{-i\mathbf{K}_h\cdot\mathbf{r}+i\mathbf{K}_{h'}\cdot\mathbf{r}'} \delta_{h,h'}$$

$$= \sum_h e^{i\mathbf{K}_h\cdot(\mathbf{r}'-\mathbf{r})} = \delta(\mathbf{r}'-\mathbf{r}) \quad (4.22)$$

where we used the Dirac δ-function (3.37).

DIRAC BRACKETS

Thus the set of functions $u_k^n(\mathbf{r})$ is orthonormal and complete. This invites us to introduce a Dirac notation where a bra $\langle nk|$ and a ket $|nk\rangle$ denote the state described by the function $u_k^n(\mathbf{r})$. Then[1]

$$\langle \mathbf{r}|nk\rangle = u_k^n(\mathbf{r}) \quad \text{and} \quad \langle nk|\mathbf{r}\rangle = u_k^{n*}(\mathbf{r}) \quad (4.23)$$

---

[1] Here we use the strict notation as introduced by Dirac (see section 16 of ref. [15]). While a ket $|\,\rangle$ denotes a well-defined vector in Hilbert space, its components depend on the choice of basis vectors. In the position representation the eigenstates $|\mathbf{r}\rangle$ of the position operator $\hat{\mathbf{r}}$ are chosen as basis vectors. Then the set of components $\langle \mathbf{r}|\,\rangle$ of a ket corresponds to the wave function represented by the ket.

We now consider an operator $Q(\mathbf{r})$ obeying the translational symmetry of the crystal lattice. Then, in this notation we may write its matrix element unambiguously as

$$\langle nk|Q(\mathbf{r})|mk'\rangle = \frac{1}{V_C}\int_{V_C} d\mathbf{r}\langle nk|\mathbf{r}\rangle Q(\mathbf{r})\langle \mathbf{r}|mk'\rangle$$

$$= \frac{1}{V_C}\int_{V_C} d\mathbf{r} u_k^{n*}(\mathbf{r})Q(\mathbf{r})u_{k'}^{m}(\mathbf{r}) \qquad (4.24)$$

Using this notation, the orthonormality (4.21) of the set of basis functions $u_k^n(\mathbf{r})$ now reads as

$$\langle nk|mk\rangle = \frac{1}{V_C}\int_{V_C} d\mathbf{r}\,\langle nk|\mathbf{r}\rangle\langle \mathbf{r}|nk\rangle$$

$$= \frac{1}{V_C}\int_{V_C} d\mathbf{r}\, u_k^{n*}(\mathbf{r})u_k^{m}(\mathbf{r}) = \delta_{n,m} \qquad (4.25)$$

Furthermore, the closure relation (4.22) leads to

$$\sum_n |nk\rangle\langle nk| = 1 \qquad (4.26)$$

MATRIX REPRESENTATION OF THE k·p-HAMILTONIAN

Finally, now we have developed a convenient notation, we consider the matrix representation of the **k·p**-Hamiltonian. To calculate the matrix elements of the unperturbed Hamiltonian $\mathcal{H}^{(0)}$, we rewrite eq. (4.11) using Dirac brackets,

$$\mathcal{H}^{(0)}|n0\rangle = E_0^n|n0\rangle$$

and multiply on the left-hand side with $\langle m0|$. Then,

$$\langle m0|\mathcal{H}^{(0)}|n0\rangle = \langle m0|E_0^n|n0\rangle = E_0^n\langle m0|n0\rangle = E_0^n\delta_{n,m} \qquad (4.27)$$

The perturbation $\mathcal{H}^{(1)}$ as given by eq. (4.8) has matrix elements:

$$\langle m0|\mathcal{H}^{(1)}|n0\rangle = \left\langle m0\left|\frac{\hbar^2}{im_e}(\mathbf{k}\cdot\nabla)\right|n0\right\rangle + \left\langle m0\left|\frac{\hbar^2}{2m_e}|\mathbf{k}|^2\right|n0\right\rangle \qquad (4.28)$$

The operator in the second term is a number, so its matrix representation is diagonal:

$$\left\langle m0\left|\frac{\hbar^2}{2m_e}|\mathbf{k}|^2\right|n0\right\rangle = \frac{\hbar^2}{2m_e}|\mathbf{k}|^2\langle m0|n0\rangle = \frac{\hbar^2}{2m_e}|\mathbf{k}|^2\,\delta_{n,m} \qquad (4.29)$$

However, the first term of eq. (4.28),

$$\left\langle m0\left|\frac{\hbar^2}{im_e}(\mathbf{k}\cdot\nabla)\right|n0\right\rangle = \frac{\hbar^2}{im_e}\langle m0|(\mathbf{k}\cdot\nabla)|n0\rangle \qquad (4.30)$$

is imaginary, because the basis states $|n0\rangle$ and the operator ∇ are real. Because the operator $(\hbar^2/im_e)(\mathbf{k}\cdot\nabla)$ is also Hermitian, we find

$$\frac{\hbar^2}{im_e}\langle n0|(\mathbf{k}\cdot\nabla)|m0\rangle = \left[\frac{\hbar^2}{im_e}\langle m0|(\mathbf{k}\cdot\nabla)|n0\rangle\right]^* = -\frac{\hbar^2}{im_e}\langle m0|(\mathbf{k}\cdot\nabla)|n0\rangle \qquad (4.31)$$

4.1 GENERAL FORMALISM

which for $n = m$ implies

$$\frac{\hbar^2}{im_e}\langle n0|(\mathbf{k}\cdot\nabla)|n0\rangle = 0 \qquad (4.32)$$

As a result the first term in eq. (4.28) is essentially non-diagonal. Thus, the diagonal matrix elements of $\mathcal{H}^{(1)}$ originate in the second term of eq. (4.8), while its non-diagonal matrix elements originate in the first term.

4.1.3 Non-Degenerate Energy Bands

In our $\mathbf{k}\cdot\mathbf{p}$-treatment we calculate the influence of $\mathcal{H}^{(1)}$ using time independent perturbation theory up to second order. In the present section we consider the special case that the unperturbed energy levels are non-degenerate. We note that such a treatment is not sufficient. This is clearly seen in Fig. 4.1 showing the part of the band structure of GaAs containing the top of the valence band and the bottom of the conduction band. Though the bottom of the conduction band is non-degenerate at the Γ-point, the top of the valence band is degenerate. Therefore, in the next section we continue using perturbation theory for degenerate levels.

Non-degenerate second-order perturbation theory is treated in section C.1.1 of appendix C. We apply eq. (C.15) to the Schrödinger equation (4.6) and use the basis states $|n0\rangle$ introduced in the previous section. Then, up to second order, the energy is found to be given by

$$E_k^n = E_0^n + \langle n0|\mathcal{H}^{(1)}|n0\rangle + \sum_{m\neq n}\frac{\langle n0|\mathcal{H}^{(1)}|m0\rangle\langle m0|\mathcal{H}^{(1)}|n0\rangle}{E_0^n - E_0^m} \qquad (4.33)$$

This result can be directly evaluated using expressions (4.29)–(4.31) for the matrix elements of $\mathcal{H}^{(1)}$. Then

$$\begin{aligned}
E_k^n &= E_0^n + \frac{\hbar^2}{2m_e}|\mathbf{k}|^2 - \frac{\hbar^4}{m_e^2}\sum_{m\neq n}\frac{\langle n0|(\mathbf{k}\cdot\nabla)|m0\rangle\langle m0|(\mathbf{k}\cdot\nabla)|n0\rangle}{E_0^n - E_0^m} \\
&= E_0^n + \frac{\hbar^2}{2m_e}|\mathbf{k}|^2 - \frac{\hbar^4}{m_e^2}\sum_{m\neq n}\sum_{\mu,\nu}k_\mu k_\nu\frac{\langle n0|\nabla_\mu|m0\rangle\langle m0|\nabla_\nu|n0\rangle}{E_0^n - E_0^m} \\
&= E_0^n + \frac{\hbar^2}{2m_e}\sum_{\mu,\nu}\frac{k_\mu k_\nu}{m_{\mu\nu}^*}
\end{aligned} \qquad (4.34)$$

Here $\mu, \nu = x, y, z$ and the so-called effective mass tensor $m_{\mu\nu}^*$ is defined by

$$\begin{aligned}
\frac{1}{m_{\mu\nu}^*} &= \delta_{\mu,\nu} - \frac{2\hbar^2}{m_e}\sum_{m\neq n}\frac{\langle n0|\nabla_\mu|m0\rangle\langle m0|\nabla_\nu|n0\rangle}{E_0^n - E_0^m} \\
&= \delta_{\mu,\nu} - \frac{2\hbar^2}{m_e}\sum_{m\neq n}\frac{P_\mu^{nm}P_\nu^{mn}}{E_0^n - E_0^m}
\end{aligned} \qquad (4.35)$$

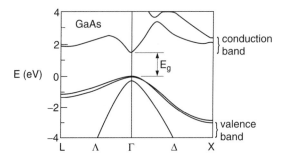

Figure 4.1 Top of the valence band and bottom of the conduction band of GaAs (adapted from Fig. 8.21, p. 103, ref. [10], © Springer Verlag).

where

$$P_\mu^{nm} = \langle n0 | \nabla_\mu | m0 \rangle \tag{4.36}$$

Comparing this result with eq. (3.70) we see that the expression for the energy of an electron in a band is very similar to the energy of a free electron. However, the mass of a band electron is anisotropic and differs from that of a free electron by a factor determined by the effective mass tensor $m_{\mu\nu}^*$. Below, in chapter 5 we will see that this effective mass has a real physical meaning: if external fields are applied, the electron behaves as though it were a free particle having this effective mass.

As we found in the previous section, the first term in $\mathcal{H}^{(1)}$ as given by eq. (4.8) yields exclusively non-diagonal matrix elements. Hence, it contributes to the second order correction to the energy only. On the other hand the second term in eq. (4.8) yields exclusively diagonal matrix elements. Thus, it contributes to the first-order correction on the energy only. Furthermore, this first-order correction yields simply the free electron mass. Clearly the second-order corrections are needed to obtain an effective mass which differs from the free electron mass.

4.1.4 Degenerate Energy Bands

As discussed above the energy levels are degenerate at the top of the valence band, so the unperturbed energy levels are degenerate. Therefore we have to apply the extended perturbation treatment of section C.1.2 in appendix C. We consider the general case that the unperturbed problem yields degenerate energy levels, each described by a multiplet consisting of g_n degenerate states $|na0\rangle$ where $a = 1, 2, \ldots, g_n$. As described in detail in section C.1.2 of appendix C, the perturbation $\mathcal{H}^1$ has two effects. First, its matrix elements between *different* multiplets modify the states $|na0\rangle$ in a similar way as described in the previous section, so they change them by a small amount into $|nak\rangle$. Second, it lifts part or all of the degeneracy by mixing these modified states $|nak\rangle$ within each multiplet. We include this latter effect by writing the exact eigenstates $|\psi\rangle$ as linear combinations of the states $|nak\rangle$ within a multiplet

$$|\psi\rangle = \sum_{a=1}^{g_n} |nak\rangle \langle nak | \psi \rangle \tag{4.37}$$

4.1 GENERAL FORMALISM

According to section C.1.2 of appendix C the coefficients $\langle nak|\psi\rangle$ and energy levels E of the full Hamiltonian are solutions of matrix equations

$$\sum_{b=1}^{g_n} (H_k^n)_{ab}\langle nbk|\psi\rangle = E\langle nak|\psi\rangle \tag{4.38}$$

while up to second order the matrix elements $(H_k^n)_{ab}$ are given by

$$(H_k^n)_{cb} = (H_k^n)_{cb}^{(0)} + (H_k^n)_{cb}^{(1)} + (H_k^n)_{cb}^{(2)}$$

$$= E_0^n \delta_{c,b} + \langle nc0|\mathcal{H}^{(1)}|nb0\rangle + \sum_{m\neq n}\sum_{a=1}^{g_m} \frac{\langle nc0|\mathcal{H}^{(1)}|ma0\rangle\langle ma0|\mathcal{H}^{(1)}|nb0\rangle}{E_0^n - E_0^m} \tag{4.39}$$

Here, $(H_k^n)_{cb}$ is evaluated using expressions (4.28) to (4.32) for the matrix elements of $\mathcal{H}^{(1)}$. Then

$$(H_k^n)_{cb} = E_0^n \delta_{c,b} + \frac{\hbar^2}{2m_e}|\mathbf{k}|^2 \delta_{c,b} + \frac{\hbar^2}{im_e}\langle nc0|(\mathbf{k}\cdot\nabla)|nb0\rangle$$

$$- \frac{\hbar^4}{m_e^2}\sum_{m\neq n}\sum_{a=1}^{g_m} \frac{\langle nc0|(\mathbf{k}\cdot\nabla)|ma0\rangle\langle ma0|(\mathbf{k}\cdot\nabla)|nb0\rangle}{E_0^n - E_0^m}$$

$$= E_0^n \delta_{c,b} + \frac{\hbar^2}{2m_e}|\mathbf{k}|^2 \delta_{c,b} + \frac{\hbar^2}{im_e}\sum_{\mu} k_\mu \langle nc0|\nabla_\mu|nb0\rangle$$

$$- \frac{\hbar^4}{m_e^2}\sum_{\mu,\nu}\sum_{m\neq n}\sum_{a=1}^{g_m} k_\mu k_\nu \frac{\langle nc0|\nabla_\mu|ma0\rangle\langle ma0|\nabla_\nu|nb0\rangle}{E_0^n - E_0^m}$$

$$= E_0^n \delta_{c,b} + \frac{\hbar^2}{im_e}\sum_{\mu} k_\mu (P_\mu)_{cb}^{nn} + \frac{\hbar^2}{2m_e}\sum_{\mu,\nu}\frac{k_\mu k_\nu}{m_{\mu\nu}^{cb}} \tag{4.40}$$

where $\mu, \nu = x, y, z$,

$$\frac{1}{m_{\mu\nu}^{cb}} = \left[\delta_{c,b}\delta_{\mu,\nu} - \frac{2\hbar^2}{m_e}\sum_{m\neq n}\sum_{a=1}^{g_m}\frac{\langle nc0|\nabla_\mu|ma0\rangle\langle ma0|\nabla_\nu|nb0\rangle}{E_0^n - E_0^m}\right]$$

$$= \left[\delta_{c,b}\delta_{\mu,\nu} - \frac{2\hbar^2}{m_e}\sum_{m\neq n}\sum_{a=1}^{g_m}\frac{(P_\mu)_{ca}^{nm}(P_\nu)_{ab}^{mn}}{E_0^n - E_0^m}\right] \tag{4.41}$$

and

$$(P_\mu)_{cb}^{nm} = \langle nc0|\nabla_\mu|mb0\rangle \tag{4.42}$$

According to eq. (4.42) $(P_\mu)_{cc}^{nn} = 0$, so the contribution of the linear term in k_μ to the diagonal matrix elements $(H_k^n)_{cc}$ is zero. In many cases, in particular the case of the top of the valence band to be treated in the next section, also for $c \neq b$, $(P_\mu)_{cb}^{nn} = 0$. Then the linear term in k_μ vanishes completely.

Now the matrix elements $(H_k^n)_{cb}$ are known up to second order, the energy levels E are obtained up to second order by solving the secular equation of the matrix consisting of these elements. Thus the energy levels E are the eigenvalues of the $g_n \times g_n$ matrix $\mathbf{H}_k^n$ with elements $(H_k^n)_{cb}$ and are obtained by solving

$$\text{Det} \mid \mathbf{H}_k^n - E\mathbf{1} \mid = 0 \qquad (4.43)$$

where $\mathbf{1}$ is the $g_n \times g_n$ unit matrix with elements $\delta_{c,b}$. Upon comparing eqs (4.8) and (4.40) it is clearly seen that the second term in the perturbation $\mathcal{H}^{(1)}$ just shifts all levels by a constant amount. The first term in $\mathcal{H}^{(1)}$ is needed to mix the levels within a multiplet and thus lift its degeneracy. In the next section we will treat a simple model for the unperturbed eigen states of semiconductors and use it for an actual calculation of the energy levels E.

4.2 Band Structure of Semiconductors

Now we have developed a method to calculate the band structure at band extrema near the Γ point, we proceed to perform such a calculation for an actual semiconductor. The first problem to attack consists of determining the basis states for the perturbation treatment of sections 4.1.3 and 4.1.4, i.e. the eigen states $|n0\rangle$ of the unperturbed Hamiltonian $\mathcal{H}^{(0)}$ given by eq. (4.7). It is important to note that the Hamiltonian $\mathcal{H}^{(0)}$ and its eigenfunctions $|n0\rangle$ show the periodicity of the crystal lattice. Therefore, this problem can be solved by considering a single primitive cell. As discussed in chapter 3 several different choices can be made for the shape of the primitive cell. For the present calculation we choose the Wigner–Seitz cell shown in Fig. 3.1b. Then the space lattice points $\mathbf{R}_l$ are located at the centre of these primitive cells. Furthermore, as was also discussed in chapter 3, one is free to choose the origin of normal space at any arbitrary position in the crystal lattice. Here we choose this origin to coincide with an atomic site. As a result the symmetry of the surroundings of an atomic site is reflected in the symmetry of the Wigner–Seitz cell. The resulting primitive cell is shown in Fig. 4.2. The black dots in the centre and at four vertices of the dodecahedron correspond to the atomic sites. We note that the latter four sites form a tetrahedron.

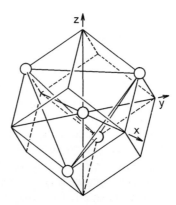

Figure 4.2 Wigner–Seitz primitive cell used for band structure calculations.

4.2 BAND STRUCTURE OF SEMICONDUCTORS

An elegant way to obtain the basis states $|n0\rangle$ is provided by group theory which allows to derive the angular dependent part of these basis states and their degeneracy. The power of group theory is that it only uses the symmetry of the potential in the primitive cell, which is the same as the symmetry of the atomic sites. Hence, one only uses this potential as invariant for those reflections and rotations that also leave the tetrahedron of atomic sites at the vertices of the Wigner–Seitz cell invariant.

However, without resorting to group theory, one may still deduce the angular dependence and degeneracy of the basis states $|n0\rangle$, using an extremely simple model which we present below.

4.2.1 The Eigenstates at k = 0

CENTRAL FORCE MODEL FOR A PRIMITIVE CELL

In our simple model we consider the valence electrons only and assume the core electrons to reside very near the atomic nuclei. Thus, we assume the core electrons to yield a simple shielding of the nuclei, just as represented by eq. (2.29). We furthermore neglect the electrostatic repulsion between the valence electrons, reducing the periodic potential experienced by a valence electron to the simple shape

$$\mathcal{V}'(\mathbf{r}) = -\sum_p \frac{\alpha_p Z_p e^2}{4\pi\varepsilon_0 r_p} \tag{4.44}$$

where

$$r_p = \sqrt{(x - X_p)^2 + (y - Y_p)^2 + (z - Z_p)^2} \tag{4.45}$$

is the distance between this valence electron and the pth core. For the various semiconductors,

IV	Si, Ge	$\alpha_p Z_p = 4$
III–V	GaAs, …	$\alpha_p Z_p = 3$ or 5
II–VI	ZnSe, …	$\alpha_p Z_p = 2$ or 6

We choose the centre of the primitive cell to coincide with the core yielding the strongest potential, and start by considering this core only. Clearly this approach is best for II–VI semiconductors where this potential is three times stronger than that of the four other cores in the Wigner–Seitz cell and worst for the elemental semiconductors where all core potentials are equal.

As we consider one single primitive cell only, our problem is now reduced to that of one single central potential. Hence, the problem is the same as that of an atom, allowing us to follow the treatment of section 2.1.1 and use the central force model. Then the angular dependent parts of the eigenstates of the valence electrons in the primitive cell are those of the atomic orbitals given by table A.3. However, the radial parts may be different from those given in table A.1 because of the boundary conditions at the surface of the primitive cell, while atomic wave functions may extend to infinity. Then within this cell the eigenstates of the valence electrons can be written as

n	$\|n0\rangle$
ν, s	$\|\nu s0\rangle = \rho_{\nu s}(r)$
ν, x	$\|\nu x0\rangle = \rho_{\nu p}(r)x$
ν, y	$\|\nu y0\rangle = \rho_{\nu p}(r)y$
ν, z	$\|\nu z0\rangle = \rho_{\nu p}(r)z$
	etc.

(4.46)

where the quantum number n is replaced by (ν, s), (ν, x), etc., and $\nu = 1, 2, 3, \ldots$ is the principal quantum number of the atomic orbitals.

When filling these states with electrons we note that valence electrons are outer electrons so they typically fill states

$$|\nu s0\rangle, |\nu x0\rangle, |\nu y0\rangle, |\nu z0\rangle, \ldots$$
$$|(\nu+1)s0\rangle, |(\nu+1)x0\rangle, |(\nu+1)y0\rangle, |(\nu+1)z0\rangle, \ldots$$

where $\nu > 1$. Each of these states corresponds to the point $\mathbf{k} = 0$ of a valence or conduction band. We now refer to chapter 3, Figs 3.7a–3.10a, showing these bands. We see that the lower four levels

$$|\nu s0\rangle, |\nu x0\rangle, |\nu y0\rangle, |\nu z0\rangle$$

correspond to the points $\mathbf{k} = 0$ of the four valence bands while all higher levels correspond to these points in the conduction bands.

CORRECTIONS ON THE CENTRAL FORCE MODEL

In the next step we consider the electrostatic potential of the cores at the vertices of the primitive cell. In first-order perturbation theory this extra potential simply shifts the energy levels without changing the eigenfunctions. Evidently, the extra potential is not spherically symmetric around the centre of the primitive cell, so the degeneracy of the atomic levels may be lifted. On the other hand, not all degeneracy will be lifted. In particular the $|\nu x0\rangle$, $|\nu y0\rangle$ and $|\nu z0\rangle$ orbitals remain degenerate, as can be seen by symmetry arguments. This is illustrated in Fig. 4.3 which shows an artist's impression of the three wave functions in the primitive cell. It is seen that the wave functions transform into each other by rotating over 120° about the trigonal axis indicated by T. Now such a rotation about T by 120° transforms the primitive cell into itself, so the periodic potential $\mathcal{V}'(\mathbf{r})$ also remains the same. As a result, the expectation value of $\mathcal{V}'(\mathbf{r})$ is the same in each of the three states $|\nu x0\rangle$, $|\nu y0\rangle$ and $|\nu z0\rangle$:

$$\langle \nu x0|\mathcal{V}'(\mathbf{r})|\nu x0\rangle = \langle \nu y0|\mathcal{V}'(\mathbf{r})|\nu y0\rangle = \langle \nu z0|\mathcal{V}'(\mathbf{r})|\nu z0\rangle \quad (4.47)$$

Otherwise the energy could be changed via a simple rotation of the frame of reference.

In most cases the extra potential of the cores at the vertices of the primitive cells raises the energy of the p-states $|\nu x0\rangle$, etc., above the s-states $|\nu s0\rangle$. This is illustrated

4.2 BAND STRUCTURE OF SEMICONDUCTORS

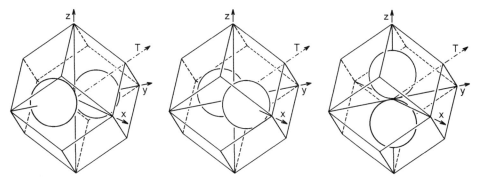

Figure 4.3 Artist's impression of the functions $|\nu y0\rangle$, $|\nu x0\rangle$ and $|\nu z0\rangle$ in a primitive cell.

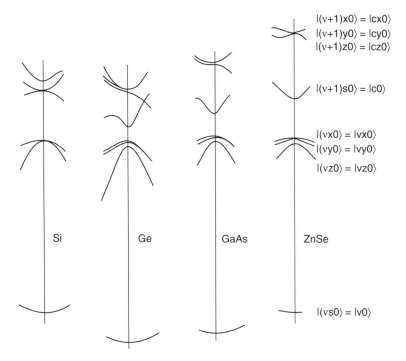

Figure 4.4 Band structures of Si, Ge, GaAs and ZnSe near **k** = 0 (adapted from Figs 8.2, 8.11, 8.21 and 8.30, pp. 81, 92, 103 and 113, ref. [10], © Springer Verlag).

in Fig. 4.4 showing the part of the band structure near **k** = 0 for Si, Ge, GaAs and ZnSe together with the associated atomic orbitals. Thus always the top of the valence band can be described by a multiplet consisting of three degenerate states

$$|\nu x0\rangle = |vx0\rangle$$
$$|\nu y0\rangle = |vy0\rangle \quad (4.48)$$
$$|\nu z0\rangle = |vz0\rangle$$

while in most cases the bottom of the conduction band at the Γ-point is described by a non-degenerate state

$$|(\nu+1)s0\rangle = |c0\rangle \tag{4.49}$$

Only in silicon, the states at the bottom of the conduction band are inverted and at the Γ-point the three states $|cx0\rangle$, $|cy0\rangle$ and $|cz0\rangle$ are lowest.

It is very important to note that *here* we discuss the Bloch functions at $\mathbf{k} = 0$ and *not* the full wave functions of the valence electrons. Thus, the atomic orbitals shown in Fig. 4.3 are *not* directly related to the atomic orbitals treated in chapter 2. The atomic orbitals in chapter 2 are obtained by adding *all* Bloch functions at *all* values of $\mathbf{k}$ and hence they do *not* correspond to just those at $\mathbf{k} = 0$.

4.2.2 k·p Treatment of the Conduction Band

In almost all semiconductors the bottom of the conduction band is non-degenerate at $\mathbf{k} = 0$. So we can treat it with the results obtained in section 4.1.3 for the band structure near $\mathbf{k} = 0$ for non-degenerate bands. Then, the energy E_k^c in the conduction band is given by eqs (4.34) and (4.35):

$$E_k^c = E_0^c + \frac{\hbar^2}{2m_e} \sum_{\mu,\nu} \frac{k_\mu k_\nu}{m^*_{\mu\nu}} \tag{4.50}$$

where $\mu, \nu = x, y, z$ and

$$\frac{1}{m^*_{\mu\nu}} = \left[\delta_{\mu,\nu} - \frac{2\hbar^2}{m_e} \sum_{m \neq c} \frac{\langle c0 | \nabla_\mu | m0 \rangle \langle m0 | \nabla_\nu | c0 \rangle}{E_0^c - E_0^m} \right] \tag{4.51}$$

where the functions $|m0\rangle$ are the eigenstates at $\mathbf{k} = 0$ corresponding to the other bands. We see from eq. (4.51) that the largest contribution to the energy is due to mixing with those other bands that are nearest to the conduction band, i.e. due to mixing with those bands for which $|E_0^c - E_0^m|$ is smallest. As can be seen from Fig. 4.4, in all cases except silicon the top of the valence band is nearest to the bottom of the conduction band. Then a good approximation is obtained by taking into account the top of the valence band only and neglecting all other bands in the sum over m in eq. (4.51).

We evaluate the matrix elements in this equation using the real basis states (4.48) and (4.49)

$$|c0\rangle = \rho_c(r)$$

$$|vx0\rangle = \rho_v(r)x$$

$$|vy0\rangle = \rho_v(r)y$$

$$|vz0\rangle = \rho_v(r)z$$

4.2 BAND STRUCTURE OF SEMICONDUCTORS

derived in the previous section. Then, because the momentum operator $(\hbar/i)\nabla_x$ is Hermitian,

$$-\frac{\hbar}{i}\langle c0|\nabla_x|vx0\rangle = \frac{\hbar}{i}\langle vx0|\nabla_x|c0\rangle$$

$$= \frac{\hbar}{i}\frac{1}{V_C}\int_{V_C} d\mathbf{r}\rho_v(r)x\frac{\partial\rho_c(r)}{\partial x}$$

$$= \frac{\hbar}{i}\frac{1}{V_C}\int_{V_C} d\mathbf{r}\rho_v(r)x\frac{\partial\rho_c(r)}{\partial r}\frac{\partial r}{\partial x}$$

$$= \frac{\hbar}{i}\frac{1}{V_C}\int_{V_C} d\mathbf{r}\rho_v(r)x\frac{\partial\rho_c(r)}{\partial r}\frac{x}{r} \equiv \frac{\hbar}{i}P \quad (4.52)$$

thus defining the real quantity P which differs from zero because the integrand is symmetric with respect to inverting $\mathbf{r}$ to $-\mathbf{r}$. In this derivation we furthermore used

$$\frac{\partial r}{\partial x} = \frac{\partial}{\partial x}\sqrt{x^2+y^2+z^2} = \frac{1}{2\sqrt{x^2+y^2+z^2}}\frac{\partial x^2}{\partial x} = \frac{2x}{2\sqrt{x^2+y^2+z^2}} = \frac{x}{r} \quad (4.53)$$

Similarly, we find

$$-\frac{\hbar}{i}\langle c0|\nabla_y|vy0\rangle = \frac{\hbar}{i}\langle vy0|\nabla_y|c0\rangle = \frac{\hbar}{i}P$$

$$-\frac{\hbar}{i}\langle c0|\nabla_z|vz0\rangle = \frac{\hbar}{i}\langle vz0|\nabla_z|c0\rangle = \frac{\hbar}{i}P \quad (4.54)$$

while all other matrix elements are found to be equal to zero. As a result the effective mass in the conduction band is found to be isotropic and given by

$$\frac{1}{m^*_{\mu\nu}} = \frac{1}{m^*_c}\delta_{\mu,\nu} = \left(1+\frac{2\hbar^2 P^2}{m_e E^0_g}\right)\delta_{\mu,\nu} \quad (4.55)$$

where

$$E^0_g = E^c_0 - E^v_0 \quad (4.56)$$

is the gap between the conduction band and the valence band.

It should be noted that $\hbar P$ has the dimension of momentum, so $2\hbar^2 P^2/m_e$ has the dimension of energy. P can be estimated from its definition (4.52). Then $P \approx 1/x_0$ where x_0 is the average radius of the wave function. Let us approximate $1/x_0$ by the length $|\mathbf{b}_\beta| = 2\sqrt{3}\pi/c$ of a vector spanning a Brillouin zone, where c is the length of the cubic unit cell. For GaAs $c = 0.5869$ nm, so $1/x_0 = 1.854 \cdot 10^{10}$ m^{-1}. Then, taking

$$\hbar = 1.055 \cdot 10^{-34} \text{ J s}$$

$$m_e = 9.110 \cdot 10^{-31} \text{ kg}$$

$$e = 1.602 \cdot 10^{-19} \text{ C}$$

we find $2\hbar^2 P^2/m_e e = 52.43$ eV. As in GaAs $E_g/e = 1.519$ eV, we estimate $m^*_c = 0.0290$, which is of the right order of magnitude considering its actual value ($m^*_c = 0.0665$).

4.2.3 k·p Treatment of the Valence Band

In all semiconductors the top of the valence band occurs at $\mathbf{k} = 0$, where it is threefold degenerate. So we have to treat it with the results obtained in section 4.1.4 for the band structure near $\mathbf{k} = 0$ for degenerate bands. Then, the energy levels E_k^v in the valence band are the eigenvalues of the matrix $\mathbf{H}_k^v$ with elements

$$(H_k^v)_{\pi\rho} = E_0^v \delta_{\pi,\rho} + \frac{\hbar^2}{2m_e} \sum_{\mu,\nu} \frac{k_\mu k_\nu}{m_{\mu\nu}^{\pi\rho}} \qquad (4.57)$$

where μ, ν, π and $\rho = x, y, z$ and

$$\frac{1}{m_{\mu\nu}^{\pi\rho}} = \delta_{\pi,\rho}\delta_{\mu,\nu} - \frac{2\hbar^2}{m_e} \sum_{m \neq v} \sum_{a=1}^{g_m} \frac{\langle v\pi 0 | \nabla_\mu | ma0 \rangle \langle ma0 | \nabla_\nu | v\rho 0 \rangle}{E_0^v - E_0^m} \qquad (4.58)$$

where the functions $|ma0\rangle$ are the eigenstates at $\mathbf{k} = 0$ corresponding to the other bands. Note that, as discussed in the previous section, the matrix elements of $(\hbar/i)\nabla$ between two p-states are equal to zero. So the term linear in k_μ vanishes.

Again the largest contribution originates in those other bands for which $|E_v - E_m|$ is smallest. Hence, a good approximation is again obtained by taking into account the conduction band only and neglecting all other bands in the sum over m. Using eqs (4.52) and (4.54) for the evaluation of the matrix elements in eq. (4.58) we find

$$\frac{1}{m_{\mu\nu}^{\pi\rho}} = \delta_{\pi,\rho}\delta_{\mu,\nu} - \frac{2\hbar^2}{m_e} \frac{\langle v\pi 0 | \nabla_\mu | c0 \rangle \langle c0 | \nabla_\nu | v\rho 0 \rangle}{E_0^v - E_0^c}$$

$$= \delta_{\pi,\rho}\delta_{\mu,\nu} - \frac{2\hbar^2 P^2}{m_e E_g} \delta_{\pi,\mu}\delta_{\rho,\nu} \qquad (4.59)$$

Thus the energy levels in the valence band are the eigenvalues of the matrix

$$\mathbf{H}_k^v = \left(E_0^v + \frac{\hbar^2 k^2}{2m_e}\right)\mathbf{1} - \frac{\hbar^4 P^2}{m_e^2 E_g} \begin{pmatrix} k_x^2 & k_x k_y & k_x k_z \\ k_y k_x & k_y^2 & k_y k_z \\ k_z k_x & k_z k_y & k_z^2 \end{pmatrix} \qquad (4.60)$$

where $\mathbf{1}$ is the 3×3 unit matrix and

$$k^2 = |\mathbf{k}|^2 = k_x^2 + k_y^2 + k_z^2$$

4.2.4 Spin-Orbit Coupling

THE ELECTRON SPIN

Unfortunately the above given description of the valence band at $\mathbf{k} = 0$ does not yield the correct structure of this band. This is caused by the fact that we have neglected the magnetic dipole moment

$$\boldsymbol{\mu}_s = -g_e \mu_B \mathbf{s} \qquad (4.61)$$

4.2 BAND STRUCTURE OF SEMICONDUCTORS

associated with the electron spin $s = \frac{1}{2}$, where

$$\mu_B = \frac{\hbar e}{2m_e} \qquad (4.62)$$

is the Bohr magneton and

$$g_e = 2.0023 \approx 2$$

is the electronic g-value. This magnetic moment lifts part of the degeneracy at the top of the valence band at $\mathbf{k} = 0$ through its interaction with the magnetic dipole moment

$$\mu_l = -\mu_B \mathbf{l} \qquad (4.63)$$

associated with the orbit of the electron, where

$$\hbar \mathbf{l} = \mathbf{r} \times \frac{\hbar}{i} \nabla \qquad (4.64)$$

is its angular momentum. The order of magnitude of this so-called spin-orbit coupling is obtained by estimating the magnetic dipole interaction between a magnetic point dipole μ_l at the centre of the orbit and a magnetic point dipole μ_s at a distance a' corresponding to the radius of the orbital and ignoring angular dependencies. Then

$$\mathcal{H}_{SO} = -\frac{\mu_0}{4\pi} \frac{\mu_s \cdot \mu_l}{(a')^3} = -\frac{\mu_0}{4\pi} \frac{(2\mu_B \mathbf{s}) \cdot (\mu_B \mathbf{l})}{(a')^3} = -\frac{\mu_0}{4\pi} \frac{\hbar^2 e^2}{m_e^2 (a')^3} (\mathbf{s} \cdot \mathbf{l}) \qquad (4.65)$$

By nature this spin-orbit coupling is a relativistic effect and should be treated by a quantum electrodynamic treatment. Fortunately a correct treatment yields the same result except for a constant factor 2 and a more accurate 'distance' a'. In the following we simply write

$$\mathcal{H}_{SO} = \Lambda (\mathbf{s} \cdot \mathbf{l}) \qquad (4.66)$$

where Λ is the spin-orbit coupling constant.

MATRIX ELEMENTS OF THE SPIN-ORBIT COUPLING

We use time-independent perturbation theory to calculate how the spin-orbit coupling $\mathcal{H}_{SO}$ lifts the degeneracy at the top of the valence band. In particular we add the spin-orbit coupling $\mathcal{H}_{SO}$ to the perturbation (4.8), so the total perturbation is given by

$$\mathcal{H}^{(1)} = \frac{\hbar^2}{im_e} (\mathbf{k} \cdot \nabla) + \frac{\hbar^2}{2m_e} |\mathbf{k}|^2 - i\Lambda \mathbf{s} \cdot (\mathbf{r} \times \nabla) \qquad (4.67)$$

and apply the results of section 4.1.4. In order to be able to do this we need to calculate the matrix elements of $\mathcal{H}_{SO}$ on the basis states (4.48) and (4.49). As an example we calculate

$$
\begin{aligned}
\langle vx0|\Lambda s_z l_z|vy0\rangle &= -i\Lambda s_z \frac{1}{V_C}\int_{V_C} d\mathbf{r}\rho_v(r)x(\mathbf{r}\times\nabla)_z\rho_v(r)y \\
&= -i\Lambda s_z \frac{1}{V_C}\int_{V_C} d\mathbf{r}\rho_v(r)x\left(x\frac{\partial}{\partial y}-y\frac{\partial}{\partial x}\right)\rho_v(r)y \\
&= -i\Lambda s_z \frac{1}{V_C}\int_{V_C} d\mathbf{r}\rho_v(r)x\left[x\frac{\partial\rho_v(r)}{\partial y}y+x\rho_v(r)-y\frac{\partial\rho_v(r)}{\partial x}y\right] \\
&= -i\Lambda s_z \frac{1}{V_C}\int_{V_C} d\mathbf{r}\rho_v(r)x\left[x\frac{\partial\rho_v(r)}{\partial r}\frac{y}{r}y+x\rho_v(r)-y\frac{\partial\rho_v(r)}{\partial r}\frac{x}{r}y\right] \\
&= -i\Lambda s_z \frac{1}{V_C}\int_{V_C} d\mathbf{r}\rho_v(r)x[x\rho_v(r)] \\
&= -i\Lambda s_z \frac{1}{V_C}\int_{V_C} d\mathbf{r}x^2\rho_v^2(r) = -2i\lambda s_z
\end{aligned}
\quad (4.68)
$$

defining the spin-orbit splitting λ. From similar evaluations we find that all matrix elements of $\mathcal{H}_{SO}$ between conduction band and valence band states are equal to zero:

$$\langle c0|\mathcal{H}_{SO}|vx0\rangle = \langle c0|\mathcal{H}_{SO}|vy0\rangle = \langle c0|\mathcal{H}_{SO}|vz0\rangle = 0 \quad (4.69)$$

Thus, according to eq. (4.39) the spin-orbit coupling does not contribute to the second order term in the energy. Furthermore, we find also all relevant diagonal matrix elements of $\mathcal{H}_{SO}$ to be equal to zero:

$$
\begin{aligned}
\langle c0|\mathcal{H}_{SO}|c0\rangle &= \langle vx0|\mathcal{H}_{SO}|vx0\rangle \\
\langle vy0|\mathcal{H}_{SO}|vy0\rangle &= \langle vz0|\mathcal{H}_{SO}|vz0\rangle = 0
\end{aligned}
\quad (4.70)
$$

Only the non-diagonal elements between valence band states yield a finite result. Using the fact that $\mathcal{H}_{SO}$ is Hermitian we find

$$
\begin{aligned}
-\langle vx0|\mathcal{H}_{SO}|vy0\rangle &= \langle vy0|\mathcal{H}_{SO}|vx0\rangle = 2i\lambda s_z \\
-\langle vy0|\mathcal{H}_{SO}|vz0\rangle &= \langle vz0|\mathcal{H}_{SO}|vy0\rangle = 2i\lambda s_x \\
-\langle vz0|\mathcal{H}_{SO}|vx0\rangle &= \langle vx0|\mathcal{H}_{SO}|vz0\rangle = 2i\lambda s_y
\end{aligned}
\quad (4.71)
$$

Hence, according to eq. (4.39) the spin-orbit coupling is expected to contribute to the energy levels in the valence band via the first-order contribution of perturbation theory. Therefore its influence must be taken into account by adding a matrix

$$
2i\lambda\begin{pmatrix} 0 & -s_z & s_y \\ s_z & 0 & -s_x \\ -s_y & s_x & 0 \end{pmatrix}
\quad (4.72)
$$

to the matrix (4.60), so it reads as

$$
\left(E_0^v+\frac{\hbar^2 k^2}{2m_e}\right)\mathbf{1} - \frac{\hbar^4 P^2}{m_e^2 E_g}\begin{pmatrix} k_x^2 & k_xk_y & k_xk_z \\ k_yk_x & k_y^2 & k_yk_z \\ k_zk_x & k_zk_y & k_z^2 \end{pmatrix} + 2i\lambda\begin{pmatrix} 0 & -s_z & s_y \\ s_z & 0 & -s_x \\ -s_y & s_x & 0 \end{pmatrix}
\quad (4.73)
$$

4.2.5 The Valence Band near k = 0

ANALYTICAL SOLUTIONS FOR THE EIGENVALUES

Solving the eigenvalues of the matrix (4.73) is not as straightforward as it may seem at first sight because it contains operators s_x, s_y and s_z that do not commute. To illustrate an analytical method to obtain the eigenvalues we consider the special case that $\mathbf{k} \parallel z'$-axis, so

$$k_x = k_y = 0 \quad \text{and} \quad k_z = k$$

The extension of this method to arbitrary directions of $\mathbf{k}$ and to the even more general case to be discussed in the next section is straightforward, however. The only reason not to present it here is the extensive algebra involved.

We start by considering the matrix equation from which the eigenvalues and eigenvectors of the matrix (4.73) are solved:

$$\begin{pmatrix} -\varepsilon & -i\lambda\sigma_z & i\lambda\sigma_y \\ i\lambda\sigma_z & -\varepsilon & -i\lambda\sigma_x \\ -i\lambda\sigma_y & i\lambda\sigma_x & -(3Bk^2+\varepsilon) \end{pmatrix} \begin{pmatrix} \alpha \\ \beta \\ \gamma \end{pmatrix} = 0 \tag{4.74}$$

corresponding to the set of linear equations,

$$\begin{aligned} -\varepsilon\alpha &- i\lambda\sigma_z\beta &+ i\lambda\sigma_y\gamma &= 0 \\ i\lambda\sigma_z\alpha &- \varepsilon\beta &- i\lambda\sigma_x\gamma &= 0 \\ -i\lambda\sigma_y\alpha &+ i\lambda\sigma_x\beta &- (3Bk^2+\varepsilon)\gamma &= 0 \end{aligned} \tag{4.75}$$

Here, we introduced the abbreviations

$$B = \frac{\hbar^4 P^2}{3m_e^2 E_g^0}, \quad \text{and} \quad \varepsilon = E - E_0^v - \frac{\hbar^2 k^2}{2m_e} \tag{4.76}$$

and introduced the Pauli operators

$$\sigma_x = 2s_x = \begin{pmatrix} 0 & 1 \\ 1 & 0 \end{pmatrix}, \quad \sigma_y = 2s_y = \begin{pmatrix} 0 & -i \\ i & 0 \end{pmatrix}, \quad \sigma_z = 2s_z = \begin{pmatrix} 1 & 0 \\ 0 & -1 \end{pmatrix} \tag{4.77}$$

Hence, the components α, β and γ of the three-dimensional eigenvectors of the matrix (4.73) are two-dimensional vectors themselves, with components α_+ and α_-, etc., on the basis of the spin states. On the basis of these spin states all spin-independent quantities are of course proportional to the 2×2 unit matrix

$$\mathbf{1} = \begin{pmatrix} 1 & 0 \\ 0 & 1 \end{pmatrix} \tag{4.78}$$

Still, it is not necessary to insert eqs (4.77) and (4.78) in the matrix in eq. (4.74), so it becomes a 6×6 matrix. The reason is that it is still possible to solve the set of equations (4.75) in the usual way, provided one keeps good track of the order of the operators involved. We start by eliminating α from the first and second equations in this set. For this purpose we multiply the first equation on the left by $+i\lambda\sigma_z$ and the second equation on the left by $-\varepsilon$. Then we obtain, respectively,

$$-i\lambda\sigma_z\varepsilon\alpha - i^2\lambda^2\sigma_z^2\beta + i^2\lambda^2\sigma_z\sigma_y\gamma = -i\lambda\sigma_z\varepsilon\alpha + \lambda^2\beta + i\lambda^2\sigma_x\gamma = 0 \quad (4.79)$$

$$-i\lambda\sigma_z\varepsilon\alpha + \varepsilon^2\beta + i\lambda\sigma_x\varepsilon\gamma = 0 \quad (4.80)$$

where we used

$$\sigma_z^2 = 1 \quad \text{and} \quad \sigma_z\sigma_y = -i\sigma_x \quad (4.81)$$

Next we subtract eqs. (4.79) and (4.80) in order to obtain

$$(\lambda^2 - \varepsilon^2)\beta + i\lambda(\lambda - \varepsilon)\sigma_x\gamma = 0 \quad (4.82)$$

In a completely similar way we eliminate α from the first and the third equations of the set (4.75). For this purpose we multiply the first equation on the left-hand side by $-i\lambda\sigma_y$ and the third equation by $-\varepsilon$. Noting that

$$\sigma_y^2 = 1 \quad \text{and} \quad \sigma_y\sigma_z = i\sigma_x \quad (4.83)$$

we obtain

$$i\lambda\sigma_y\varepsilon\alpha + i^2\lambda^2\sigma_y\sigma_z\beta - i^2\lambda^2\sigma_y^2\gamma = i\lambda\sigma_y\varepsilon\alpha - i\lambda^2\sigma_x\beta + \lambda^2\gamma = 0 \quad (4.84)$$

$$i\lambda\sigma_y\varepsilon\alpha - i\lambda\sigma_x\varepsilon\beta + \varepsilon(3Bk^2 + \varepsilon)\gamma = 0 \quad (4.85)$$

which upon subtraction yields

$$-i\lambda(\lambda - \varepsilon)\sigma_x\beta + (\lambda^2 - \varepsilon^2 - 3\varepsilon Bk^2)\gamma = 0 \quad (4.86)$$

Thus we have obtained a set of two linear equations (4.82) and (4.86) for β and γ. In the final step we eliminate β from these two equations. For this purpose we multiply eq. (4.82) on the left-hand side with $-i\lambda(\lambda - \varepsilon)\sigma_x$ and eq. (4.86) on the left-hand side with $(\lambda^2 - \varepsilon^2)$. Noting that

$$\sigma_x^2 = 1 \quad (4.87)$$

we find

$$-i\lambda(\lambda - \varepsilon)(\lambda^2 - \varepsilon^2)\sigma_x\beta - i^2\lambda^2(\lambda - \varepsilon)^2\sigma_x^2\gamma$$
$$= -i\lambda(\lambda - \varepsilon)(\lambda^2 - \varepsilon^2)\sigma_x\beta + \lambda^2(\lambda - \varepsilon)^2\gamma = 0 \quad (4.88)$$

$$-i\lambda(\lambda - \varepsilon)(\lambda^2 - \varepsilon^2)\sigma_x\beta + (\lambda^2 - \varepsilon^2)(\lambda^2 - \varepsilon^2 - 3\varepsilon Bk^2)\gamma = 0 \quad (4.89)$$

which upon subtraction yields

$$[\lambda^2(\lambda - \varepsilon)^2 - (\lambda^2 - \varepsilon^2)(\lambda^2 - \varepsilon^2 - 3\varepsilon Bk^2)]\gamma = 0 \quad (4.90)$$

Note that this equation has the striking property that all spin operators have been eliminated. Hence, we can now continue without bothering about the sequence of the various parameters. After dividing out γ and rearranging the various terms, we obtain a third order equation,

$$\varepsilon^3 + (3Bk^2)\varepsilon^2 - (3\lambda^2)\varepsilon + (2\lambda - 3Bk^2)\lambda^2 = 0 \quad (4.91)$$

This equation has three solutions which can be analytically found. The fact that we only find three solutions corresponds to the disappearance of the spin operators from

4.2 BAND STRUCTURE OF SEMICONDUCTORS

the set of equations (4.75) upon eliminating α and β. As a result we cannot determine the spin state, while moreover the eigenvalues are degenerate with respect to this spin state. This degeneracy is commonly denoted by Kramer's degeneracy.

THE THREE VALENCE BANDS

Though eq. (4.91) is analytically solvable, we will restrict ourselves to the simple approximation that $3Bk^2 \ll \lambda$. Then the solutions are readily found to be

$$\varepsilon = +\lambda \tag{4.92}$$

$$\varepsilon = +\lambda - 2Bk^2 \tag{4.93}$$

$$\varepsilon = -2\lambda - Bk^2 \tag{4.94}$$

Thus, inserting the definitions (4.76) for B and ε we find three bands

$$E_{hh} = E_0^v + \lambda + \frac{\hbar^2}{2m_e} k^2 \tag{4.95}$$

$$E_{lh} = E_0^v + \lambda + \frac{\hbar^2}{2m_e}\left(1 - \frac{4\hbar^2 P^2}{3m_e E_g^0}\right) k^2 \tag{4.96}$$

$$E_{\lambda h} = E_0^v - 2\lambda + \frac{\hbar^2}{2m_e}\left(1 - \frac{2\hbar^2 P^2}{3m_e E_g^0}\right) k^2 \tag{4.97}$$

The result is plotted in Fig. 4.5. One band, for which $E = E_{\lambda h}$, is shifted to lower energy by $\delta E = 3\lambda$ compared to the other two bands. The effective mass is negative and of the same order of magnitude as the effective mass of electrons in the conduction band. This band is called the *spin-orbit band*. The two other bands are degenerate at $\mathbf{k} = 0$. Electrons in the so-called *light holes band*, for which $E = E_{lh}$ also have a negative mass, which also is of the same order of magnitude as that of conduction band electrons. The last band, the *heavy holes band*, is characterized by a much larger effective mass, because in lowest order, the $\mathbf{k} \cdot \mathbf{p}$ approximation yields no mixing with

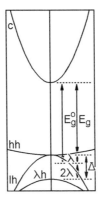

Figure 4.5 The band structure near $\mathbf{k} = 0$.

the conduction band. In the present rough treatment we find a positive effective mass. As we will see below in a more refined treatment also the heavy holes band is found to be characterized by a negative effective mass [19].

The fact that the effective mass of an electron in the valence band is negative yields no problems. Conductivity in this band is due to holes or 'absent' electrons. These are in turn characterized by a positive effective mass, so they behave 'properly' in external fields.

4.2.6 Further Refinements

THE CONDUCTION BAND

Though the above results were obtained with an extremely simple model, they still give a good impression of the band structure at the Γ-point. More accurate results may sometimes be required however. Therefore the present section presents some refinements without giving full derivations.

When calculating the effective mass of conduction electrons at the Γ-point two refinements are generally made to improve the result (4.55) obtained in section 4.2.2. First, the spin-orbit splitting of the valence band is included in the calculation. The following argument, though not exact, may be used to deduce the general shape of the effective mass when this refinement is made. We recall that the simple result (4.55) originates in the more general expression (4.51), where the second term represents mixing of the conduction band and the three valence bands due to the term $(\hbar^2/im_e)(\mathbf{k}\cdot\nabla)$ in the $\mathbf{k}\cdot\mathbf{p}$-Hamiltonian given by eqs (4.7) and (4.8). Let us now suppose that in eq. (4.51) the index m represents the heavy holes, the light holes and the spin-orbit band, instead of the three basis states $|vx0\rangle$, $|vy0\rangle$ and $|vz0\rangle$. Then, the denominator in eq. (4.51) should represent the splitting between the conduction band and each of the three valence bands. Let us furthermore consider eqs (4.95)–(4.97), showing that the heavy holes band does not mix with the conduction band, while mixing between the light holes band and the conduction band is twice as large as mixing between the spin-orbit band and the conduction band. Then, the numerator in eq. (4.51) should be zero for the heavy holes band and twice as large for the light holes band as for the spin-orbit band. As a result, we expect the conduction band effective mass to have the general shape,

$$\frac{1}{m_c^*} = 1 + \frac{2}{m_e}\left[\frac{2\hbar^2 P^2}{3E_g} + \frac{\hbar^2 P^2}{3(E_g + \Delta)}\right] \tag{4.98}$$

where the gap energy

$$E_g = E_0^c - E_0^v - \lambda \tag{4.99}$$

is the energy separation from the bottom of the conduction band to the top of the light—and heavy holes—bands and

$$E_g + \Delta = E_g + 3\lambda = E_0^c - E_0^v + 2\lambda \tag{4.100}$$

is the energy separation from the bottom of the conduction band to the top of the spin-orbit band.

4.2 BAND STRUCTURE OF SEMICONDUCTORS

Table 4.1 Typical parameters for the conduction band at the Γ-point of some semiconductors.

	E_g/e (eV)	Δ/e (eV)	m_c^*	$2P^2/em_e$ (eV) simple model	$2P^2/em_e$ (eV) 3 level k·p	$2P^2/em_e$ (eV) 5 level k·p
Ge	0.898	0.296	0.037	25.9	25.5	26.3
AlAs	3.13	0.275	0.124	22.7	23.1	21.1
GaAs	1.519	0.35	0.067	22.8	22.6	25.7
GaSb	0.811	0.76	0.041	24.9	22.6	22.4
InP	1.424	0.21	0.077	17.9	17.8	20.4
InAs	0.418	0.43	0.024	22.8	20.5	22.2
InSb	0.235	0.82	0.014	35.8	22.3	23.1

A second correction on the result of section 4.2.2 consists of including the higher conduction bands—in particular the three *p*-type levels above the lowest conduction band—in the **k·p**-approximation. One generally calls this a five level **k·p**-approximation, counting one level for the conduction band, one level for the spin orbit band, one level for the degenerate heavy and light holes bands and two levels for the higher *p*-type conduction bands.

Table 4.1 summarizes typical parameters for the conduction band minimum at the Γ-point of some semiconductors [24]. Except for the effective mass which is dimensionless, all quantities in that table have the dimension of energy and are given in electron volts. The values of E_g/e, Δ/e and m_c^* result from experiments. The first column for $2P^2/em_e$ is calculated from these experimental values using our simple model, i.e. eqs (4.55) and (4.56). The second column for $2P^2/em_e$ is obtained in the same way, but incorporating a correction for the spin-orbit splitting in the valence band, i.e. using eqs (4.98) to (4.100). The third column for $2P^2/em_e$ gives the results of a five level **k·p** approximation as found in ref. [24].

It is seen that the correction due to the inclusion of the spin-orbit splitting of the valence band is almost negligible in the case of Ge, AlAs, GaAs and InP where the spin-orbit splitting is much smaller than the band gap. Even in the case of GaSb and InAs, where the spin-orbit splitting is equal to the band gap, the resulting correction is only about 10%. Only in the case of InSb, where the spin-orbit splitting is much larger than the band gap, we find a significant difference between our extremely simple model and a refined model including the spin-orbit splitting of the valence band. On the other hand corrections due to the inclusion of higher conduction bands do not follow such a simple rule and are almost always typically 10%. It is interesting to note that the values for $2P^2/em_e$ are nearly the same for all semiconductors. Hence, in all these semiconductors the matrix elements iP representing the mixing of the conduction band and the valence band are apparently nearly equal. This suggests that also the basis functions $|c0\rangle$, $|vx0\rangle$, etc., are very similar in all these substances.

THE VALENCE BAND

As the spin-orbit coupling has already been taken into account in the previous section, the only important correction on the results (4.95) to (4.97) for the valence bands is due to the inclusion of higher conduction bands in the **k·p** approximation.

Unfortunately, this results in a much more complicated matrix $\mathbf{H}_k^v$ from which the energy levels in the valence band must be solved. The reason for this more complicated shape is seen by returning to section 4.2.3. The extremely simple form

$$\frac{\hbar^4 P^2}{m_e^2 E_g} \begin{pmatrix} k_x^2 & k_x k_y & k_x k_z \\ k_y k_x & k_y^2 & k_y k_z \\ k_z k_x & k_z k_y & k_z^2 \end{pmatrix}$$

of the second matrix in eq. (4.60) originates in the many δ-functions in expression (4.59) for $1/m_{\mu\nu}^{\pi\rho}$. However, these δ-functions are specific for the simple case of mixing with the lowest conduction band only. If we include more conduction bands in the $\mathbf{k}\cdot\mathbf{p}$-approximation, we need to return to the more general expression (4.58) for $1/m_{\mu\nu}^{\pi\rho}$ which does not contain such δ-functions. Then eq. (4.60) for the matrix $\mathbf{H}_k^v$ becomes

$$\mathbf{H}_k^v = E_0^v \mathbf{1} + \frac{\hbar^2}{2m_e} \sum_{\mu,\nu} \begin{pmatrix} (m_{\mu\nu}^{xx})^{-1} & (m_{\mu\nu}^{xy})^{-1} & (m_{\mu\nu}^{xz})^{-1} \\ (m_{\mu\nu}^{yx})^{-1} & (m_{\mu\nu}^{yy})^{-1} & (m_{\mu\nu}^{yz})^{-1} \\ (m_{\mu\nu}^{zx})^{-1} & (m_{\mu\nu}^{zy})^{-1} & (m_{\mu\nu}^{zz})^{-1} \end{pmatrix} k_\mu k_\nu \quad (4.101)$$

where again $\mathbf{1}$ is the 3×3 unit matrix.

At this point it is very profitable to invoke the symmetry properties of the crystal. This can be done very elegantly by means of group theory [4], but such a treatment is beyond the scope of this book. Here, we will simply use some results of group theory as compiled in appendix D for cubic crystals. To be able to use those results, we start by noting that the quantities $(m_{\mu\nu}^{\pi\rho})^{-1}$ in eq. (4.101) constitute a tensor of rank 4. Next, we consider Fig. 4.2 showing the Wigner–Seitz primitive cell used to calculate the band structure. The basis states of the tensor in eq. (4.101) are p-orbitals given by eq. (4.48) and centred at the *origin* of this Wigner–Seitz primitive cell. Now, as seen in fig. 4.2 this origin is surrounded by a regular tetrahedron of nearest neighbour atoms, so the symmetry at the origin of this Wigner–Seitz cell is tetrahedral. Hence, the tensor of rank 4 consisting of the quantities $(m_{\mu\nu}^{\pi\rho})^{-1}$ should obey tetrahedral symmetry. As discussed in Appendix D, it then has three independent non-zero components (D.10) to (D.12) only. Here, we denote them as

$$\frac{1}{m_{xx}^{xx}} = \frac{1}{m_{yy}^{yy}} = \frac{1}{m_{zz}^{zz}} = 1 - \frac{2\hbar^2}{m_e} \sum_{m \neq v} \sum_{a=1}^{g_m} \frac{\langle vz0|\nabla_z|ma0\rangle\langle ma0|\nabla_z|vz0\rangle}{E_0^v - E_0^m} = L \quad (4.102)$$

$$\frac{1}{m_{xx}^{yy}} = \frac{1}{m_{yy}^{zz}} = \frac{1}{m_{zz}^{xx}} = \frac{1}{m_{yy}^{xx}} = \frac{1}{m_{zz}^{yy}} = \frac{1}{m_{xx}^{zz}}$$

$$= 1 - \frac{2\hbar^2}{m_e} \sum_{m \neq v} \sum_{a=1}^{g_m} \frac{\langle vz0|\nabla_x|ma0\rangle\langle ma0|\nabla_x|vz0\rangle}{E_0^v - E_0^m} = M \quad (4.103)$$

$$\frac{1}{m_{xy}^{xy}} = \frac{1}{m_{yz}^{yz}} = \frac{1}{m_{zx}^{zx}} = \frac{1}{m_{yx}^{yx}} = \frac{1}{m_{zy}^{zy}} = \frac{1}{m_{xz}^{xz}}$$

$$= -\frac{2\hbar^2}{m_e} \sum_{m \neq v} \sum_{a=1}^{g_m} \frac{\langle v\pi 0|\nabla_\mu|ma0\rangle\langle ma0|\nabla_\nu|v\rho 0\rangle}{E_0^v - E_0^m} = N \quad (4.104)$$

4.2 BAND STRUCTURE OF SEMICONDUCTORS

As a result, after summing over μ and ν, the matrix in eq. (4.101) reduces to

$$\frac{\hbar^2}{2m_e} \begin{pmatrix} Lk_x^2 + M(k_y^2 + k_z^2) & Nk_xk_y & Nk_xk_z \\ Nk_yk_x & Lk_y^2 + M(k_z^2 + k_x^2) & Nk_yk_z \\ Nk_zk_x & Nk_zk_y & Lk_z^2 + M(k_x^2 + k_y^2) \end{pmatrix} \quad (4.105)$$

After this reduction a set of equations similar to eq. (4.74) is obtained which can still be solved analytically [18]. However, the resulting expressions are cumbersome and here we will only give results for the case that the matrix (4.105) is small compared to the spin-orbit coupling, comprised in the third matrix in eq. (4.73). For the energy of the light and heavy hole bands one obtains

$$E_{hh} = E_0^v + \lambda + \frac{\hbar^2}{2m_e} f_{hh} k^2 \quad (4.106)$$

$$E_{lh} = E_0^v + \lambda + \frac{\hbar^2}{2m_e} f_{lh} k^2 \quad (4.107)$$

where

$$f_{hh} = A + \left(B^2 + C^2 \frac{k_x^2 k_y^2 + k_y^2 k_z^2 + k_z^2 k_x^2}{k^4} \right)^{1/2} \quad (4.108)$$

$$f_{lh} = A - \left(B^2 + C^2 \frac{k_x^2 k_y^2 + k_y^2 k_z^2 + k_z^2 k_x^2}{k^4} \right)^{1/2} \quad (4.109)$$

and

$$A = \frac{1}{3}[L + 2M] \quad (4.110)$$

$$B = \frac{1}{3}[L - M] \quad (4.111)$$

$$C^2 = \frac{1}{3}[N^2 - (L - M)^2] \quad (4.112)$$

These bands have thus lost their isotropy and have obtained a complicated orientation dependence in reciprocal space. This can be seen in Fig. 4.6 showing a surface of constant energy in reciprocal space. This complicated orientational dependence of the bands is called 'warped'.

The first three columns of Table 4.2 summarize the quantities L, M and N for some semiconductors as calculated from the values of A, B and $|C|$ given in ref. [24]. It is interesting to compare these numbers with the extremely simple model presented in the previous section. Then,

$$M - 1 = 0 \quad (4.113)$$

$$L - 1 = N = -\frac{2\hbar^2 P^2}{m_e E_g^0} = \left(1 - \frac{1}{m_c^*} \right) \quad (4.114)$$

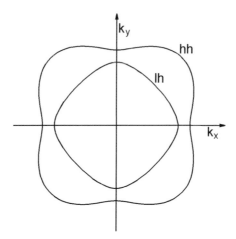

Figure 4.6 Surface of constant energy in reciprocal space for a warped band. The figure is obtained using the band structure parameters for silicon given in Table 4.2.

Table 4.2 Parameters describing the heavy and light hole bands.

	L	M	N	$1 - (m_c^*)^{-1}$
Si	−3.6	−6.5	−8.7	
Ge	−30.4	−5.7	−33.9	−26.0
AlAs	−8.2	−2.9	−10.2	−7.1
GaAs	−16.0	−3.5	−17.3	−13.9
GaSb	−28.1	−4.5	−31.2	−23.4
InP	−14.6	−3.1	−16.5	−12.0
InAs	−53	−4	−55	−41
InSb	−98	−5	−101	−70

These expressions are easily evaluated using the experimental values for m^* at the Γ-point as listed in Table 4.1. The results are tabulated in the last column of Table 4.2. Note that the case of Si is omitted because at the Γ-point its conduction band does not behave normally. It is seen that our extremely simple model is less good than in the case of the conduction band. For L and N the simple model yields values that are about 20% lower than those found in ref. [24]. However, the deviation is very much the same in all cases, so the relative values are still quite good.

5 Effective Mass Theory

5.1 Externally Applied Fields
5.1.1 Introduction
WAVE PACKETS

In real life, i.e. in practical applications, electric fields and sometimes also magnetic fields are applied to a semiconductor. As a result electrons in the conduction band and holes in the valence band are accelerated and start to move through it. This is observed as the conductivity of the semiconductor. Our present purpose is to describe this motion of electrons and holes quantum mechanically, i.e. as the evolution of their wave functions. Above we found that Bloch functions describe the eigenstates of electrons and holes and therefore it seems logical to use these functions for the description of electron and hole motion. Unfortunately however, pure Bloch functions cannot be directly used for this purpose. The reason is that Bloch functions are time independent and extend over the crystal as a whole, while a moving electron can only be represented by a wave packet which is more localized, so it has a well defined centre of gravity that may move in time. On the other hand, because the set of Bloch functions

$$b_k^n(\mathbf{r}) = u_k^n(\mathbf{r})e^{i\mathbf{k}\cdot\mathbf{r}} = e^{i\mathbf{k}\cdot\mathbf{r}}|nk\rangle \tag{5.1}$$

is orthogonal and complete, we may construct such a wave packet from this set. Then such a wave packet is described by the wave function

$$\Psi(\mathbf{r},t) = \sum_n \int_{V_B} d\mathbf{k}\, G^n(\mathbf{k},t) b_k^n(\mathbf{r})$$

$$= \sum_n \int_{V_B} d\mathbf{k}\, G^n(\mathbf{k},t) u_k^n(\mathbf{r}) e^{i\mathbf{k}\cdot\mathbf{r}}$$

$$= \sum_n \int_{V_B} d\mathbf{k}\, G^n(\mathbf{k},t) e^{i\mathbf{k}\cdot\mathbf{r}}|nk\rangle \tag{5.2}$$

where $G^n(\mathbf{k},t)$ is time dependent to allow for the motion of the wave packet. Note that in eq. (5.2) the sum over bands $\sum_n$ combined with the integral over a single Brillouin zone $\int_{V_B} d\mathbf{k}$ corresponds to an integral over full reciprocal space.

Furthermore, while Bloch functions extend over infinite space and are generally normalized in a unit volume, wave packets are localized and should be normalized in infinite space. Now, using eq. (5.2)

$$\int_\infty d\mathbf{r}\, |\Psi(\mathbf{r},t)|^2$$

$$= \int_\infty d\mathbf{r} \left[\sum_m \int_{V_B} d\mathbf{k}'\, G^m(\mathbf{k}',t) b_{k'}^m(\mathbf{r})\right]^* \left[\sum_n \int_{V_B} d\mathbf{k}\, G^n(\mathbf{k},t) b_k^n(\mathbf{r})\right]$$

$$= \sum_m \sum_n \int_{V_B} d\mathbf{k}' \int_{V_B} d\mathbf{k}\, G^{m*}(\mathbf{k}',t) G^n(\mathbf{k},t) \int_\infty d\mathbf{r}\, b_{k'}^{m*}(\mathbf{r}) b_k^n(\mathbf{r})$$

$$= \sum_m \sum_n \int_{V_B} d\mathbf{k}' \int_{V_B} d\mathbf{k}\, G^{m*}(\mathbf{k}',t) G^n(\mathbf{k},t) \delta_{m,n} \delta(\mathbf{k}' - \mathbf{k})$$

$$= \sum_n \int_{V_B} d\mathbf{k}\, |G^n(\mathbf{k},t)|^2 \tag{5.3}$$

Hence, when $\Psi(\mathbf{r},t)$ is normalized in infinite normal space, then $G^n(\mathbf{k},t)$ is normalized in infinite reciprocal space. Thus, eq. (5.2) retains the normalization convention given by eqs (3.23)–(3.25) for non-periodic functions.

EFFECTIVE MASS THEORY

The present chapter is devoted to the derivation of a formalism allowing insight in the shape of $\Psi(\mathbf{r},t)$ and its evolution under the influence of externally applied fields, electron–phonon interaction and impurity scattering. For this purpose we derive a powerful tool: effective mass theory. It is based on the following idea. The previous chapter described the band structure near the Γ-point as a series development in $\mathbf{k}$. Thus, for a non-degenerate band, as e.g. the conduction band, we find up to second order in $\mathbf{k}$:

$$E_k^n = E_0^n + \frac{\hbar^2}{2m_e} \sum_{\mu,\nu} \frac{k_\mu k_\nu}{m_{\mu\nu}^*} \tag{5.4}$$

As $\hbar\mathbf{k}$ has the dimension of momentum, the $\mathbf{k}$-dependent term in E_k^n has the shape of the kinetic energy of a *free* particle with an anisotropic mass $m_e m_{\mu\nu}^*$. Effective mass theory shows how the motion of an electron in a non-degenerate band like the conduction band can *indeed* be described as that of such a *free* particle with a mass $m_e m_{\mu\nu}^*$. Thus we obtain an interesting simplification: the complicated periodic potential $\mathcal{V}'$ is replaced by a simple adaptation of the electron mass.

In effective mass theory $\hbar\mathbf{k}$ represents the momentum of this free particle, explaining its name, crystal momentum. This allows a more precise argument showing that wave packets consisting of several Bloch functions are needed to describe motion of electrons and holes. Of course, also in effective mass theory the uncertainty relation, $\Delta|\hbar\mathbf{k}|\Delta|\mathbf{r}| \approx \hbar$ should hold. So a spread in the value of $\mathbf{k}$ is needed to construct a wave packet with a finite spatial extension $\Delta|\mathbf{r}|$.

5.1.2 The Hamiltonian with Externally Applied Fields

INSERTION OF THE SCALAR AND VECTOR POTENTIAL

We begin our treatment by considering the extra terms in the single electron Hamiltonian that arise when electric and magnetic fields are applied. To determine these terms we go all the way back to the Hamiltonian (2.20) of a complete crystal and add the influence of an externally applied electric field with a strength $\mathbf{E}_0(\mathbf{r}_i, t)$ and a magnetic field $\mathbf{B}_0(\mathbf{r}_i, t)$ at the position $\mathbf{r}_i = (x_i, y_i, z_i)$ of the ith electron. Time dependence of these fields is indicated explicitly. The index 0 indicates that these are the vacuum values of these fields. Because the Hamiltonian (2.20) gives a microscopic description of the crystal, corrections on these fields due to electrical polarization and magnetization of the crystal have not yet been taken into account. As we will see below, these corrections appear during the derivation of the single electron Hamiltonian. At this stage it is worthwhile estimating the influence of such polarization and magnetization, however. For this purpose we consider Table 5.1 giving the values of the relative permittivity and the magnetic susceptibility of some semiconductors. As can be seen from this table, the relative permittivity is quite large, so effects due to polarization of the crystal cannot be neglected. On the other hand, the magnetic susceptibility is always very small, so we may neglect the influence of magnetization of the medium. Therefore in the following we will consider the former effect only.

We start by writing $\mathbf{E}_0(\mathbf{r}_i, t)$ and $\mathbf{B}_0(\mathbf{r}_i, t)$ in the standard way as the derivatives of a scalar potential $U_0(\mathbf{r}_i, t)$ and a vector potential $\mathbf{A}_0(\mathbf{r}_i, t)$,

$$\mathbf{E}_0(\mathbf{r}_i, t) = -\nabla_i U_0(\mathbf{r}_i, t) - \frac{\partial \mathbf{A}_0(\mathbf{r}_i, t)}{\partial t}$$

$$\mathbf{B}_0(\mathbf{r}_i, t) = \nabla_i \times \mathbf{A}_0(\mathbf{r}_i, t) \tag{5.5}$$

Next, we insert these potentials in the general Hamiltonian (2.20). Then,

$$\mathcal{H} = \sum_i \left\{ \frac{1}{2m_e} \left[\frac{\hbar}{i} \nabla_i + e\mathbf{A}_0(\mathbf{r}_i, t) \right] \cdot \left[\frac{\hbar}{i} \nabla_i + e\mathbf{A}_0(\mathbf{r}_i, t) \right] - eU_0(\mathbf{r}_i, t) \right\} + \mathcal{V} \tag{5.6}$$

where $\mathcal{V}$ contains all electrostatic potentials given by eqs (2.21), (2.22) and (2.23).

As in chapter 2 we construct a single electron Hamiltonian to calculate the eigenfunctions of the individual electrons. Because of the externally applied fields, this single electron Hamiltonian is now given by

$$\mathcal{H}'(\mathbf{r}_i, t) = \frac{1}{2m_e} \left[\frac{\hbar}{i} \nabla_i + e\mathbf{A}_0(\mathbf{r}_i, t) \right] \cdot \left[\frac{\hbar}{i} \nabla_i + e\mathbf{A}_0(\mathbf{r}_i, t) \right] - eU_0(\mathbf{r}_i, t) + \mathcal{V}''(\mathbf{r}_i, t) \tag{5.7}$$

Table 5.1 Relative permittivities and susceptibilities of some semiconductors.

	ε_r	$\chi_m = \mu_r - 1$
Si	11.9	$-0.42 \cdot 10^{-7}$
Ge	16.2	$+0.92 \cdot 10^{-7}$
GaAs	12.91	$-0.98 \cdot 10^{-7}$

Again we use a self-consistent method to determine the potential $V''(\mathbf{r}_i, t)$. However, the result may be different from eq. (2.28), because of the externally applied fields. This is denoted by adding double primes to this potential. So now,

$$V''(\mathbf{r}_i, t) = -\sum_l \frac{Z_l e^2}{4\pi\varepsilon_0 r_{il}} + \int d\mathbf{r}_1 \int d\mathbf{r}_j \Psi'^*(t) \frac{1}{2} \sum_{j \neq i} \frac{e^2}{4\pi\varepsilon_0 r_{ij}} \Psi'(t) \qquad (5.8)$$

where r_{il} is the distance of the ith electron to the lth nucleus and r_{ij} its distance to the jth electron. Furthermore, $\Psi'(t)$ is the eigenstate of the crystal in the presence of externally applied fields. To distinguish it from its counterpart in the absence of external fields we add a prime. Finally, $\Psi'(t)$ will generally be time dependent because the externally applied fields are time dependent. As a result the self-consistent potential will in general be time dependent as well.

ELECTRICAL POLARIZATION

Thus, in eq. (5.7) the externally applied fields act in two different ways. First, they appear directly in the single electron Hamiltonian. Second, they change the self-consistent potential by adding a term

$$\delta V''(\mathbf{r}_i, t) = V''(\mathbf{r}_i, t) - V'(\mathbf{r}_i), \qquad (5.9)$$

where $V'(\mathbf{r}_i)$ is the original potential given by eq. (2.28). The character of this extra potential energy can be understood as follows. The externally applied electric field polarizes the crystal, i.e. it changes the electronic state in such a way that the electric dipole moment has a finite expectation value. The thus created polarization yields an additional electric field. The change (5.9) of the self-consistent potential can then be interpreted as the extra energy of the ith electron due to this additional electric field.

The way to handle this effect can be intuitively understood by resorting to classical electromagnetism of continuous media. In externally applied fields that are not too strong, the polarization grows linearly with these fields and it can be accounted for by introducing a relative permittivity ε_r in the potentials. Thus, while in vacuum the scalar potential due to a charge q at a position $\mathbf{R} = 0$ is given by

$$U_0(\mathbf{r}) = \frac{q}{4\pi\varepsilon_0 |\mathbf{r}|}$$

in a continuous medium it changes to

$$U(\mathbf{r}) = \frac{q}{4\pi\varepsilon_0 \varepsilon_r |\mathbf{r}|} = U_0(\mathbf{r}) + \delta U(\mathbf{r})$$

where

$$\delta U(\mathbf{r}) = \frac{q}{4\pi\varepsilon_0} \left(\frac{1 - \varepsilon_r}{\varepsilon_r} \right) \frac{1}{|\mathbf{r}|}$$

is the additional scalar potential due to the polarization of the medium. As a result an electron at position $\mathbf{r}_i$ in a polarized continuous medium experiences an extra potential energy $e\delta U(\mathbf{r})$.

As discussed above, $\delta V''(\mathbf{r}_i, t)$ given by eq. (5.9) also represents the extra potential energy of an electron due to polarization of the crystal. Therefore, we could venture to

5.1 EXTERNALLY APPLIED FIELDS

take care of its influence by simply introducing a relative permittivity ε_r in the scalar and vector potentials. However, before doing so, we need to realize that on a microscopic scale a semiconductor is not a continuous medium and the electron and nuclear charge densities vary widely over a primitive cell. Hence, also the polarization induced by an external electric field varies strongly as a function of the position in a this cell. These microscopic effects are accounted for in $\delta \mathcal{V}''(\mathbf{r}_i, t)$ and would be lost if we would just insert the macroscopic permittivity ε_r given in Table 5.1.

We solve the problem by simply introducing a position-dependent relative permittivity $\varepsilon_r(\mathbf{r})$. Its relation to the macroscopic relative permittivity ε_r is not trivial. In section 5.2.2 we will discuss this matter in more detail. We enter this position dependent relative permittivity in the scalar and vector potentials in the single electron Hamiltonian, so *all* microscopic and macroscopic effects due to polarization of the medium are incorporated in these potentials. Of course, simultaneously the term $\delta \mathcal{V}''(\mathbf{r}_i, t)$ is omitted from the single electron Hamiltonian (5.7).

We note that the position dependent relative permittivity $\varepsilon_r(\mathbf{r})$ is an intrinsic property of the crystal and should obey the translational symmetry of the crystal lattice. Therefore, similar to eq. (3.33) we may write it as

$$\varepsilon_r(\mathbf{r}) = \sum_h \varepsilon_h e^{i\mathbf{K}_h \cdot \mathbf{r}} \qquad (5.10)$$

where ε_h is called the dielectric function of the crystal.

As a result, in the presence of externally applied electric and magnetic fields that are not too strong, the single electron Hamiltonian reduces to

$$\mathcal{H}'(\mathbf{r}_i, t) = \frac{1}{2m_e} \left[\frac{\hbar}{i} \nabla_i + e\mathbf{A}(\mathbf{r}_i, t) \right] \cdot \left[\frac{\hbar}{i} \nabla_i + e\mathbf{A}(\mathbf{r}_i, t) \right] - eU(\mathbf{r}_i, t) + \mathcal{V}'(\mathbf{r}_i) \qquad (5.11)$$

where $\mathcal{V}'(\mathbf{r}_i)$ is the self-consistent potential in the absence of externally applied electric and magnetic fields and $\mathbf{A}(\mathbf{r}_i, t)$ and $U(\mathbf{r}_i, t)$ incorporate the position dependent relative permittivity. For example, to account for a time independent electric field, we add a scalar potential

$$U(\mathbf{r}_i) = \frac{1}{\varepsilon_r(\mathbf{r}_i)} U_0(\mathbf{r}_i) \qquad (5.12)$$

where $U_0(\mathbf{r}_i)$ would have been this potential in vacuum.

FIELD TERMS IN THE HAMILTONIAN

We rewrite eq. (5.11) in order to identify the extra term due to the applied external fields. First, we assume the gauge of the vector potential such that

$$\nabla \cdot \mathbf{A}(\mathbf{r}_i, t) = 0 \qquad (5.13)$$

Then, we split the Hamiltonian into two terms

$$\mathcal{H}'(\mathbf{r}_i, t) = \mathcal{H}_0(\mathbf{r}_i) + \mathcal{H}_F(\mathbf{r}_i, t) \qquad (5.14)$$

where

$$\mathcal{H}_0(\mathbf{r}_i) = \frac{1}{2m_e} \left(\frac{\hbar}{i} \nabla_i \right) \cdot \left(\frac{\hbar}{i} \nabla_i \right) + \mathcal{V}'(\mathbf{r}_i) \qquad (5.15)$$

is the Hamiltonian in the absence of externally applied fields and

$$\mathcal{H}_F(\mathbf{r}_i, t) = \frac{e}{m_e}\left[\mathbf{A}(\mathbf{r}_i, t) \cdot \frac{\hbar}{i}\nabla_i\right] + \frac{e^2}{m_e}|\mathbf{A}(\mathbf{r}_i, t)|^2 - eU(\mathbf{r}_i, t) \qquad (5.16)$$

is the additional term due to these fields.

Note that the vector potential $\mathbf{A}(\mathbf{r}_i, t)$, the scalar potential $U(\mathbf{r}_i, t)$ and hence $\mathcal{H}_F(\mathbf{r}_i, t)$ are not translationally symmetric. Hence, the eigenfunctions of $\mathcal{H}'(\mathbf{r}_i, t)$ *cannot* be Bloch functions. Thus we obtain a still different argument why we have to resort to wave packets rather than Bloch functions to describe motion of electrons and holes in semiconductors.

5.1.3 Transitions Induced by Externally Applied Fields

TRANSITION MATRIX ELEMENTS

We now continue by investigating the effect of an externally applied electromagnetic field on a particular band electron. As we concentrate on a single electron, there is no need to retain the index i distinguishing it from the other electrons, and we will drop this index henceforward. Thus, we will calculate the matrix elements of $\mathcal{H}_F(\mathbf{r}, t)$ between eigenstates of the unperturbed Hamiltonian (5.15), i.e. between two Bloch states. Before performing the actual calculation we consider the spatial dependence of this term in more detail. We note that the potentials $\mathbf{A}(\mathbf{r}, t)$ and $U(\mathbf{r}, t)$ are due to externally applied fields that vary very little on a microscopic scale. This is even valid for ultraviolet light with a wavelength of 100 nm, because the primitive cells of semiconductors have dimensions of typically 0.5 nm only. Hence the Fourier transforms of the vacuum values $\mathbf{A}_0(\mathbf{r}, t)$ and $U_0(\mathbf{r}, t)$ of the vector and scalar potentials contain very low values of $\mathbf{k}$ only. On the other hand, as discussed above, the Fourier transform of $\varepsilon_r(\mathbf{r})$ contains reciprocal lattice vectors $\mathbf{K}_h$ only. Because of these properties of $\mathbf{A}_0(\mathbf{r}, t)$, $U(\mathbf{r}, t)$ and $\varepsilon_r(\mathbf{r})$, one expects that the Fourier transform of the field term $\mathcal{H}_F(\mathbf{r}, t)$ can be decomposed in Fourier components with wave vector $\mathbf{k}_q$,

$$\mathcal{H}_F(\mathbf{r}, t) = \sum_q \mathcal{H}_F^q \quad \text{where} \quad \mathcal{H}_F^q = h_q(\mathbf{r})e^{i\mathbf{k}_q \cdot \mathbf{r}} \qquad (5.17)$$

where $\mathbf{k}_q$ is small and $h_q(\mathbf{r})$ obeys the periodic symmetry of the crystal. Thus we assume that $\mathcal{H}_F^q$ has the same shape as a Bloch function.

Note that we have assumed a finite crystal with periodic boundary conditions as discussed in section 3.1.5 and shown in Fig. 3.5. It consists of a block of $N = L_1 L_2 L_3$ primitive cells numbered by integer indices

$$0 \leq q_\alpha < L_\alpha$$

Then $\mathbf{k}$ may only take discrete values $\mathbf{k}_q$ as given by eq. (3.52). Here q is a shorthand notation for the triple index (q_1, q_2, q_3).

We now consider the matrix elements of $\mathcal{H}_F(\mathbf{r}, t)$ between two Bloch states

$$b_i^n(\mathbf{r}) = u_i^n(\mathbf{r})e^{i\mathbf{k}_i \cdot \mathbf{r}} \qquad (5.18)$$

and

$$b_f^m(\mathbf{r}) = u_f^m(\mathbf{r})e^{i\mathbf{k}_f \cdot \mathbf{r}} \qquad (5.19)$$

where the indices i and f are shorthand notations for the triple indices (i_1, i_2, i_3) and (f_1, f_2, f_3) identifying the discrete values $\mathbf{k}_i$ and $\mathbf{k}_f$ where these two Bloch functions are defined. Using this notation, the energy of an electron in these two Bloch states is given by E_i and E_f, respectively. For our finite crystal containing N primitive cells, so it has a volume NV_C, direct evaluation yields

$$\begin{aligned}(\mathcal{H}_F^q)_{ni,mf} &= \frac{1}{NV_C}\int_{NV_C} d\mathbf{r}\, b_i^{n*}(\mathbf{r})\mathcal{H}_F^q b_f^m(\mathbf{r}) \\ &= \frac{1}{NV_C}\int_{NV_C} d\mathbf{r}\, u_i^{n*}(\mathbf{r})e^{-i\mathbf{k}_i\cdot\mathbf{r}} h_q(\mathbf{r})e^{i\mathbf{k}_q\cdot\mathbf{r}} u_f^m(\mathbf{r})e^{i\mathbf{k}_f\cdot\mathbf{r}} \\ &= \frac{1}{NV_C}\int_{NV_C} d\mathbf{r}\, u_i^{n*}(\mathbf{r})h_q(\mathbf{r})u_f^m(\mathbf{r})e^{i(\mathbf{k}_q-\mathbf{k}_i+\mathbf{k}_f)\cdot\mathbf{r}} \end{aligned} \quad (5.20)$$

We continue by splitting the integral over the crystal in a sum of integrals over the individual primitive cells. Next, we replace $\mathbf{r}$ by $(\mathbf{r}-\mathbf{R}_l)+\mathbf{R}_l$ in the exponential function. Then,

$$\begin{aligned}(\mathcal{H}_F^q)_{ni,mf} &= \frac{1}{N}\sum_l^N \frac{1}{V_C}\int_{V_C^l} d\mathbf{r}\, u_i^{n*}(\mathbf{r})h_q(\mathbf{r})u_f^m(\mathbf{r})e^{i(\mathbf{k}_q-\mathbf{k}_i+\mathbf{k}_f)\cdot\mathbf{r}} \\ &= \frac{1}{N}\sum_l^N e^{i(\mathbf{k}_q-\mathbf{k}_i+\mathbf{k}_f)\cdot\mathbf{R}_l} \times \frac{1}{V_C}\int_{V_C^l} d\mathbf{r}\, u_i^{n*}(\mathbf{r})h_q(\mathbf{r})u_f^m(\mathbf{r})e^{i(\mathbf{k}_q-\mathbf{k}_i+\mathbf{k}_f)\cdot(\mathbf{r}-\mathbf{R}_l)} \end{aligned} \quad (5.21)$$

where $\int_{V_C^l}$ denotes integration over the lth primitive cell. Now, the three functions $u_i^n(\mathbf{r})$, $h_q(\mathbf{r})$ and $u_f^m(\mathbf{r})$ obey the translational symmetry of the crystal lattice. Furthermore, in each integral the exponential function yields the same contribution. Therefore the integral is independent of the primitive cell. Hence, we may replace all integrals by the integral over the primitive cell at the origin of our frame of reference, where $\mathbf{R}_l = 0$. Using the Dirac notation defined in eqs (4.23)–(4.26), we find

$$\begin{aligned}(\mathcal{H}_F^q)_{ni,mf} &= \frac{1}{N}\sum_l^N e^{i(\mathbf{k}_q-\mathbf{k}_i+\mathbf{k}_f)\cdot\mathbf{R}_l} \times \frac{1}{V_C}\int_{V_C^0} d\mathbf{r}\, u_i^{n*}(\mathbf{r})h_q(\mathbf{r})u_f^m(\mathbf{r})\, e^{i(\mathbf{k}_q-\mathbf{k}_i+\mathbf{k}_f)\cdot\mathbf{r}} \\ &= \langle n\mathbf{k}_i | h_q(\mathbf{r})e^{i(\mathbf{k}_q-\mathbf{k}_i+\mathbf{k}_f)\cdot\mathbf{r}} | m\mathbf{k}_f\rangle \frac{1}{N}\sum_l^N e^{i(\mathbf{k}_q-\mathbf{k}_i+\mathbf{k}_f)\cdot\mathbf{R}_l} \\ &= \langle n\mathbf{k}_i | h_q(\mathbf{r}) | m\mathbf{k}_f\rangle\, \delta_{q,q_{if}} \end{aligned} \quad (5.22)$$

Here we also inserted the Kronecker δ-function defined by eqs (3.55)–(3.57) and subsequently used this δ-function to evaluate the exponential function within the Dirac brackets. Note that for simplicity we introduced the index q_{if} to denote the vector $\mathbf{k}_i - \mathbf{k}_f$.

INTRABAND AND INTERBAND TRANSITIONS

Eq. (5.22) allows us to consider the kind of mixing that can be induced by $\mathcal{H}_F(\mathbf{r}_i, t)$. As discussed above, even for near ultraviolet light the wave vector $\mathbf{k}_q$ is small, i.e. much smaller than the size of a Brillouin zone. Hence, eq. (5.22) yields finite matrix elements only between Bloch states for which $\mathbf{k}_i$ is very near to $\mathbf{k}_f$ in the Brillouin zone. Let us

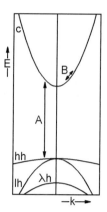

Figure 5.1 Examples of Block states mixed by externally applied fields: (A) interband mixing; (B) intraband mixing.

consider a typical band structure as shown in Fig 5.1. Matrix elements between Bloch states belonging to two different bands, $n \neq m$, cause so-called interband mixing which is indicated in Fig. 5.1 by arrow 'A'. Then the requirement $\mathbf{k}_i \approx \mathbf{k}_f$ implies that these Bloch states must be almost vertically above each other. On the other hand matrix elements between two Bloch states belonging to the same band, so $n = m$, cause intraband mixing indicated by arrow 'B'. Then the same argument restricts mixing to states that are very near to each other in the Brillouin zone.

Mixing of type 'A' may result in interband transitions of band electrons. According to first-order time-dependent perturbation theory, the probability of these interband transitions is given by

$$W_{ni,mf} = \frac{2\pi}{\hbar} \left| (\mathcal{H}_F^q)_{ni,mf} \right|^2 \delta(E_f^m - E_i^n \pm \hbar\omega)$$

$$= \frac{2\pi}{\hbar} |\langle nk_i | h_q(\mathbf{r}) \delta_{q,q_{if}} | mk_f \rangle|^2 \, \delta(E_f^m - E_i^n \pm \hbar\omega) \quad (5.23)$$

where ω is the frequency of the electromagnetic field. The latter Dirac δ-function corresponds to conservation of energy and requires large quanta, $(E_f^m - E_i^n)$, to be transferred from the electromagnetic field to the band electrons or vice versa. Therefore such transitions can be induced only if the external fields vary rapidly in time with a frequency,

$$\omega = (E_f^m - E_i^n)/\hbar \quad (5.24)$$

If we take a rapid look at the band structures shown in Figs. 3.7a–3.10a, such transitions bridge energy splittings of typically 1 eV, corresponding to optical frequencies in the near infrared to visible regime.

On the other hand mixing of type 'B' may induce intraband transitions that require very small energy quanta only. Therefore such transitions are induced only by slowly varying or constant externally applied fields. In chapter 9 we will consider interband transitions of type 'A' in more detail. Here we discuss the effects of constant or slowly varying external fields inducing intraband mixing of type 'B'.

5.2 The Effective Hamiltonian
5.2.1 Envelope Functions

We now concentrate on the development of the wave function of an electron in the conduction band under the influence of slowly varying electric and magnetic fields. Then, as discussed in section 5.1.3, interband mixing is small and we may write the wave packet (5.2) as a linear combination of Bloch functions belonging to a *single* band:

$$\Psi^n(\mathbf{r}, t) = \int_{V_B} d\mathbf{k}\, G^n(\mathbf{k}, t) \tilde{b}_k^n(\mathbf{r}) \tag{5.25}$$

where the Bloch functions

$$\tilde{b}_k^n(\mathbf{r}) = e^{i\mathbf{k}\cdot\mathbf{r}} \widetilde{|nk\rangle} \tag{5.26}$$

are marked with a tilde because they may be weakly modified by interband mixing. The basic idea behind effective mass theory is that we do not need to calculate the evolution of the full wave function to obtain the trajectory of an electron through the crystal. In particular the microscopic structure of this full wave function is not relevant. It is sufficient to calculate the evolution of the so-called envelope function, a smooth function which is obtained by replacing $\Psi^n(\mathbf{r}, t)$ by its average in each primitive cell.

We will use the Dirac notation developed in the previous chapter to describe the method by which the envelope function is defined. Before we can use this notation, we need to modify it slightly however. The reason is as follows. In chapter 4 all operators $Q(\mathbf{r}, t)$ obeyed the translational symmetry of the crystal, so the magnitude of their matrix elements (4.24) did not depend on the choice of the primitive cell and we did not need to indicate this choice in the notation. Now we have to consider quantities like the externally applied electric and magnetic fields, that vary from primitive cell to primitive cell. Therefore we need to indicate the primitive cell for which the integral applies. We do this by adding an index l denoting that we integrate over the lth primitive cell V_C^l. Thus, matrix elements are written as

$$\langle_l nk | Q(\mathbf{r}, t) | mk'\rangle = \frac{1}{V_C} \int_{V_C^l} d\mathbf{r}\, u_k^{n*}(\mathbf{r}) Q(\mathbf{r}, t) u_{k'}^m(\mathbf{r}) \tag{5.27}$$

Using this notation we define the discrete envelope function $F_l^n(t)$ of the wave function $\Psi^n(\mathbf{r}, t)$ as

$$\begin{aligned} F_l^n(t) &\equiv \frac{1}{(2\pi)^{3/2}} \langle_l n0 | \Psi^n(\mathbf{r}, t)\rangle \\ &= \frac{1}{(2\pi)^{3/2}} \langle_l n0 | \int_{V_B} d\mathbf{k}\, G^n(\mathbf{k}, t) e^{i\mathbf{k}\cdot\mathbf{r}} \widetilde{|nk\rangle} \\ &= \frac{1}{(2\pi)^{3/2}} \int_{V_B} d\mathbf{k}\, G^n(\mathbf{k}, t) \langle_l n0 | e^{i\mathbf{k}\cdot\mathbf{r}} \widetilde{|nk\rangle} \end{aligned} \tag{5.28}$$

Below we will see that the addition of the factor $1/(2\pi)^{3/2}$ is needed to retain the normalization convention of chapter 3. To study the relation between the envelope function in normal space $F_l^n(t)$ and its counterpart in reciprocal space, we write

$$\langle {}_l n 0 | e^{i\mathbf{k}\cdot\mathbf{r}} | \widetilde{nk} \rangle = \langle {}_l n 0 | e^{i\mathbf{k}\cdot(\mathbf{r}-\mathbf{R}_l)} | \widetilde{nk} \rangle e^{i\mathbf{k}\cdot\mathbf{R}_l} \tag{5.29}$$

Subsequently we use the same argument as above in the derivation of eq. (5.22): because the expression between Dirac brackets is independent of the choice of the primitive cell, we may replace the integral over the lth primitive cell by an integral over the 0th primitive cell for which we choose $\mathbf{R}_0 = 0$. So

$$\langle {}_l n 0 | e^{i\mathbf{k}\cdot\mathbf{r}} | \widetilde{nk} \rangle = \langle {}_0 n 0 | e^{i\mathbf{k}\cdot\mathbf{r}} | \widetilde{nk} \rangle e^{i\mathbf{k}\cdot\mathbf{R}_l} \tag{5.30}$$

We now define

$$\widetilde{G}^n(\mathbf{k}, t) \equiv G^n(\mathbf{k}, t) \langle {}_0 n 0 | e^{i\mathbf{k}\cdot\mathbf{r}} | \widetilde{nk} \rangle \tag{5.31}$$

which is a function of $\mathbf{k}$ only. Then,

$$F_l^n(t) = \frac{1}{(2\pi)^{3/2}} \int_{V_B} d\mathbf{k} \, \widetilde{G}^n(\mathbf{k}, t) e^{i\mathbf{k}\cdot\mathbf{R}_l} \tag{5.32}$$

We note that the factor $\langle {}_0 n 0 | \exp(i\mathbf{k}\cdot\mathbf{r}) | \widetilde{nk} \rangle$ is very close to 1 near the Γ-point where $\mathbf{k}$ is much smaller than the size $\mathbf{b}_\alpha$ of a Brillouin zone. This is seen as follows. We are calculating the average over a single primitive cell around $\mathbf{r} = 0$, so $\mathbf{r}$ is smaller than the size $\mathbf{a}_\alpha$ of this primitive cell. Thus $\mathbf{k}\cdot\mathbf{r}$ is much smaller than $\mathbf{a}_\alpha \cdot \mathbf{b}_\alpha = 2\pi$ and the factor $\exp(i\mathbf{k}\cdot\mathbf{r}) \approx 1$. Therefore, for $\mathbf{k}$ close to the Γ-point, $\widetilde{G}^n(\mathbf{k}, t) \approx G^n(\mathbf{k}, t)$ and $F_l^n(t)$ is approximately equal to the Fourier transform of $G^n(\mathbf{k}, t)$. Thus, we only need to know the averaged values $F_l^n(t)$ to obtain $G^n(\mathbf{k}, t)$ up to high precision and, using eq. (5.25) to obtain the full wave function $\Psi^n(\mathbf{r}, t)$.

Eq. (5.32) defines the envelope function $F_l^n(t)$ at discrete positions $\mathbf{R}_l$ only. A continuous envelope function $F^n(\mathbf{r}, t)$ may be defined by extending eq. (5.32) to

$$F^n(\mathbf{r}, t) = \frac{1}{(2\pi)^{3/2}} \int_{V_B} d\mathbf{k} \, \widetilde{G}^n(\mathbf{k}, t) e^{i\mathbf{k}\cdot\mathbf{r}} \tag{5.33}$$

This definition ensures that

$$F^n(\mathbf{r}, t) = F^n(\mathbf{R}_l, t) = F_l^n(t) \quad \text{for} \quad \mathbf{r} = \mathbf{R}_l \tag{5.34}$$

so $F^n(\mathbf{r}, t)$ is simply the continuous limit of $F_l^n(t)$. We recall that we are concerned with wave packets constructed of Bloch functions belonging to a single band with index n. Hence, $\widetilde{G}^n(\mathbf{k}, t)$ is defined in the first Brillouin zone only and may be set to zero in all other Brillouin zones. Thus, in eq. (5.33) the integral over V_B may be extended to full reciprocal space. As a consequence, eq. (5.33) is completely equivalent to the Fourier transform (3.17) and its inverse is given by

$$\widetilde{G}^n(\mathbf{k}, t) = \frac{1}{(2\pi)^{3/2}} \int_\infty d\mathbf{r} \, F^n(\mathbf{r}, t) e^{-i\mathbf{k}\cdot\mathbf{r}} \tag{5.35}$$

As a further consequence the transformation pair (5.33) and (5.35) obeys the normalization convention of eqs. (3.23)–(3.25). Hence, if the envelope function $F^n(\mathbf{r}, t)$ is normalized in infinite normal space, then its transform $\widetilde{G}^n(\mathbf{k}, t)$ is normalized in infinite reciprocal space. Now, according to eq. (5.3), we have defined the function $G^n(\mathbf{k}, t)$ in such a way, that if the full wave packet $\Psi(\mathbf{r}, t)$ is normalized in infinite space, then its counterpart $G^n(\mathbf{k}, t)$ is normalized in infinite reciprocal space. Hence, apart from a factor $|\langle_0 n0|\exp(i\mathbf{k}\cdot\mathbf{r})|\widetilde{nk}\rangle|^2 \approx 1$, the normalization convention for the envelope function is the same as the one for the full wave packet.

5.2.2 The P·π Method

AVERAGING THE SCHRÖDINGER EQUATIONS OVER PRIMITIVE CELLS

We continue by studying the evolution of the envelope function under the influence of externally applied electric and magnetic fields. The behaviour of the full wave function $\Psi^n(\mathbf{r}, t)$ is obtained by solving the time-independent and time-dependent Schrödinger equations

$$\mathcal{H}'\Psi^n(\mathbf{r}, t) = E^n \Psi^n(\mathbf{r}, t) \tag{5.36}$$

$$\mathcal{H}'\Psi^n(\mathbf{r}, t) = i\hbar \frac{\partial}{\partial t} \Psi^n(\mathbf{r}, t) \tag{5.37}$$

where

$$\mathcal{H}' = \frac{1}{2m_e}\left[\frac{\hbar}{i}\nabla + e\mathbf{A}(\mathbf{r}, t)\right] \cdot \left[\frac{\hbar}{i}\nabla + e\mathbf{A}(\mathbf{r}, t)\right] - eU(\mathbf{r}, t) + \mathcal{V}'$$

is the full single electron Hamiltonian given by eq. (5.11). As discussed above, we would be quite happy knowing the shape and time evolution of the envelope function $F_l^n(t)$ only. We now propose that we can construct a simple effective Hamiltonian $\mathcal{H}_{eff}$ that can be inserted in simple effective Schrödinger equations from which the shape and time evolution of $F_l^n(t)$ can be solved.

To obtain such effective Schrödinger equations we start with the full Schrödinger equations (5.36) and (5.37) and average following the procedure (5.28) for $F_l^n(t)$. Then

$$\frac{1}{(2\pi)^{3/2}} \int_{V_B} d\mathbf{k}\, G^n(\mathbf{k}, t)\langle_l n0|\mathcal{H}' e^{i\mathbf{k}\cdot\mathbf{r}}|\widetilde{nk}\rangle = E^n \frac{\langle_l n0|\Psi^n(\mathbf{r}, t)\rangle}{(2\pi)^{3/2}} = E^n F_l^n(t) \tag{5.38}$$

$$\frac{1}{(2\pi)^{3/2}} \int_{V_B} d\mathbf{k}\, G^n(\mathbf{k}, t)\langle_l n0|\mathcal{H}' e^{i\mathbf{k}\cdot\mathbf{r}}|\widetilde{nk}\rangle = i\hbar \frac{\partial}{\partial t}\frac{\langle_l n0|\Psi^n(\mathbf{r}, t)\rangle}{(2\pi)^{3/2}} = i\hbar \frac{\partial F_l^n(t)}{\partial t} \tag{5.39}$$

k·p TREATMENT

Next we calculate the matrix element on the left hand side of these two equations. For this purpose we follow the procedure of the **k·p** method treated in chapter 4. First, we derive the **k·p** Hamiltonian for the case of externally applied fields. We note that similar to eq. (4.4),

$$\left[\frac{\hbar}{i}\nabla + e\mathbf{A}(\mathbf{r}, t)\right] e^{i\mathbf{k}\cdot\mathbf{r}}|\widetilde{nk}\rangle = e^{i\mathbf{k}\cdot\mathbf{r}}\left[\frac{\hbar}{i}\nabla + \hbar\mathbf{k} + e\mathbf{A}(\mathbf{r}, t)\right]|\widetilde{nk}\rangle \tag{5.40}$$

Hence, choosing the gauge of the vector potential such that

$$\nabla \cdot \mathbf{A}(\mathbf{r}, t) = 0$$

we find

$$\left[\frac{\hbar}{i}\nabla + e\mathbf{A}(\mathbf{r},t)\right] \cdot \left[\frac{\hbar}{i}\nabla + e\mathbf{A}(\mathbf{r},t)\right] e^{i\mathbf{k}\cdot\mathbf{r}} |\widetilde{nk}\rangle$$

$$= e^{i\mathbf{k}\cdot\mathbf{r}}\left[\frac{\hbar}{i}\nabla + \hbar\mathbf{k} + e\mathbf{A}(\mathbf{r},t)\right] \cdot \left[\frac{\hbar}{i}\nabla + \hbar\mathbf{k} + e\mathbf{A}(\mathbf{r},t)\right] |\widetilde{nk}\rangle$$

$$= e^{i\mathbf{k}\cdot\mathbf{r}}\left\{\left(\frac{\hbar}{i}\nabla\right) \cdot \left(\frac{\hbar}{i}\nabla\right) + 2[\hbar\mathbf{k} + e\mathbf{A}(\mathbf{r},t)]\cdot\left(\frac{\hbar}{i}\nabla\right) + |\hbar\mathbf{k} + e\mathbf{A}(\mathbf{r},t)|^2\right\}|\widetilde{nk}\rangle \quad (5.41)$$

AVERAGE EXTERNAL FIELDS

The externally applied fields are macroscopic. Hence, as discussed in section 5.1.3 the vacuum values $\mathbf{A}_0(\mathbf{r},t)$ and $U_0(\mathbf{r},t)$ of the vector and scalar potentials vary little over a primitive cell. As discussed in section 5.1.2, the externally applied fields change the periodic potential in the crystal. This was accounted for in the single electron Hamiltonian (5.11) by incorporating a position dependent relative permittivity $\varepsilon_r(\mathbf{r})$ in the vector potential $\mathbf{A}(\mathbf{r},t)$ and the scalar potential $U(\mathbf{r},t)$. Thus, like the periodic potential, this position dependent relative permittivity obeys the translational symmetry of the crystal lattice. To evaluate eqs (5.38) and (5.39) we need to calculate weighed averages of these potentials over a primitive cell. For example, the scalar potential is averaged as

$$\langle_l n0|U(\mathbf{r},t)|\widetilde{nk}\rangle = \langle_l n0|\frac{1}{\varepsilon_r(\mathbf{r})}U_0(\mathbf{r},t)|\widetilde{nk}\rangle$$

Because the fields are long range we may confidently approximate the vacuum value $U_0(\mathbf{r},t)$ by $U_0(\mathbf{R}_l,t)$. We furthermore assume that the weighed average of $1/\varepsilon_r(\mathbf{r})$ corresponds simply to the macroscopic relative permittivity $1/\varepsilon_r$. This latter assumption is not trivial. However, it is corroborated by experimental results like the energy levels of a bound donor electron to be presented in the next section. Thus we write

$$\langle_l n0|U(\mathbf{r},t)|\widetilde{nk}\rangle = \langle_l n0|\frac{1}{\varepsilon_r(\mathbf{r})}|\widetilde{nk}\rangle U_0(\mathbf{R}_l,t)$$

$$= \langle_l n0|\frac{1}{\varepsilon_r(\mathbf{r})}|\widetilde{nk}\rangle U_0(\mathbf{R}_l,t)$$

$$= \frac{1}{\varepsilon_r}U_0(\mathbf{R}_l,t) = \bar{U}(\mathbf{R}_l,t) \quad (5.42)$$

where we put a bar over $\bar{U}(\mathbf{R}_l,t)$ to distinguish it from the value of the unaveraged potential $U(\mathbf{R}_l,t)$ at the position $\mathbf{R}_l$. Thus $\bar{U}(\mathbf{R}_l,t)$ is simply equal to the macroscopic value of the scalar potential, i.e. its value in a continuous medium with relative permittivity ε_r. Similarly, we assume the weighed average of $\mathbf{A}(\mathbf{r},t)$ to be equal to $\bar{\mathbf{A}}(\mathbf{R}_l,t)$, being its value in such a continuous medium. As a result

$$\mathcal{H}'e^{i\mathbf{k}\cdot\mathbf{r}}|\widetilde{nk}\rangle = e^{i\mathbf{k}\cdot\mathbf{r}}(\mathcal{H}^{(0)} + \mathcal{H}^{(1)})|\widetilde{nk}\rangle \quad (5.43)$$

5.2 THE EFFECTIVE HAMILTONIAN

where $\mathcal{H}^{(0)}$ is the same as in the absence of externally applied fields,

$$\mathcal{H}^{(0)} = \frac{1}{2m_e}\left(\frac{\hbar}{i}\nabla\right)\cdot\left(\frac{\hbar}{i}\nabla\right) + \mathcal{V}' \tag{5.44}$$

and

$$\mathcal{H}^{(1)} = \frac{1}{m_e}[\hbar\mathbf{k} + e\bar{\mathbf{A}}(\mathbf{R}_l,t)]\cdot\left(\frac{\hbar}{i}\nabla\right) + \frac{1}{2m_e}|\hbar\mathbf{k} + e\bar{\mathbf{A}}(\mathbf{R}_l,t)|^2 - e\bar{U}(\mathbf{R}_l,t) \tag{5.45}$$

represents the extra term due the externally applied fields and to the fact that $\mathbf{k} \neq 0$. It seen that $\mathcal{H}^{(1)}$ is very similar to its counterpart (4.8) derived in section 4.1.1 as the first step in the $\mathbf{k}\cdot\mathbf{p}$-approximation. Here, a term for the scalar potential is added while furthermore $\hbar\mathbf{k}$ is replaced by $\hbar\mathbf{k} + e\bar{\mathbf{A}}(\mathbf{R}_l,t)$. In literature one often writes

$$\mathbf{P} = \hbar\mathbf{k} + e\bar{\mathbf{A}}(\mathbf{R}_l,t) \tag{5.46}$$

$$\pi \approx \left(\frac{\hbar}{i}\nabla\right) \tag{5.47}$$

Then $(\mathcal{H}^{(0)} + \mathcal{H}^{(1)})$ is called the $\mathbf{P}\cdot\pi$-Hamiltonian and the perturbation treatment based on this Hamiltonian is named the $\mathbf{P}\cdot\pi$ method. Inserting eq. (5.43) in the averaged Schrödinger equations (5.38) and (5.39) finally yields

$$\frac{1}{(2\pi)^{3/2}}\int_{V_B} d\mathbf{k}\, G^n(\mathbf{k},t)\langle_l n0|e^{i\mathbf{k}\cdot\mathbf{r}}(\mathcal{H}^{(0)} + \mathcal{H}^{(1)})|\widetilde{n\mathbf{k}}\rangle = E^n F_l^n(t) \tag{5.48}$$

$$\frac{1}{(2\pi)^{3/2}}\int_{V_B} d\mathbf{k}\, G^n(\mathbf{k},t)\langle_l n0|e^{i\mathbf{k}\cdot\mathbf{r}}(\mathcal{H}^{(0)} + \mathcal{H}^{(1)})|\widetilde{n\mathbf{k}}\rangle = i\hbar\frac{\partial F_l^n(t)}{\partial t} \tag{5.49}$$

5.2.3 Effective Schrödinger Equations

PERTURBATION EXPANSION

We proceed by evaluating the left-hand sides of eqs (5.48) and (5.49) using the results of second-order time-independent perturbation theory obtained in appendix C. Using the perturbation expansion (C.2), we write

$$(\mathcal{H}^{(0)} + \mathcal{H}^{(1)})|\widetilde{n\mathbf{k}}\rangle = (E_0^n + E_k^{n(1)} + E_k^{n(2)} + \cdots)|\widetilde{n\mathbf{k}}\rangle \tag{5.50}$$

We furthermore split the exponential function as in eq. (5.29),

$$e^{i\mathbf{k}\cdot\mathbf{r}} = e^{i\mathbf{k}\cdot\mathbf{R}_l}\cdot e^{i\mathbf{k}\cdot(\mathbf{r}-\mathbf{R}_l)}$$

to find

$$\frac{1}{(2\pi)^{3/2}} \int_{V_B} d\mathbf{k}\, G^n(\mathbf{k},t) \langle {}_l n0 | e^{i\mathbf{k}\cdot\mathbf{r}} (\mathcal{H}^{(0)} + \mathcal{H}^{(1)}) | \widetilde{n\mathbf{k}} \rangle$$

$$= \frac{1}{(2\pi)^{3/2}} \int_{V_B} d\mathbf{k}\, G^n(\mathbf{k},t) \langle {}_l n0 | e^{i\mathbf{k}\cdot\mathbf{R}_l} \cdot e^{i\mathbf{k}\cdot(\mathbf{r}-\mathbf{R}_l)} \times (E_0^n + E_k^{n(1)} + E_k^{n(2)} + \cdots) | \widetilde{n\mathbf{k}} \rangle$$

$$= \frac{1}{(2\pi)^{3/2}} \int_{V_B} d\mathbf{k}\, G^n(\mathbf{k},t) \langle {}_l n0 | e^{i\mathbf{k}\cdot(\mathbf{r}-\mathbf{R}_l)} | \widetilde{n\mathbf{k}} \rangle e^{i\mathbf{k}\cdot\mathbf{R}_l} \times (E_0^n + E_k^{n(1)} + E_k^{n(2)} + \cdots)$$

$$= \frac{1}{(2\pi)^{3/2}} \int_{V_B} d\mathbf{k}\, G^n(\mathbf{k},t) \langle {}_0 n0 | e^{i\mathbf{k}\cdot\mathbf{r}} | \widetilde{n\mathbf{k}} \rangle e^{i\mathbf{k}\cdot\mathbf{R}_l} (E_0^n + E_k^{n(1)} + E_k^{n(2)} + \cdots)$$

$$= \frac{1}{(2\pi)^{3/2}} \int_{V_B} d\mathbf{k}\, \widetilde{G}^n(\mathbf{k},t) e^{i\mathbf{k}\cdot\mathbf{R}_l} (E_0^n + E_k^{n(1)} + E_k^{n(2)} + \cdots) \quad (5.51)$$

where, as before in eq. (5.30), we replace the integral over the lth primitive cell by an integral over the primitive cell at the origin of our frame of reference where $\mathbf{R}_l = \mathbf{R}_0 = 0$. Subsequently we develop according to eq. (C.15),

$$E_0^n + E_k^{n(1)} + E_k^{n(2)} + \cdots = E_0^n + \langle {}_0 n0 | \mathcal{H}^{(1)} | n0 \rangle$$
$$- \sum_{m \neq n} \frac{\langle {}_0 n0 | \mathcal{H}^{(1)} | m0 \rangle \langle {}_0 m0 | \mathcal{H}^{(1)} | n0 \rangle}{E_0^m - E_0^n} + \cdots \quad (5.52)$$

Then we insert eq. (5.45) for $\mathcal{H}^{(1)}$. We note that $e\bar{U}(\mathbf{R}_l, t)$ and $e\bar{\mathbf{A}}(\mathbf{R}_l, t)$ are constant in the integration interval defined by the Dirac brackets, so e.g.,

$$\langle {}_0 n0 | e\bar{U}(\mathbf{R}_l, t) | m0 \rangle = e\bar{U}(\mathbf{R}_l, t) \langle {}_0 n0 | m0 \rangle = e\bar{U}(\mathbf{R}_l, t) \delta_{n,m} \quad (5.53)$$

Thus we obtain

$$E_0^n + E_k^{n(1)} + E_k^{n(2)} + \cdots \simeq E_0^n + \frac{1}{2m_e} \sum_{\mu,\nu} \frac{[\hbar k_\mu + eA_\mu(\mathbf{R}_l, t)][\hbar k_\nu + eA_\nu(\mathbf{R}_l, t)]}{m^*_{\mu\nu}}$$
$$- e\bar{U}(\mathbf{R}_l, t) \quad (5.54)$$

where $\mu, \nu = x, y, z$ and

$$\frac{1}{m^*_{\mu\nu}} = \left[\delta_{\mu,\nu} + \frac{2\hbar^2}{m_e} \sum_{m \neq n} \frac{\langle {}_0 n0 | \nabla_\mu | m0 \rangle \langle {}_0 m0 | \nabla_\nu | n0 \rangle}{E_0^m - E_0^n} \right]$$

$$= \left[\delta_{\mu,\nu} + \frac{2\hbar^2}{m_e} \sum_{m \neq n} \frac{P_\mu^{nm} P_\nu^{mn}}{E_0^m - E_0^n} \right] \quad (5.55)$$

is the effective mass tensor (4.35) derived earlier in section 4.1.3. After inserting this expression in eq. (5.51) and the result in the left-hand sides of eqs (5.48) and (5.49), the

5.2 THE EFFECTIVE HAMILTONIAN

averaged Schrödinger equations become

$$\frac{1}{(2\pi)^{3/2}}\int_{V_B} d\mathbf{k}\, \widetilde{G}^n(\mathbf{k},t)e^{i\mathbf{k}\cdot\mathbf{R}_l}$$

$$\times \left[E_0^n + \frac{1}{2m_e}\sum_{\mu,\nu}\frac{[\hbar k_\mu + e\bar{A}_\mu(\mathbf{R}_l,t)][\hbar k_\nu + e\bar{A}_\nu(\mathbf{R}_l,t)]}{m^*_{\mu\nu}} - e\bar{U}(\mathbf{R}_l,t) \right] = E^n F_l^n \quad (5.56)$$

$$\frac{1}{(2\pi)^{3/2}}\int_{V_B} d\mathbf{k}\, \widetilde{G}^n(\mathbf{k},t)e^{i\mathbf{k}\cdot\mathbf{R}_l}$$

$$\times \left[E_0^n + \frac{1}{2m_e}\sum_{\mu,\nu}\frac{[\hbar k_\mu + e\bar{A}_\mu(\mathbf{R}_l,t)][\hbar k_\nu + e\bar{A}_\nu(\mathbf{R}_l,t)]}{m^*_{\mu\nu}} - e\bar{U}(\mathbf{R}_l,t) \right] = i\hbar \frac{\partial F_l^n}{\partial t} \quad (5.57)$$

THE CONTINUOUS LIMIT

These are not yet the desired effective Schrödinger equations because the envelope function F_l^n is defined at discrete positions $\mathbf{R}_l$ only. To bring eqs (5.56) and (5.57) in the shape of effective Schrödinger equations we take the continuous limit and replace F_l^n by the continuous function $F^n(\mathbf{r},t)$ defined in eq. (5.33). Because $F_l^n(t)$ is the solution of eqs (5.56) and (5.57), its continuous limit $F^n(\mathbf{r},t)$ can be defined as the solution of the same equations, where we replace $\mathbf{R}_l$ by $\mathbf{r}$. Hence, it obeys

$$\frac{1}{(2\pi)^{3/2}}\int_{V_B} d\mathbf{k}\, \widetilde{G}^n(\mathbf{k},t)e^{i\mathbf{k}\cdot\mathbf{r}}$$

$$\times \left[E_0^n + \frac{1}{2m_e}\sum_{\mu,\nu}\frac{[\hbar k_\mu + e\bar{A}_\mu(\mathbf{r},t)][\hbar k_\nu + e\bar{A}_\nu(\mathbf{r},t)]}{m^*_{\mu\nu}} - e\bar{U}(\mathbf{r},t) \right] = E^n F^n(\mathbf{r},t) \quad (5.58)$$

$$\frac{1}{(2\pi)^{3/2}}\int_{V_B} d\mathbf{k}\, \widetilde{G}^n(\mathbf{k},t)e^{i\mathbf{k}\cdot\mathbf{r}}$$

$$\times \left[E_0^n + \frac{1}{2m_e}\sum_{\mu,\nu}\frac{[\hbar k_\mu + e\bar{A}_\mu(\mathbf{r},t)][\hbar k_\nu + e\bar{A}_\nu(\mathbf{r},t)]}{m^*_{\mu\nu}} - e\bar{U}(\mathbf{r},t) \right] = i\hbar \frac{\partial F^n(\mathbf{r},t)}{\partial t} \quad (5.59)$$

INVERSE k·p TREATMENT

We now reverse the procedure of eq. (5.41). However, such a reversal has to be performed with care because now the mass is anisotropic. First, we note that

$$\left\{ -\hbar^2 \nabla_\mu \nabla_\nu + \frac{\hbar e}{i}[\bar{A}_\mu(\mathbf{r},t)\nabla_\nu + \bar{A}_\nu(\mathbf{r},t)\nabla_\mu] + e^2 \bar{A}_\mu(\mathbf{r},t)\bar{A}_\nu(\mathbf{r},t) \right\} e^{i\mathbf{k}\cdot\mathbf{r}}$$

$$= e^{i\mathbf{k}\cdot\mathbf{r}}\left[\hbar^2 k_\mu k_\nu + e\bar{A}_\mu(\mathbf{r},t)\,\hbar k_\nu + \hbar k_\mu\, e\bar{A}_\nu(\mathbf{r},t) + e^2 \bar{A}_\mu(\mathbf{r},t)\bar{A}_\nu(\mathbf{r},t) \right]$$

$$= e^{i\mathbf{k}\cdot\mathbf{r}}[\hbar k_\mu + e\bar{A}_\mu(\mathbf{r},t)][\hbar k_\nu + e\bar{A}_\nu(\mathbf{r},t)] \quad (5.60)$$

So, the left hand side of eqs (5.58) and (5.59) can be written as

$$\frac{1}{(2\pi)^{3/2}} \int_{V_B} d\mathbf{k}\, \tilde{G}^n(\mathbf{k}, t)$$

$$\times \left(E_0^n + \frac{1}{2m_e} \sum_{\mu,\nu} \left\{ -\frac{\hbar^2}{m_{\mu\nu}^*} \nabla_\mu \nabla_\nu + \frac{\hbar e}{im_{\mu\nu}^*} [\bar{A}_\mu(\mathbf{r}, t)\nabla_\nu + \bar{A}_\nu(\mathbf{r}, t)\nabla_\mu] \right. \right.$$

$$\left. \left. + \frac{e^2}{m_{\mu\nu}^*} \bar{A}_\mu(\mathbf{r}, t)\bar{A}_\nu(\mathbf{r}, t) \right\} - e\bar{U}(\mathbf{r}, t) \right) e^{i\mathbf{k}\cdot\mathbf{r}}$$

$$= \left(E_0^n + \frac{1}{2m_e} \sum_{\mu,\nu} \left\{ -\frac{\hbar^2}{m_{\mu\nu}^*} \nabla_\mu \nabla_\nu + \frac{\hbar e}{im_{\mu\nu}^*} [\bar{A}_\mu(\mathbf{r}, t)\nabla_\nu + \bar{A}_\nu(\mathbf{r}, t)\nabla_\mu] \right. \right.$$

$$\left. \left. + \frac{e^2}{m_{\mu\nu}^*} \bar{A}_\mu(\mathbf{r}, t)\bar{A}_\nu(\mathbf{r}, t) \right\} - e\bar{U}(\mathbf{r}, t) \right) F^n(\mathbf{r}, t) \qquad (5.61)$$

In the second line we could finally integrate over $\mathbf{k}$ because none of the remaining operators and quantities is a function of $\mathbf{k}$.

THE EFFECTIVE HAMILTONIAN

Thus we obtain the desired effective Schrödinger equations from which $F^n(\mathbf{r}, t)$ can be solved:

$$\mathcal{H}_{eff} F^n(\mathbf{r}, t) = E^n F^n(\mathbf{r}, t) \qquad (5.62)$$

$$\mathcal{H}_{eff} F^n(\mathbf{r}, t) = i\hbar \frac{\partial F^n(\mathbf{r}, t)}{\partial t} \qquad (5.63)$$

Here the effective Hamiltonian is given by

$$\mathcal{H}_{eff} = E_0^n + \frac{1}{2m_e} \sum_{\mu,\nu} \left\{ -\frac{\hbar^2}{m_{\mu\nu}^*} \nabla_\mu \nabla_\nu + \frac{\hbar e}{im_{\mu\nu}^*} [\bar{A}_\mu(\mathbf{r}, t)\nabla_\nu + \bar{A}_\nu(\mathbf{r}, t)\nabla_\mu] \right.$$

$$\left. + \frac{e^2}{m_{\mu\nu}^*} \bar{A}_\mu(\mathbf{r}, t)\bar{A}_\nu(\mathbf{r}, t) \right\} - e\bar{U}(\mathbf{r}, t)$$

$$= E_0^n + \frac{1}{2m_e} \sum_{\mu,\nu} \frac{\left[\frac{\hbar}{i}\nabla_\mu + e\bar{A}_\mu(\mathbf{r}, t)\right] \left[\frac{\hbar}{i}\nabla_\nu + e\bar{A}_\nu(\mathbf{r}, t)\right]}{m_{\mu\nu}^*}$$

$$- \frac{\hbar e}{2im_e} \sum_{\mu,\nu} \frac{[\nabla_\mu \bar{A}_\nu(\mathbf{r}, t)]}{m_{\mu\nu}^*} - e\bar{U}(\mathbf{r}, t) \qquad (5.64)$$

Except for the term

$$-\frac{\hbar e}{2im_e} \sum_{\mu,\nu} \frac{[\nabla_\mu \bar{A}_\nu(\mathbf{r}, t)]}{m_{\mu\nu}^*} \qquad (5.65)$$

5.2 THE EFFECTIVE HAMILTONIAN

the effective Hamiltonian (5.64) has precisely the shape of the Hamiltonian of a free electron with an anisotropic mass $m_e m^*_{\mu\nu}$ influenced by external fields described by potentials $\bar{\mathbf{A}}(\mathbf{r},t)$ and $\bar{U}(\mathbf{r},t)$. Note that now these potentials are simply the macroscopic potentials as they contain the macroscopic relative permittivity ε_r, while all microscopic information is averaged out.

The beauty of the effective Hamiltonian (5.64) is that it allows to solve $F^n(\mathbf{r},t)$ and as discussed in section 5.2.1, once $F^n(\mathbf{r},t)$ is known, we can use eq. (5.33) to obtain $\widetilde{G}^n(\mathbf{k},t)$ and hence $G^n(\mathbf{k},t)$ up to high precision. As a result, using eq. (5.25), once the envelope function $F^n(\mathbf{r},t)$ is solved from eqs (5.62) and (5.63), also the full wave function $\Psi^n(\mathbf{r},t)$ can be derived. Thus, the behaviour of an electron in a crystal can be very well understood using the effective Hamiltonian only. Moreover, to obtain this effective Hamiltonian only the effective mass needs to be known, without requiring further detailed information about the complicated periodic potential $\mathcal{V}'$.

The extra term (5.65) is completely due to anisotropy of the effective mass. For an isotropic effective mass,

$$\frac{1}{m^*_{\mu\nu}} = \frac{1}{m^*}\delta_{\mu,\nu} \tag{5.66}$$

and

$$-\frac{\hbar e}{2im_e}\sum_{\mu,\nu}\frac{[\nabla_\mu \bar{A}_\nu(\mathbf{r},t)]}{m^*_{\mu\nu}} = -\frac{\hbar e}{2im_e}\sum_{\mu,\nu}\frac{[\nabla_\mu \bar{A}_\nu(\mathbf{r},t)]}{m^*}\delta_{\mu,\nu}$$

$$= -\frac{\hbar e}{2im_e}\sum_\mu \frac{[\nabla_\mu \bar{A}_\mu(\mathbf{r},t)]}{m^*} = -\frac{\hbar e}{2im_e m^*}[\nabla \cdot \bar{\mathbf{A}}(\mathbf{r},t)] = 0 \tag{5.67}$$

because of the gauge $\nabla \cdot \mathbf{A}(\mathbf{r},t) = 0$ chosen earlier.

CONSTANT MAGNETIC FIELDS

Moreover, the extra term (5.65) vanishes in the special case that the magnetic field $\mathbf{B}$ is constant in space while moreover the effective mass tensor is symmetric. Then, we may choose the vector potential to obey,

$$\bar{\mathbf{A}}(\mathbf{r},t) = \frac{1}{2}(\mathbf{r} \times \mathbf{B}) \tag{5.68}$$

To show that the term (5.65) vanishes, we insert the vector potential (5.68). Furthermore, in order to facilitate the calculation, we introduce a cyclic notation, μ, $\mu+1$ and $\mu+2$, for the three possible values that ν may take. Thus, if $\mu = x$ then, $\mu+1 = y$ and $\mu+2 = z$. However, when $\mu = y$, then $\mu+1 = z$ and $\mu+2 = x$, while when $\mu = z$, then $\mu+1 = x$ and $\mu+2 = y$. As a result,

$$-\frac{\hbar e}{2im_e}\sum_{\mu,\nu}\frac{[\nabla_\mu \bar{A}_\nu(\mathbf{r},t)]}{m^*_{\mu\nu}} = -\frac{\hbar e}{4im_e}\sum_{\mu,\nu}\frac{[\nabla_\mu(\mathbf{r}\times\mathbf{B})_\nu]}{m^*_{\mu\nu}}$$

$$= -\frac{\hbar e}{4im_e}\sum_{\mu}\left\{\frac{[\nabla_\mu(\mathbf{r}\times\mathbf{B})_\mu]}{m^*_{\mu\mu}} + \frac{[\nabla_\mu(\mathbf{r}\times\mathbf{B})_{(\mu+1)}]}{m^*_{\mu(\mu+1)}} + \frac{[\nabla_\mu(\mathbf{r}\times\mathbf{B})_{(\mu+2)}]}{m^*_{\mu(\mu+2)}}\right\} \quad (5.69)$$

Now,

$$\nabla_\mu(\mathbf{r}\times\mathbf{B})_\mu = \nabla_\mu(r_{\mu+1}B_{\mu+2} - r_{\mu+2}B_{\mu+1}) = 0 \quad (5.70)$$

Note that then also

$$\nabla\cdot\mathbf{A}(\mathbf{r},t) = 0$$

so the vector potential (5.68) does not contradict this previous choice for the gauge. Furthermore,

$$\nabla_\mu(\mathbf{r}\times\mathbf{B})_{(\mu+1)} = \nabla_\mu(r_{\mu+2}B_\mu - r_\mu B_{\mu+2}) = -B_{\mu+2} \quad (5.71)$$

$$\nabla_\mu(\mathbf{r}\times\mathbf{B})_{(\mu+2)} = \nabla_\mu(r_\mu B_{\mu+1} - r_{\mu+1}B_\mu) = B_{\mu+1} \quad (5.72)$$

So,

$$-\frac{\hbar e}{2im_e}\sum_{\mu,\nu}\frac{[\nabla_\mu \bar{A}_\nu(\mathbf{r},t)]}{m^*_{\mu\nu}} = -\frac{\hbar e}{4im_e}\sum_{\mu}\left(-\frac{B_{\mu+2}}{m^*_{\mu(\mu+1)}} + \frac{B_{\mu+1}}{m^*_{\mu(\mu+2)}}\right) \quad (5.73)$$

Then, by renumbering the second sum, so μ is replaced by $\mu+1$, etc.,

$$-\frac{\hbar e}{2im_e}\sum_{\mu,\nu}\frac{[\nabla_\mu \bar{A}_\nu(\mathbf{r},t)]}{m^*_{\mu\nu}} = \frac{\hbar e}{4im_e}\sum_{\mu}\left(\frac{1}{m^*_{\mu(\mu+1)}} - \frac{1}{m^*_{(\mu+1)\mu}}\right)B_{\mu+2} \quad (5.74)$$

which indeed vanishes when the effective mass tensor is symmetric, so $m^*_{\mu\nu} = m^*_{\nu\mu}$.

5.2.4 Plane Waves and Bloch Functions

A special case of the effective Schrödinger equations considers the limit that the externally applied fields vanish, so the potentials $\bar{\mathbf{A}}(\mathbf{r},t)$ and $\bar{U}(\mathbf{r},t)$ may be set to zero. This case is seemingly trivial but still deserves attention, because of the insight one obtains in the nature of the envelope function. We first consider the effective Hamiltonian (5.64) which for this special case reduces to the Hamiltonian of a free particle with an anisotropic effective mass,

$$\mathcal{H}_{eff} = E_0^n + \frac{1}{2m_e}\sum_{\mu,\nu}\frac{\left(\frac{\hbar}{i}\nabla_\mu\right)\left(\frac{\hbar}{i}\nabla_\nu\right)}{m^*_{\mu\nu}} = E_0^n - \frac{\hbar^2}{2m_e}\sum_{\mu,\nu}\frac{\nabla_\mu\nabla_\nu}{m^*_{\mu\nu}} \quad (5.75)$$

The eigenvalues of $\mathcal{H}_{eff}$ are given by

$$E^n = E_0^n + \frac{\hbar^2}{2m_e}\sum_{\mu,\nu}\frac{k_\mu k_\nu}{m^*_{\mu\nu}} \quad (5.76)$$

where **k** is the quantum number identifying the corresponding eigenstate,

$$F^n(\mathbf{r}) = e^{i\mathbf{k}\cdot\mathbf{r}} \qquad (5.77)$$

which extends to infinity so we need to normalize it in a unit volume, contrary to earlier envelope functions. Clearly, in the absence of externally applied fields the envelope function corresponds to a plane wave with wave vector **k**.

On the other hand, we also know from chapter 3 that the full eigenfunctions of an electron in a given energy band, n, are Bloch functions,

$$b_k^n(\mathbf{r}) = e^{i\mathbf{k}\cdot\mathbf{r}} |nk\rangle \qquad (5.78)$$

which are also normalized in a unit volume. Moreover, we know from chapter 4, eq. (4.34) that the $\mathbf{k}\cdot\mathbf{p}$-approximation yields precisely expression (5.76) for the corresponding eigenvalues. Hence, in the absence of externally applied fields, the envelope function is just the exponential part of the Bloch function describing the state of the band electron.

Then, however, the envelope functions solved from eqs (5.62)–(5.64) can be considered to be linear combinations of these plane waves, i.e. linear combinations of the exponential parts of Bloch functions. Hence, using the effective Hamiltonian assumes that the averaged externally applied fields do not mix the periodic parts $|nk\rangle$ of Bloch functions, but only these exponential parts. In section 5.1.3 we already made clear that this restriction does not always hold. In particular interband optical transitions—type A in Fig. 5.1—cannot be described using effective mass theory. Thus, a full description of electrons in a band demands a hybrid approach, it can partly be given using the effective mass approximation, but it partly also needs the solution of the full Schrödinger equations.

5.3 Quantization in Effective Mass Theory

Upon solution, the time-independent effective Schrödinger equation (5.62) will generally yield quantum states described by wave functions $F^n(\mathbf{r}, t)$ and energy levels E^n. On the one hand this provides us with the information needed to determine the full wave function $\Psi^n(\mathbf{r}, t)$ of a wave packet. However, on the other hand this may also lead to further quantization effects: as the energy levels E^n may be discrete, the originally 'continuous' energy band may be split into a number of discrete subbands or even into discrete energy levels. The present section is devoted to some examples of these further quantization effects. In particular we will consider Landau subbands induced by an externally applied magnetic field, discrete energy levels of electrons near impurities and subbands in artificially grown layer structures.

For simplicity we will restrict our discussion to the simple case of the conduction band minimum at the Γ-point. There, as we saw in sections 4.2.2 and 4.2.6, the effective mass tensor is isotropic and diagonal, so

$$\frac{1}{m_{\mu\nu}} = \frac{1}{m^*}\delta_{\mu,\nu} \qquad (5.79)$$

Hence the effective Hamiltonian (5.64) reduces to the simple shape

$$\mathcal{H}_{eff} = \frac{\left[\frac{\hbar}{i}\nabla + e\bar{\mathbf{A}}(\mathbf{r})\right] \cdot \left[\frac{\hbar}{i}\nabla + e\bar{\mathbf{A}}(\mathbf{r})\right]}{2m_e m^*} - e\bar{U}(\mathbf{r}) \tag{5.80}$$

where we also use eq. (5.67) allowing us to skip the term (5.65).

5.3.1 Landau Levels

THE EFFECTIVE HAMILTONIAN IN A CONSTANT MAGNETIC FIELD

In this section we study the subbands induced by a constant magnetic field **B**. As we have simplified the calculations by restricting ourselves to an isotropic effective mass tensor, we are in the position to make an optimal choice for the direction of the axes of our frame of reference. We choose the z-axis parallel to the direction of the magnetic field. Furthermore, we choose a gauge where the scalar potential

$$\bar{U}(\mathbf{r}) = 0 \tag{5.81}$$

and the vector potential

$$\bar{\mathbf{A}}(\mathbf{r}) = \frac{1}{2}(\mathbf{r} \times \mathbf{B}) = \frac{1}{2}(yB, -xB, 0) \tag{5.82}$$

where $B = |\mathbf{B}|$. This choice is not only consistent with skipping the term (6.65) in the Hamiltonian, but, according to eq. (5.70) it also ensures that

$$\nabla \cdot \bar{\mathbf{A}}(\mathbf{r}) = 0 \tag{5.83}$$

Eq. (5.80) for $\mathcal{H}_{eff}$ is now readily evaluated by calculating

$$\left[\frac{\hbar}{i}\nabla + e\bar{\mathbf{A}}(\mathbf{r})\right] \cdot \left[\frac{\hbar}{i}\nabla + e\bar{\mathbf{A}}(\mathbf{r})\right] = -\hbar^2 \nabla \cdot \nabla + \frac{2\hbar e}{i}\bar{\mathbf{A}}(\mathbf{r}) \cdot \nabla + e^2 \bar{\mathbf{A}}(\mathbf{r}) \cdot \bar{\mathbf{A}}(\mathbf{r})$$

$$= -\hbar^2 \nabla \cdot \nabla + \frac{\hbar e}{i}(\mathbf{r} \times \mathbf{B}) \cdot \nabla + \frac{e^2}{4}(\mathbf{r} \times \mathbf{B}) \cdot (\mathbf{r} \times \mathbf{B})$$

$$= -\hbar^2 \nabla \cdot \nabla - \frac{\hbar e}{i}(\mathbf{r} \times \nabla) \cdot \mathbf{B} + \frac{e^2}{4}(\mathbf{r} \times \mathbf{B}) \cdot (\mathbf{r} \times \mathbf{B})$$

$$= -\hbar^2 \nabla \cdot \nabla - e\hbar l_z B + \frac{e^2}{4}(x^2 + y^2)B^2 \tag{5.84}$$

where

$$\hbar l_z \equiv \frac{\hbar}{i}(\mathbf{r} \times \nabla)_z \tag{5.85}$$

is the z-component of angular momentum.

As a result $\mathcal{H}_{eff}$ breaks down in four terms

$$\mathcal{H}_{eff} = \mathcal{H}_{eff}^z + \mathcal{H}_{eff}^x + \mathcal{H}_{eff}^y + \mathcal{H}_{eff}^m \tag{5.86}$$

5.3 QUANTIZATION IN EFFECTIVE MASS THEORY

where

$$\mathcal{H}_{eff}^z = -\frac{\hbar^2}{2m_e m^*} \frac{\partial^2}{\partial z^2} \tag{5.87}$$

corresponds to the Hamiltonian of a free electron with mass $m_e m^*$ moving in the z-direction. The two following terms

$$\mathcal{H}_{eff}^x = -\frac{\hbar^2}{2m_e m^*} \frac{\partial^2}{\partial x^2} + \frac{e^2 B^2}{8m_e m^*} x^2 = -\frac{\hbar^2}{2m_e m^*} \frac{\partial^2}{\partial x^2} + \frac{m_e m^*}{2} \left(\frac{\omega_c}{2}\right)^2 x^2 \tag{5.88}$$

$$\mathcal{H}_{eff}^y = -\frac{\hbar^2}{2m_e m^*} \frac{\partial^2}{\partial y^2} + \frac{e^2 B^2}{8m_e m^*} y^2 = -\frac{\hbar^2}{2m_e m^*} \frac{\partial^2}{\partial y^2} + \frac{m_e m^*}{2} \left(\frac{\omega_c}{2}\right)^2 y^2 \tag{5.89}$$

represent one-dimensional harmonic oscillators in the x- and y-directions and the last term

$$\mathcal{H}_{eff}^m = -\frac{eB}{2m_e m^*} \hbar l_z = -\frac{\omega_c}{2} \hbar l_z \tag{5.90}$$

is the Zeeman interaction of an electron with mass $m_e m^*$ and angular momentum $\hbar l_z$ parallel to the magnetic field. In these equations the so-called cyclotron resonance frequency,

$$\omega_c = \frac{eB}{m_e m^*} \tag{5.91}$$

was introduced to shorten the notation.

The first term $\mathcal{H}_{eff}^z$ is a function of z only, while the other three terms are functions of x and y only. Hence this first term is completely independent and we can calculate its energy levels separately. We find a continuous spectrum of eigenvalues

$$E_{\vec{k}}^z = \frac{\hbar^2 k_z^2}{2m_e m^*} \tag{5.92}$$

where $\hbar k_z$ is crystal momentum in the direction of the magnetic field. As we will see below, the remaining part of $\mathcal{H}_{eff}$,

$$\mathcal{H}_{eff}^L = \mathcal{H}_{eff}^x + \mathcal{H}_{eff}^y + \mathcal{H}_{eff}^m \tag{5.93}$$

yields discrete energy levels, called Landau levels. Therefore, we call $\mathcal{H}_{eff}^L$ the Landau Hamiltonian. Thus, the spectrum of eigenvalues of the full effective Hamiltonian $\mathcal{H}_{eff}$ consists of a set of bands, the so-called Landau subbands, each corresponding to a discrete energy level of $\mathcal{H}_{eff}^L$ and broadened by $\mathcal{H}_{eff}^z$.

To calculate the Landau levels, i.e. the eigenvalues of $\mathcal{H}_{eff}^L$, we start with the well known method to solve the quantum mechanical problem of a one-dimensional harmonic oscillator. A concise treatment is given in appendix B. For further details the reader is referred to ref. [11]. Following eqs (B.7) and (B.8) we introduce annihilation and creation operators

$$a_x = \frac{1}{\sqrt{2}}\left(\sqrt{\frac{m_e m^* \omega_c}{2\hbar}}\, x + \sqrt{\frac{2\hbar}{m_e m^* \omega_c}}\, \frac{\partial}{\partial x}\right) \tag{5.94}$$

$$a_x^\dagger = \frac{1}{\sqrt{2}}\left(\sqrt{\frac{m_e m^* \omega_c}{2\hbar}}\, x - \sqrt{\frac{2\hbar}{m_e m^* \omega_c}}\, \frac{\partial}{\partial x}\right) \tag{5.95}$$

$$a_y = \frac{1}{\sqrt{2}}\left(\sqrt{\frac{m_e m^* \omega_c}{2\hbar}}\, y + \sqrt{\frac{2\hbar}{m_e m^* \omega_c}}\, \frac{\partial}{\partial y}\right) \tag{5.96}$$

$$a_y^\dagger = \frac{1}{\sqrt{2}}\left(\sqrt{\frac{m_e m^* \omega_c}{2\hbar}}\, y - \sqrt{\frac{2\hbar}{m_e m^* \omega_c}}\, \frac{\partial}{\partial y}\right) \tag{5.97}$$

Hence, equivalently to eq. (B.11)

$$\mathcal{H}_{eff}^x = \frac{\hbar \omega_c}{2}\left(a_x^\dagger a_x + \frac{1}{2}\right) \tag{5.98}$$

and

$$\mathcal{H}_{eff}^y = \frac{\hbar \omega_c}{2}\left(a_y^\dagger a_y + \frac{1}{2}\right) \tag{5.99}$$

The eigenstates of a one-dimensional harmonic oscillators are given by eq. (B.18). Hence, the eigenstates of $\mathcal{H}_{eff}^x$ and $\mathcal{H}_{eff}^y$ can be written as

$$|0\rangle, |1\rangle, \ldots, |n_x\rangle, \ldots \quad \text{where} \quad |n_x\rangle = \frac{1}{\sqrt{n_x!}}(a_x^\dagger)^{n_x}|0\rangle \tag{5.100}$$

$$|0\rangle, |1\rangle, \ldots, |n_y\rangle, \ldots \quad \text{where} \quad |n_y\rangle = \frac{1}{\sqrt{n_y!}}(a_y^\dagger)^{n_y}|0\rangle \tag{5.101}$$

According to eq. (B.14), the corresponding eigenvalues are

$$E^{n_x} = \frac{\hbar \omega_c}{2}\left(n_x + \frac{1}{2}\right) \tag{5.102}$$

$$E^{n_y} = \frac{\hbar \omega_c}{2}\left(n_y + \frac{1}{2}\right) \tag{5.103}$$

where n_x and n_y are positive integers. As $\mathcal{H}_{eff}^x$ is a function of x only, while $\mathcal{H}_{eff}^y$ is a function of y only, these two terms in the Landau Hamiltonian are independent. Hence the eigenstates of their sum $\mathcal{H}_{eff}^x + \mathcal{H}_{eff}^y$ are simply products of the eigenstates given by eqs (5.100) and (5.101),

$$|n_x\rangle|n_y\rangle \tag{5.104}$$

while the eigenvalues of $\mathcal{H}_{eff}^x + \mathcal{H}_{eff}^y$ are simply sums of the eigenvalues given by eqs (5.102) and (5.103),

$$E^{n_x} + E^{n_y} = \frac{\hbar \omega_c}{2}(n_x + n_y + 1) = \frac{\hbar \omega_c}{2}(2j + 1) \tag{5.105}$$

5.3 QUANTIZATION IN EFFECTIVE MASS THEORY

We see that all states corresponding to the same value of $2j = n_x + n_y$ are degenerate. Thus the eigenstates of $\mathcal{H}_{eff}^x + \mathcal{H}_{eff}^y$ consist of multiplets separated by splittings $\hbar\omega_c/2$, each multiplet consisting of $2j + 1$ states, that can be denoted by

$$|n_x\rangle|n_y\rangle = |2j\rangle|0\rangle, |2j-1\rangle|1\rangle, \ldots, |1\rangle|2j-1\rangle, |0\rangle|2j\rangle \quad (5.106)$$

We now consider the Zeeman term $\mathcal{H}_{eff}^m$. This term can also be expressed in the creation and annihilation operators (5.94)–(5.97) [42]. To show this we consider

$$\frac{\hbar\omega_c}{2i}(a_x a_y^\dagger - a_x^\dagger a_y)$$

$$= \frac{\hbar\omega_c}{4i}\left(\sqrt{\frac{m_e m^* \omega_c}{2\hbar}} x + \sqrt{\frac{2\hbar}{m_e m^* \omega_c}} \frac{\partial}{\partial x}\right)\left(\sqrt{\frac{m_e m^* \omega_c}{2\hbar}} y - \sqrt{\frac{2\hbar}{m_e m^* \omega_c}} \frac{\partial}{\partial y}\right)$$

$$- \frac{\hbar\omega_c}{4i}\left(\sqrt{\frac{m_e m^* \omega_c}{2\hbar}} x - \sqrt{\frac{2\hbar}{m_e m^* \omega_c}} \frac{\partial}{\partial x}\right)\left(\sqrt{\frac{m_e m^* \omega_c}{2\hbar}} y + \sqrt{\frac{2\hbar}{m_e m^* \omega_c}} \frac{\partial}{\partial y}\right)$$

$$= \frac{\hbar\omega_c}{4i}\left(\frac{m_e m^* \omega_c}{2\hbar} xy - \left(x\frac{\partial}{\partial y} - y\frac{\partial}{\partial x}\right) - \frac{2\hbar}{m_e m^* \omega_c}\frac{\partial^2}{\partial x \partial y}\right)$$

$$- \frac{\hbar\omega_c}{4i}\left(\frac{m_e m^* \omega_c}{2\hbar} xy + \left(x\frac{\partial}{\partial y} - y\frac{\partial}{\partial x}\right) - \frac{2\hbar}{m_e m^* \omega_c}\frac{\partial^2}{\partial x \partial y}\right)$$

$$= -\frac{\hbar\omega_c}{2i}\left(x\frac{\partial}{\partial y} - y\frac{\partial}{\partial x}\right) = -\frac{\omega_c}{2}\hbar l_z = \mathcal{H}_{eff}^m \quad (5.107)$$

We use this result to calculate the matrix elements of $\mathcal{H}_{eff}^m$ on the basis of the harmonic oscillator states (5.104). Now, using eqs (B.16) and (B.17),

$$(a_x a_y^\dagger - a_x^\dagger a_y)|n_x\rangle|n_y\rangle$$

$$= [\sqrt{n_x(n_y+1)}|n_x-1\rangle|n_y+1\rangle - \sqrt{(n_x+1)n_y}|n_x+1\rangle|n_y-1\rangle] \quad (5.108)$$

As a result the only matrix elements of $\mathcal{H}_{eff}^m$ that differ from zero are given by

$$\langle n_x - 1|\langle n_y + 1|\frac{\hbar\omega_c}{2i}(a_x a_y^\dagger - a_x^\dagger a_y)|n_x\rangle|n_y\rangle = \frac{\hbar\omega_c}{2i}\sqrt{n_x(n_y+1)} \quad (5.109)$$

$$\langle n_x + 1|\langle n_y - 1|\frac{\hbar\omega_c}{2i}(a_x a_y^\dagger - a_x^\dagger a_y)|n_x\rangle|n_y\rangle = -\frac{\hbar\omega_c}{2i}\sqrt{(n_x+1)n_y} \quad (5.110)$$

We may immediately draw the following conclusion from this result: the matrix elements of the Zeeman term $\mathcal{H}_{eff}^m$ do not mix states with different values of $2j = n_x + n_y$. Hence j is a good quantum number. As we saw above, the terms $\mathcal{H}_{eff}^x + \mathcal{H}_{eff}^y$ leave states with the same value of j degenerate. Hence, the Zeeman term $\mathcal{H}_{eff}^m$ may resolve the degeneracy of these $(2j+1)$-fold multiplets, but it will do nothing more.

We now consider an interesting equivalence between the present problem and that of the quantum mechanical description of angular momentum. For this purpose, we introduce a new quantum number

$$\tilde{m} = \frac{1}{2}(n_x - n_y) = -j, -j+1, \ldots, j-1, j \qquad (5.111)$$

denoting the states within the multiplets with degeneracy $(2j+1)$. Thus, these states are written as

$$|n_x\rangle|n_y\rangle = |j, \tilde{m}\rangle$$
$$|2j\rangle|0\rangle = |j, -j\rangle$$
$$|2j-1\rangle|1\rangle = |j, -j+1\rangle \qquad (5.112)$$
$$|1\rangle|2j-1\rangle = |j, j-1\rangle$$
$$|0\rangle|2j\rangle = |j, j\rangle$$

Furthermore, the non-zero matrix elements of $\mathcal{H}_{eff}^m$ are given by

$$\langle j, \tilde{m}-1| \frac{\hbar\omega_c}{2i}(a_x a_y^\dagger - a_x^\dagger a_y)|j, \tilde{m}\rangle = \frac{\hbar\omega_c}{2i}\sqrt{(j+\tilde{m})(j-\tilde{m}+1)} \qquad (5.113)$$

$$\langle j, \tilde{m}+1| \frac{\hbar\omega_c}{2i}(a_x a_y^\dagger - a_x^\dagger a_y)|j, \tilde{m}\rangle = -\frac{\hbar\omega_c}{2i}\sqrt{(j+\tilde{m}+1)(j-\tilde{m})} \qquad (5.114)$$

We now observe that, formally, the eigen states (5.112) correspond to the basis states of a fictitious angular momentum $\hbar \mathbf{J}$, such that [11]

$$\langle j, \tilde{m}|J^2|j, \tilde{m}\rangle = j(j+1) \qquad (5.115)$$

$$\langle j, \tilde{m}|J_{\tilde{z}}|j, \tilde{m}\rangle = \tilde{m} \qquad (5.116)$$

Here the $\tilde{z}$-axes is a direction which is formally defined by the basis states $|j, \tilde{m}\rangle$ and therefore *not* related to the z-axis of the frame of reference used to describe our Hamiltonian (5.86). If we now carefully consider the matrix elements (5.113) and (5.114) of the Zeeman term $\mathcal{H}_{eff}^m$, we see that this term in the Hamiltonian can be formally written as

$$\mathcal{H}_{eff}^m = \hbar\omega_c J_{\tilde{y}} \qquad (5.117)$$

because the matrix elements of the right hand side of this expression are precisely those given by eqs (5.113) and (5.114).

Now, the diagonalization of $\mathcal{H}_{eff}^m$ is straightforward. We rotate the formally defined frame of reference $(\tilde{x}, \tilde{y}, \tilde{z})$ to become $(\bar{x}, \bar{y}, \bar{z})$, so $\tilde{y} = \bar{z}$. Then,

$$\mathcal{H}_{eff}^m = \hbar\omega_c J_{\bar{z}} \qquad (5.118)$$

and the eigenvalues of the Zeeman term are given by

$$E_m^m = \hbar\omega_c m_j \qquad (5.119)$$

where

$$m_j = -j, -j+1, \ldots, j-1, j \qquad (5.120)$$

5.3 QUANTIZATION IN EFFECTIVE MASS THEORY

Note that $J_{\bar{z}}$ is formally the $\bar{z}$-component of a fictitious angular momentum. It should be clearly distinguished from the z-component l_z of the real angular momentum. Their relation is simple however, comparing eqs (5.90) and (5.118) we find,

$$l_z = -2J_{\bar{z}} \tag{5.121}$$

Combining this result with eq. (5.105) we finally obtain the Landau levels, i.e. the eigen values of the Landau Hamiltonian (5.93),

$$E_N = \hbar\omega_c\left(j + m_j + \frac{1}{2}\right) = \hbar\omega_c\left(N + \frac{1}{2}\right) \tag{5.122}$$

where we note that $N = j + m_j > 0$ is always an integer, though j may have values $0, \frac{1}{2}, 1$, etc.

The quantum numbers j and m_j used above are not only very useful to obtain the final result, but also show how the eigenstates of isotropic multi-dimensional harmonic oscillators are connected to those of angular momenta [42]. On the other hand, they are not usually encountered in literature on semiconductors. There one normally uses $N = j + m_j$ introduced above which denotes the Landau level. Furthermore one uses $m = -2m_j$, corresponding to the z-component of the *real* angular momentum $\hbar\mathbf{l}$ instead of m_j corresponding to the $\bar{z}$-component of the fictitious angular momentum $\hbar\mathbf{J}$. Finally one often encounters a quantum number $n = j - |m_j|$. In this latter notation

$$E_N = \frac{\hbar\omega_c}{2}(2n + |m| + m + 1) = \frac{\hbar\omega_c}{2}(2N + 1) \tag{5.123}$$

We finish this section by putting some numbers in the final result (5.122). As an example we consider electrons in the conduction band minimum at the Γ-point of GaAs. Then,

$$e = 1.602 \cdot 10^{-19} \text{ C}$$

$$m_e = 9.110 \cdot 10^{-31} \text{ kg}$$

$$\hbar = 1.055 \cdot 10^{-34} \text{ J s}$$

$$m^* = 0.0665$$

and the splitting between subsequent Landau levels is $\hbar\omega_c/e = 1.74 \cdot 10^{-3}$ eV per Tesla. Thus, even at 10 Tesla this splitting is much smaller than e.g. the band gap of a semiconductor. On the other hand it is large enough to be easily observed by spectroscopy. As we will see below, splittings of a few meV are also typical for other quantization effects evolving from effective mass treatments.

5.3.2 Hydrogenic Impurities

DONOR AND ACCEPTOR IMPURITIES

Donor and acceptor impurities and the scattering of electrons and holes at such impurities present an excellent example of a case that can be perfectly understood using effective mass formalism. We start our discussion of such impurities using Table 5.2, representing the part of the periodic table containing the major semiconductors.

Table 5.2 The part of the periodic table containing the silicon-like semiconductors.

I	II	IIa	III	IV	V	VI	VII	0
H								He
Li	Be		B	C	N	O	F	Ne
Na	Mg		Al	Si	P	S	Cl	Ar
K	Ca	...Zn	Ga	Ge	As	Se	Br	Kr
Rb	Sr	...Cd	In	Sn	Sb	Te	I	Xe

First, we define various types of impurities using the example of silicon, which is found in the fourth column. We recall that in silicon the atoms form cores consisting of its nucleus surrounded by 10 core electrons while the remaining four electrons are used for homopolar binding with neighbouring atoms. These latter electrons exactly fill the valence band. We now distinguish the following types of impurities replacing a silicon atom:

Isoelectronic Impurities. These are impurities in the same column as silicon, e.g. germanium. Such impurities also form cores with a charge $+4$. The remaining four electrons are again needed for homopolar binding with neighbouring silicon atoms, so again the valence band is exactly filled. As a result the only change consists of the slightly different size and shape of the germanium core compared to the silicon it replaces. This results in a highly localized disturbance of the periodic potential experienced by the valence electrons.

Donor Impurities. These are impurities from the fifth column, e.g. phosphorus. This element differs from silicon in two respects only: its nuclear charge is $15e$ and it provides 15 electrons, both one more than silicon. When built into the silicon lattice it forms cores consisting of that nucleus and 10 electrons, so their electronic structure is the same as that of the silicon cores. Four of the remaining electrons are needed for homopolar binding with neighbouring silicon atoms. They exactly fill the valence band. For the remaining electron only conduction band states are available and in first approximation it may therefore move freely in the crystal. At the position of a phosphorus core it experiences two changes of the potential however. Because of its higher nuclear charge a phosphorus core is slightly smaller than a silicon core, yielding a highly localized change of the periodic potential, just as in the case of isoelectronic impurities. Secondly, the extra nuclear charge of the phosphorus core is not compensated by its surrounding valence electrons, resulting in a long range Coulomb potential of that charge.

Acceptor Impurities. Acceptor impurities are just the opposite of donor impurities and originate in the third column. A typical example is aluminium with a nuclear charge $13e$ and 13 electrons, both one less than silicon. Again, when built into the silicon lattice it forms cores consisting of that nucleus and 10 electrons, so it has the same electronic structure as silicon. Four more electrons would be needed to fill the valence band. However, one electron is missing, thus there remains a hole in the valence band, which in first approximation may move freely through the crystal. At the position of an aluminium core again two effects change the potential experienced by this hole. Because of its lower nuclear charge an aluminium core is slightly larger than a silicon core, yielding again a highly localized change of the periodic potential. Secondly, the lower nuclear charge of the phosphorus core is more than

SHALLOW IMPURITIES

The impurities described above can again be divided into two classes. Deep impurities are those where the local disturbance of the periodic potential consists of a deep potential well at the position of the impurity. Then the donor electron or the hole provided by the acceptor is trapped in that well and large energies are required to excite them into the conduction or valence band.

Shallow impurities are those where the local disturbance of the periodic potential is small or repulsive, so the attractive long range Coulomb potential dominates. As this potential is long range, we may use effective mass theory to describe its influence on the free electron or hole provided by that impurity. Then, for a donor yielding an electron to a non-degenerate conduction band, we simply find

$$\mathcal{H}_{eff} = \frac{1}{2m_e} \sum_{\mu,\nu} \frac{1}{m^*_{\mu\nu}} \left(\frac{\hbar}{i}\nabla_\mu\right)\left(\frac{\hbar}{i}\nabla_\nu\right) - \frac{e^2}{4\pi\varepsilon_0\varepsilon_r r} \qquad (5.124)$$

Here we use eq. (5.64) for the special case that $\bar{\mathbf{A}}(\mathbf{r}) = 0$ and the Coulomb potential (A.2) of a single positive elementary charge. The latter is modified to include dielectric effects as described in section 5.1.2, noting that $\varepsilon_r(\mathbf{r})$ must be averaged over a primitive cell, so it becomes equal to ε_r.

Eq. (5.124) shows a striking property of shallow donors. In effective mass theory the Hamiltonian of a shallow donor does not contain information on the chemical nature of the donor itself. It only contains the properties $m^*_{\mu\nu}$ and ε_r of the host semiconductor crystal. Thus, for shallow donors, the energy levels of a donor electron are found to be independent of the chemical nature of that donor. This property of shallow donors must be contrasted to the case of deep impurities. There the potential well originates in the local disturbance of the periodic potential due to the different nature of the impurity itself. Hence, for deep impurities, the energy levels depend strongly on the chemical nature of these impurities.

BOUND DONOR STATES FOR AN ISOTROPIC EFFECTIVE MASS

A shallow donor becomes extremely simple in the case of an isotropic effective mass. This situation is encountered in e.g. GaAs, where the conduction band minimum is at the Γ-point. Then,

$$\mathcal{H}_{eff} = -\frac{\hbar^2}{2m_e m^*}\Delta - \frac{e^2}{4\pi\varepsilon_0\varepsilon_r r} \qquad (5.125)$$

which, except for m_e being replaced by $m_e m^*$ and ε_0 by $\varepsilon_0\varepsilon_r$, is exactly the Hamiltonian of the hydrogen atom. Therefore such an impurity is called a hydrogenic impurity. Thus, following eqs (A.3)–(A.5), the donor electron has bound states given by

$$E_n = -\frac{Ry^*}{n^2} \qquad (5.126)$$

where

$$Ry^* = \frac{e^2}{8\pi\varepsilon_0\varepsilon_r a^*} = \frac{m^*}{\varepsilon_r^2} Ry \qquad (5.127)$$

is the effective Rydberg and

$$a^* = \frac{4\pi\varepsilon_0\varepsilon_r \hbar^2}{m_e m^* e^2} = \frac{\varepsilon_r}{m^*} a \qquad (5.128)$$

is the effective Bohr radius. In these bound states the electron provided by the donor is bound by the Coulomb potential of that donor. The binding energies correspond to those shown in Fig. A.1 except that the vertical scale is modified by a factor m^*/ε_r^2. Furthermore the wave functions are those given in Tables A.1 and A.2, except that the radii of the wave functions are larger by a factor ε_r/m^*. The energy level scheme scaled to the case of GaAs is shown in Fig. 5.2.

Though qualitatively the same, the energy levels and eigenstates are quantitatively very different from those of the hydrogen atom. For example, for a donor in GaAs we must take into account that $\varepsilon_r = 12.8$ while for the conduction band minimum at the Γ point $m^* = 0.0665$. Then $a^* = 192.5a = 10.2 \times 10^{-9}$ m and $Ry^*/e = 0.406 \times 10^{-3} Ry/e = 5.52 \times 10^{-3}$ eV. Thus the wave function of a bound donor electron extends over thousands of primitive cells while its binding energy is less than 1% of the gap between the valence and conduction bands. It is crucial to note that Ry^* corresponds to only 64 K. Hence at room temperature donor electrons will be mostly ionized into the conduction band. Thus, shallow donors contribute to the conductivity of semiconductors and can be used to create such conductivity. The same applies for shallow acceptors, that can be described in a completely equivalent way.

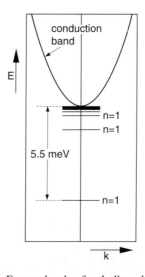

Figure 5.2 Energy levels of a shallow donor in GaAs.

5.3.3 Quantum Wells

HETEROSTRUCTURES

Semiconductor physics relies heavily on epitaxial growing techniques like Molecular Beam Epitaxy (MBE). With these methods crystals are grown by depositing single atomic layers one after another on a substrate. Thus one produces so-called heterostructures consisting of thin layers of different chemical composition while the boundary between these layers is sharp on an atomic scale. Such a boundary, e.g. between $Al_xGa_{1-x}As$ and GaAs, is called a heterojunction. At such a heterojunction the periodic potential $\mathcal{V}'$ and hence the band structure changes. In particular the gap between the valence and conduction bands and the effective masses describing the electrons and the holes in both bands change. Moreover, the absolute value of the energy at the top of their valence bands changes, contrary to what is suggested by standard band structure diagrams shown in Figs 3.7–3.10 where the zero point of the energy is chosen arbitrarily at the top of the valence band. This is seen as follows. The absolute value of the energy at the top of the valence band corresponds to the energy needed to excite a valence electron from the semiconductor into a vacuum outside the crystal. Evidently, for different semiconductors this excitation energy is different, so the absolute value of the energy at the top of the valence band is also different. The resulting situation is depicted in Fig. 5.3a, showing the energy level scheme of a heterojunction between undoped $Al_{0.2}Ga_{0.8}$ and GaAs.

A simple heterostructure is obtained when a thin layer (typically a few nm) of one semiconductor material is grown between two thicker layers of another semiconductor material. This situation is depicted in Fig. 5.3b showing the position of the conduction and valence band edges for a thin layer of GaAs embedded between two thicker layers of $Al_{0.2}Ga_{0.8}$. It is seen that the GaAs layer constitutes a square potential well for the electrons in the conduction band as well as for the holes in the valence band.

The situation shown in Fig. 5.3 must be modified when donor or acceptor impurities, and electrons or holes are present. Let us consider the influence of donor impurities on the potential of a heterojunction as shown in Fig. 5.3a. As the binding energy of the donor electrons is smaller than the height of the potential step,

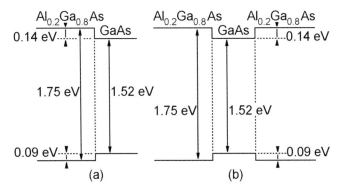

Figure 5.3 (a) Heterojunction of undoped $Al_{0.2}Ga_{0.8}$ and GaAs. (b) Heterostructure consisting of a thin layer of undoped GaAs embedded between layers of $Al_{0.2}Ga_{0.8}$.

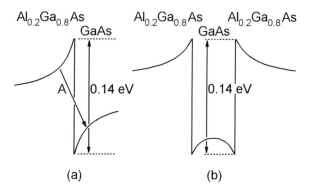

Figure 5.4 The influence of space charge. (a) Heterojunction of doped $Al_{0.2}Ga_{0.8}$ and GaAs. (b) Thin layer of doped GaAs between layers of doped $Al_{0.2}Ga_{0.8}$.

energy is gained by transferring donor electrons from their bound donor states to the conduction band down this step (see arrow A in Fig. 5.4a). This transfer creates positive space charge due to the remaining ionized donors at the 'high' side of the potential step and negative space charge due to the excess electrons at the 'low' side of that step. In equilibrium the number of transferred electrons is so large that the potential of these space charges compensates the step completely, as is depicted in Fig. 5.4a. Similarly, in the presence of donor impurities the square potential well shown in Fig. 5.3b is modified by these band bending effects in the way shown in Fig. 5.4b.

With current growing techniques narrow potential wells are obtained with widths below a few nanometres, corresponding to a few tens of atomic layers. Below we treat two simple cases, an infinitely deep square well depicted in Fig. 5.5a and approximating the wells shown in Figs. 5.3b and 5.4b, and a triangular well depicted in Fig. 5.5b and approximating the heterojunction shown in Fig. 5.4a. In both simplifications the potential barriers are approximated to be infinitely high and the electrons are completely confined to one semiconductor material: in the square well shown in Fig. 5.5a the electrons do not penetrate the barriers and the wave functions can be calculated using the properties of the material inside the well only. In the triangular well of Fig. 5.5b the electrons do not penetrate the infinitely high potential step and again the wave functions can be calculated using the properties of one material only. Moreover, as the wells are one to two orders of magnitude larger than the primitive cells their potentials may be considered as long range and the behaviour of charge carriers in these wells may be calculated using the effective mass formalism developed above. Because of the small width of these wells this treatment needs necessarily to be quantum mechanical and therefore these wells are normally called quantum wells.

As earlier we simplify further and assume an isotropic and diagonal effective mass tensor. This approximation is actually quite realistic as many experiments involving quantum wells are performed on electrons in the conduction band of GaAs. There the lowest minimum of the conduction band is found at the Γ-point, where the effective mass tensor is indeed so simple. Then the effective Hamiltonian of an electron in a

5.3 QUANTIZATION IN EFFECTIVE MASS THEORY

quantum well is given by

$$\mathcal{H}_{eff} = \mathcal{H}_{eff}^x + \mathcal{H}_{eff}^y + \mathcal{H}_{eff}^z \tag{5.129}$$

where

$$\mathcal{H}_{eff}^x = -\frac{\hbar^2}{2m_e m^*}\frac{\partial^2}{\partial x^2} \tag{5.130}$$

$$\mathcal{H}_{eff}^y = -\frac{\hbar^2}{2m_e m^*}\frac{\partial^2}{\partial y^2} \tag{5.131}$$

$$\mathcal{H}_{eff}^z = -\frac{\hbar^2}{2m_e m^*}\frac{\partial^2}{\partial z^2} - eU(z) \tag{5.132}$$

The first two terms (5.130) and (5.131) describe a free particle. The last term (5.132) corresponds to the problem of a one-dimensional potential well, yielding bound states. As electrons that are bound in these states are still free to move in the two other directions, they are said to form a two-dimensional electron gas (2-DEG).

THE SQUARE QUANTUM WELL

The infinitely deep square quantum well represents a textbook problem of Quantum Mechanics [11]. To treat it we consider Fig. 5.5a. In the interval $-z_0 < z < +z_0$ the potential is equal to zero, while outside this interval it is equal to ∞. Within the well the effective Schrödinger equation is given by

$$\mathcal{H}_{eff}^z F_n(z) = -\frac{\hbar^2}{2m_e m^*}\frac{\partial^2}{\partial z^2} F_n(z) = E_n F_n(z) \tag{5.133}$$

The effective Hamiltonian is real, so, as proven in section 4.1.2, the eigen functions can also be chosen to be real. Then eq. (5.133) has oscillating solutions

$$F_n(z) = A_n \sin(k_n(z+z_0) + \phi) \tag{5.134}$$

corresponding to energy levels

$$E_n = \frac{\hbar^2 k_n^2}{2m_e m^*} \tag{5.135}$$

while $k_n > 0$. Because the potential barriers are infinitely high, the electrons do not penetrate in them and the wave functions are equal to zero outside the well. Because these functions must also be continuous, this leads to boundary conditions

$$F_n(-z_0) = A_n \sin\phi = 0 \quad \text{and} \quad F_n(z_0) = A_n \sin(2k_n z_0 + \phi) = 0 \tag{5.136}$$

As a result k_n is quantized to values

$$k_n = \frac{n\pi}{2z_0}$$

$$n = 1, 2, 3, \ldots \tag{5.137}$$

$$\text{and} \quad \phi = 0$$

and the energy takes discrete values

$$E_n = \frac{\pi^2\hbar^2}{(2z_0)^2}\frac{n^2}{2m_e m^*} \quad (5.138)$$

Both these wave functions and energy levels are inserted in Fig. 5.5a.

As earlier, we put some numbers in this result in order to obtain insight in the magnitude of the splittings between the levels. Again we take GaAs as an example and assuming a typical well with a width of 10 nm, i.e. about 100 atomic layers. Then,

$$e = 1.602 \cdot 10^{-19} \text{ C}$$
$$m_e = 9.110 \cdot 10^{-31} \text{ kg}$$
$$\hbar = 1.055 \cdot 10^{-34} \text{ J s}$$
$$m^* = 0.0665$$

and the splitting between the lowest levels ($n = 1, 2$) is equal to $42.4 \cdot 10^{-3}$ eV.

THE TRIANGULAR QUANTUM WELL

The triangular quantum well shown in Fig. 5.5b is more complicated. For $z < 0$ the potential is infinitely high, so we only need to solve the effective Schrödinger equation for $z \geq 0$, using as a boundary condition that the envelope function $F_n(0) = 0$. For $z \geq 0$ the potential increases linearly, equivalently to the case that a constant electric field E is applied. Then the effective Schrödinger equation is given by

$$\mathcal{H}^z_{\text{eff}} F_n(z) = \left(-\frac{\hbar^2}{2m_e m^*}\frac{\partial^2}{\partial z^2} + eEz\right) F_n(z) = E_n F_n(z) \quad (5.139)$$

Though this case is less commonly treated than the square quantum well, it may be found in some textbooks on quantum mechanics [23]. To solve eq. (5.139) we introduce a reduced coordinate

$$\zeta = \left(\frac{2m_e m^*}{\hbar^2 e^2 E^2}\right)^{1/3} \cdot (eEz - E_n) \quad (5.140)$$

Then the Schrödinger equation reduces to Airy's equation

$$\left(\frac{\partial^2}{\partial \zeta^2} - \zeta\right)\phi(\zeta) = 0 \quad \text{where} \quad \phi(\zeta) = F_n(z) \quad (5.141)$$

As may be checked by direct insertion, the solution is Airy's function defined by

$$Ai(\zeta) = C \int_{\epsilon-i\infty}^{\epsilon+i\infty} dt\, e^{\zeta t - 1/3 t^3} \quad (5.142)$$

where the integral is taken in the complex plane along a line which is very close to the imaginary axis.

The infinitely high potential barrier at $z = 0$ yields the boundary condition $F_n(0) = Ai(\zeta) = 0$. Thus $z = 0$ should correspond to one of the discrete values

$$\zeta = \zeta_1, \zeta_2, \ldots, \zeta_n, \ldots$$

5.3 QUANTIZATION IN EFFECTIVE MASS THEORY

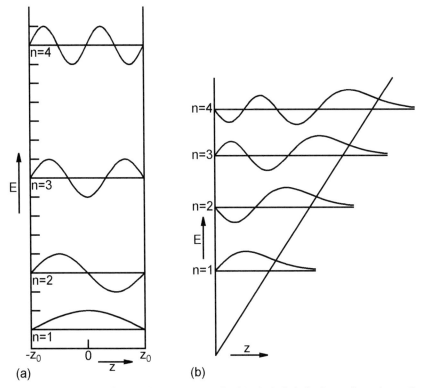

Figure 5.5 (a) The infinitely deep square well. (b) The infinitely deep triangular well.

where $Ai(\zeta) = 0$. Hence by definition (5.140) of ζ, also the energy takes discrete values

$$E_n = \zeta_n \left(\frac{\hbar^2 e^2 E^2}{2m_e m^*} \right)^{1/3} \tag{5.143}$$

The corresponding energy levels and wave functions are shown in Fig. 5.5b.

QUANTUM WELLS IN A MAGNETIC FIELD

As seen above, the potential of a quantum well yields quantized energy levels E_n for the z-component of the effective Hamiltonian, while leaving the motion of the electron in the other two directions free. At low temperatures, where $kT \ll E_2 - E_1$, only the lowest of these discrete levels is populated. Then the effective Hamiltonian can be conveniently written as

$$\mathcal{H}_{\mathit{eff}} = E_1 - \frac{\hbar^2}{2m_e m^*} \left(\frac{\partial^2}{\partial x^2} + \frac{\partial^2}{\partial y^2} \right) \tag{5.144}$$

describing a perfectly two-dimensional free electron.

We now consider the case that also a magnetic field is applied along the z-axis, i.e. in a direction perpendicular to the plane of the quantum well. According to the results of section 5.3.1, the motion of the electrons in the x- and y-directions is now restricted as well. Using eq. (5.122) we find energy levels

$$E_n + \hbar\omega_c\left(N + \frac{1}{2}\right) \tag{5.145}$$

where E_n denotes a discrete level following from the confinement in the quantum well, $\omega_c = eB/m_e m^*$ is the cyclotron resonance and N denotes the Landau level. Thus, a magnetic field applied perpendicularly to the plane of the quantum well renders all quantum states discrete.

6 The Crystal Lattice

6.1 The Lattice Hamiltonian

Up till now we considered the eigenstates of electrons in a semiconductor crystal assuming the nuclei to occupy fixed positions. However, as discussed above in chapter 2, this is only an approximation because in reality the nuclei vibrate about these fixed positions. Therefore, in the present and following chapter we consider a more complete description where such lattice vibrations and their coupling with the electrons are taken into account. Still, we will exploit many of the results obtained above, i.e. translational symmetry, the $\mathbf{k} \cdot \mathbf{p}$-approximation and effective mass theory.

We start considering the crystal lattice Hamiltonian (2.45),

$$\mathcal{H}_L = -\sum_{l,s,\mu}^{N} \frac{\hbar^2}{2M_s} \left(\frac{\partial}{\partial u_\mu^{ls}} \right)^2 + \frac{1}{2} \sum_{ll',ss',\mu\nu}^{N} \Phi_{\mu\nu}^{ll',ss'} u_\mu^{ls} u_\nu^{l's'} \tag{6.1}$$

Here, as in section 2.3.1, u_μ^{ls} is the displacement of a nucleus in the μ-direction, where μ and ν denote Cartesian coordinates. However, to be able to exploit the translational symmetry of the crystal lattice, we number the nuclei differently. Now, the index l denotes the primitive cell where this nucleus is situated, while s numbers the nuclei within these primitive cells. Note that, because of lattice periodicity the nuclear mass M_s needs to bear the index s only. Furthermore we have restricted the sums over the indices l and l' to a maximum N. Thus, eq. (6.1) represents a finite crystal containing N of those cells. This is done because $\mathcal{H}_L$ represents the total lattice energy of the crystal and hence its eigenvalues would diverge if the crystal were infinitely large.

As we already found in section 2.3.2, the lattice Hamiltonian (6.1) has the shape of a multi-dimensional harmonic oscillator, but it can always be reduced to a sum of independent one-dimensional harmonic oscillators using a unitary transformation. In the present chapter we will see that this unitary transformation can be split into four successive steps. First, we transform to reciprocal space, which leads to complex generalized coordinates. Next, we split the relative and centre of mass motion of the two nuclei in the primitive cell. This separates the so-called optical and acoustical branches of the lattice Hamiltonian. Subsequently we rotate our frame of reference, so one of its axes is parallel to the wave vector in reciprocal space, leading to what are called longitudinal and transverse modes. Finally, in the last step we introduce real generalized coordinates, allowing to use standard solutions for one-dimensional harmonic oscillators.

6.1.1 Reciprocal Space and Plane Standing Waves

NUCLEAR DISPLACEMENTS IN RECIPROCAL SPACE

Clearly, the nuclear displacements u_μ^{ls} are discrete functions of space lattice points $\mathbf{R}_l$ in normal space. Hence, the method presented in section 3.1.5 applies to transform these quantities to reciprocal space. Thus, u_μ^{ls} can be decomposed into Fourier components $\tilde{u}_\mu^s(\mathbf{k})$ using eq. (3.44). Then,

$$u_\mu^{ls} = \frac{1}{V_B} \int_{V_B} d\mathbf{k}\, \tilde{u}_\mu^s(\mathbf{k})\, e^{i\mathbf{k}\cdot\mathbf{R}_l} \qquad (6.2)$$

where $\int_{V_B}$ denotes integration over the first Brillouin zone.

As discussed above, the lattice Hamiltonian diverges unless the crystal is finite. To accommodate the finite dimensions of the crystal in the transformation to reciprocal space, we recall the finite crystal shown in Fig. 3.5. It consists of a block of $L_1 L_2 L_3 = N$ primitive cells and is assumed to obey periodic boundary conditions. Then, as discussed in section 3.1.5, the crystal as a whole can be seen as a 'maxi primitive cell' giving rise to 'mini Brillouin zones'. As all functions in normal space obey periodic boundary conditions, their transforms to reciprocal space should be discrete functions of 'mini reciprocal lattice vectors', $\mathbf{k}_q$, given by eq. (3.52),

$$\mathbf{k}_q = (q_1/L_1)\mathbf{b}_1 + (q_2/L_2)\mathbf{b}_2 + (q_3/L_3)\mathbf{b}_3 \qquad (6.3)$$

and lying in the first Brillouin zone, so the integers q_α obey

$$0 \leq q_\alpha < L_\alpha \qquad (6.4)$$

In particular, equivalently to eq. (3.53),

$$\tilde{u}_\mu^s(\mathbf{k}) = \frac{V_B}{\sqrt{N}} \sum_q^N v_\mu^{qs} \delta(\mathbf{k} - \mathbf{k}_q) \qquad (6.5)$$

where the factor $V_B/\sqrt{N}$ scales the amplitudes of the discrete Fourier components v_μ^{qs} in order to render the transformation to reciprocal space unitary. Then equivalently to eqs (3.54) and (3.58),

$$u_\mu^{ls} = \frac{1}{\sqrt{N}} \sum_q^N v_\mu^{qs}\, e^{i\mathbf{k}_q\cdot\mathbf{R}_l} \qquad (6.6)$$

$$v_\mu^{qs} = \frac{1}{\sqrt{N}} \sum_l^N u_\mu^{ls}\, e^{-i\mathbf{k}_q\cdot\mathbf{R}_l} \qquad (6.7)$$

6.1 THE LATTICE HAMILTONIAN

Note again that the $N \times N$ transformation matrix is unitary, as

$$\sum_{l}^{N} \left(\frac{1}{\sqrt{N}} e^{i\mathbf{k}_q \cdot \mathbf{R}_l}\right)^* \left(\frac{1}{\sqrt{N}} e^{i\mathbf{k}_{q'} \cdot \mathbf{R}_l}\right) = \prod_{\alpha} \delta_{q_\alpha, q'_\alpha} \equiv \delta_{q,q'} \tag{6.8}$$

RELATIONS BETWEEN THE FOURIER COMPONENTS

Before continuing we investigate the physical meaning of the Fourier components v_μ^{qs} of the nuclear displacements u_μ^{ls}. Intuitively, one interprets eq. (6.6) as resolving the nuclear motion in standing plane waves with wave vector $\mathbf{k}_q$. However, the situation is not that simple because the parameters v_μ^{qs} are complex and can hence only be indirectly connected to the amplitude of such standing waves. To find this connection, we start by using eq. (6.7) to calculate the complex conjugate of v_μ^{qs}. Noting that u_μ^{ls} is the component of the displacement of a nucleus, so it is real, we find

$$v_\mu^{qs*} = \frac{1}{\sqrt{N}} \sum_{l}^{N} u_\mu^{ls} \left(e^{-i\mathbf{k}_q \cdot \mathbf{R}_l}\right)^*$$

$$= \frac{1}{\sqrt{N}} \sum_{l}^{N} u_\mu^{ls} e^{i\mathbf{k}_q \cdot \mathbf{R}_l} = v_\mu^{-qs} \tag{6.9}$$

where the index $-q$ indicates the point $-\mathbf{k}_q$ in reciprocal space. We remark that this expression can only be used to full advantage if both $\mathbf{k}_q$ and $-\mathbf{k}_q$ lie in the first Brillouin zone. Fortunately this can be achieved by choosing this zone to have the shape shown in Fig. 3.3 with the origin located at the Γ-point. Then, as can be seen by inspecting Fig. 3.3, the Brillouin zone exhibits inversion symmetry, i.e. for every point $\mathbf{k}_q$ in this zone there is also a point $-\mathbf{k}_q$ in it.

It is also pointed out that eq. (6.9) solves a problem which is hidden in the concise notation of eqs (6.6) and (6.7). Recall that the quantities u_μ^{ls} are real. Thus, with three dimensions and two nuclei in a primitive cell one has $6N$ independent real variables u_μ^{ls}. On the other hand the Fourier transforms v_μ^{qs} defined by eq. (6.7) are complex. As the number of allowed positions $\mathbf{k}_q$ in reciprocal space is equal to the number N of primitive cells in the crystal, this transformation yields $6N$ real variables $\Re\{v_\mu^{qs}\}$ and another $6N$ real variables $\Im\{v_\mu^{qs}\}$. Clearly these $12N$ variables cannot be completely independent and $6N$ hidden relations between the Fourier transforms v_μ^{qs} should exist. Well, eq. (6.9) provides precisely these $6N$ relations.

PLANE STANDING WAVES

We now use eq. (6.9) to rewrite the Fourier transform (6.6). For this purpose we split the sum over q in two stages. First we consider all pairs of terms corresponding to $\mathbf{k}_q$ and $-\mathbf{k}_q$. Next we sum over these pairs. Thus we write,

$$u_\mu^{ls} = \frac{1}{\sqrt{N}} \sum_{q}^{N/2} \left(v_\mu^{qs} e^{i\mathbf{k}_q \cdot \mathbf{R}_l} + v_\mu^{-qs} e^{-i\mathbf{k}_q \cdot \mathbf{R}_l}\right) \tag{6.10}$$

Here, the notation $\sum_q^{N/2}$ implies that we sum over the $N/2$ pairs $\mathbf{k}_q$ and $-\mathbf{k}_q$. Inserting eq. (6.9), this expression may be rewritten as,

$$u_\mu^{ls} = \frac{1}{\sqrt{N}} \sum_q^{N/2} \left(v_\mu^{qs} e^{i\mathbf{k}_q \cdot \mathbf{R}_l} + v_\mu^{qs*} e^{-i\mathbf{k}_q \cdot \mathbf{R}_l} \right)$$

$$= \frac{1}{\sqrt{N}} \sum_q^{N/2} \left[\frac{1}{2}(v_\mu^{qs} + v_\mu^{qs*})(e^{i\mathbf{k}_q \cdot \mathbf{R}_l} + e^{-i\mathbf{k}_q \cdot \mathbf{R}_l}) + \frac{1}{2}(v_\mu^{qs} - v_\mu^{qs*})(e^{i\mathbf{k}_q \cdot \mathbf{R}_l} - e^{-i\mathbf{k}_q \cdot \mathbf{R}_l}) \right]$$

$$= \frac{1}{\sqrt{N}} \sum_q^{N/2} \left[2\Re\{v_\mu^{qs}\} \cos(\mathbf{k}_q \cdot \mathbf{R}_l) - 2\Im\{v_\mu^{qs}\} \sin(\mathbf{k}_q \cdot \mathbf{R}_l) \right] \qquad (6.11)$$

We now write

$$v_\mu^{qs} = |v_\mu^{qs}| e^{i\phi_\mu^{qs}} \qquad (6.12)$$

So,

$$\Re\{v_\mu^{qs}\} = |v_\mu^{qs}| \cos \phi_\mu^{qs} \quad \text{and} \quad \Im\{v_\mu^{qs}\} = |v_\mu^{qs}| \sin \phi_\mu^{qs} \qquad (6.13)$$

As a result we find

$$u_\mu^{ls} = \frac{1}{\sqrt{N}} \sum_q^{N/2} \left[2|v_\mu^{qs}| \cos \phi_\mu^{qs} \cos(\mathbf{k}_q \cdot \mathbf{R}_l) - 2|v_\mu^{qs}| \sin \phi_\mu^{qs} \sin(\mathbf{k}_q \cdot \mathbf{R}_l) \right]$$

$$= \frac{1}{\sqrt{N}} \sum_q^{N/2} 2|v_\mu^{qs}| \cos(\mathbf{k}_q \cdot \mathbf{R}_l + \phi_\mu^{qs}) \qquad (6.14)$$

We see that we have not only succeeded in resolving the nuclear displacements in standing plane waves with wave vector $\mathbf{k}_q$, but that we also have found the physical meaning of the Fourier components v_μ^{qs} of these nuclear displacements. Thus their absolute value $2|v_\mu^{qs}|$ is the amplitude of these plane waves and ϕ_μ^{qs} their phase. Moreover, eq. (6.10) shows that we need *both* $\mathbf{k}_q$ and $-\mathbf{k}_q$ to construct the corresponding standing plane wave. This situation is very similar to the construction of a standing wave out of *two* propagating waves with opposite wave vector. A further similarity with the construction of standing waves emerges from eqs (6.11) and (6.14). An arbitrary standing wave with amplitude $2|v_\mu^{qs}|$ and wave vector $\mathbf{k}_q$ is either obtained by introducing a phase ϕ_μ^{qs} as in eq. (6.14), or by mixing two of those waves shifted by 90° and with independent amplitudes $\Re\{v_\mu^{qs}\}$ and $\Im\{v_\mu^{qs}\}$ as in eq. (6.11).

To illustrate the present discussion the vertical axis in Fig. 6.1 shows the nuclear displacements u_μ^{ls} in the μ-direction for the case that only one plane wave with a wave vector $\mathbf{k}_q$, an amplitude $2|v_\mu^{q1}|$ and a phase $\phi_\mu^{q1} = 0$ is present. The horizontal axis corresponds to a one-dimensional row of atoms in the $\mathbf{k}_q$ direction. One should be aware that in real space the μ-axis need not be perpendicular to the wave vector $\mathbf{k}_q$, though they are represented as such in Fig. 6.1. Note that the value of s is specified, so only one of the atoms in each primitive cell is moved away from its equilibrium position, the other one remaining in its equilibrium position.

Note, that each of the formulations (6.6), (6.10), (6.11) and (6.14) involves N independent parameters, i.e. two per pair of wave vectors $(\mathbf{k}_q, -\mathbf{k}_q)$. No reduction

6.1 THE LATTICE HAMILTONIAN

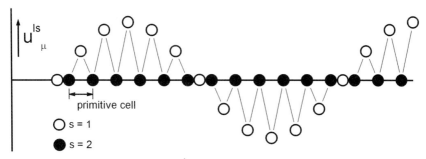

Figure 6.1 The nuclear displacements u_μ^{ls} due to a single plane wave with a wave vector $\mathbf{k}_q$, an amplitude $|v_\mu^{q1}|$ and a phase $\phi_\mu^{q1} = 0$.

of complexity is obtained by introducing real transforms (6.11) or (6.14) at this stage. On the contrary, the complex transform pair (6.6) and (6.7) is more simple to handle and therefore we continue using this particular pair.

6.1.2 The Lattice Hamiltonian in Reciprocal Space

THE POTENTIAL ENERGY IN RECIPROCAL SPACE

The potential term in the lattice Hamiltonian (6.1) is now easily transformed. We first insert the Fourier transform (6.6) for the displacements u_μ^{ls}. Then,

$$\frac{1}{2}\sum_{ll',ss',\mu\nu}^{N} \Phi_{\mu\nu}^{ll',ss'} u_\mu^{ls} u_\nu^{l's'} = \frac{1}{2}\sum_{ll',ss',\mu\nu}^{N} \Phi_{\mu\nu}^{ll',ss'} \frac{1}{\sqrt{N}}\sum_q^N v_\mu^{qs} e^{i\mathbf{k}_q\cdot\mathbf{R}_l} \frac{1}{\sqrt{N}}\sum_{q'}^N v_\nu^{q's'} e^{i\mathbf{k}_{q'}\cdot\mathbf{R}_{l'}}$$

$$= \frac{1}{2}\sum_{ss',\mu\nu} \frac{1}{N} \sum_{q,q'}^N v_\mu^{qs} v_\nu^{q's'} \sum_{l,l'}^N \Phi_{\mu\nu}^{ll',ss'} e^{i\mathbf{k}_q\cdot\mathbf{R}_l} e^{i\mathbf{k}_{q'}\cdot\mathbf{R}_{l'}} \quad (6.15)$$

We consider the sums over l and l' in more detail. They may be rewritten as

$$\sum_{l,l'}^N \Phi_{\mu\nu}^{ll',ss'} e^{i\mathbf{k}_q\cdot\mathbf{R}_l} e^{i\mathbf{k}_{q'}\cdot\mathbf{R}_{l'}} = \sum_{l,l'}^N \Phi_{\mu\nu}^{ll',ss'} e^{i(\mathbf{k}_q+\mathbf{k}_{q'})\cdot\mathbf{R}_l} e^{i\mathbf{k}_{q'}\cdot(\mathbf{R}_{l'}-\mathbf{R}_l)}$$

$$= \sum_l^N e^{i(\mathbf{k}_q+\mathbf{k}_{q'})\cdot\mathbf{R}_l} \left[\sum_{l'}^N \Phi_{\mu\nu}^{ll',ss'} e^{i\mathbf{k}_{q'}\cdot(\mathbf{R}_{l'}-\mathbf{R}_l)}\right] \quad (6.16)$$

The potential term should obey translational symmetry. This implies that the second sum depends on the difference between l and l' only. This allows us to define

$$\Psi_{\mu\nu}^{qss'} = \sum_{l'}^N \Phi_{\mu\nu}^{0l',ss'} e^{-i\mathbf{k}_q\cdot\mathbf{R}_{l'}} = \sum_{l'}^N \Phi_{\mu\nu}^{ll',ss'} e^{-i\mathbf{k}_q\cdot(\mathbf{R}_{l'}-\mathbf{R}_l)} \quad (6.17)$$

Inserting definition (6.17) in eq. (6.16) and the result in eq. (6.15), the potential term finally reduces to

$$\frac{1}{2}\sum_{ss',\mu\nu}\frac{1}{N}\sum_{q,q'}^{N} v_\mu^{qs} v_\nu^{q's'} \left(\sum_{l}^{N} e^{i(\mathbf{k}_q+\mathbf{k}_{q'})\cdot\mathbf{R}_l}\right) \Psi_{\mu\nu}^{-q'ss'} = \frac{1}{2}\sum_{ss',\mu\nu}\frac{1}{N}\sum_{q,q'}^{N} v_\mu^{qs} v_\nu^{q's'} N \delta_{q,-q'} \Psi_{\mu\nu}^{-q'ss'}$$

$$= \frac{1}{2}\sum_{ss',\mu\nu}\sum_{q}^{N} \Psi_{\mu\nu}^{qss'} v_\mu^{qs} v_\nu^{-qs'} \qquad (6.18)$$

where we insert the Kronecker δ-function (6.8).

The resulting expression (6.18) for the potential energy term in the lattice Hamiltonian contains complex quantities $\Psi_{\mu\nu}^{qss'}$ and v_μ^{qs}. On the other hand the original expression in eq. (6.1) was real, so the present expression should be real as well. To investigate whether this is true, we first consider a pair of terms corresponding to $\mathbf{k}_q$ and $-\mathbf{k}_q$ in eq. (6.18),

$$\frac{1}{2}\Psi_{\mu\nu}^{qss'} v_\mu^{qs} v_\nu^{-qs'} + \frac{1}{2}\Psi_{\mu\nu}^{-qss'} v_\mu^{-qs} v_\nu^{qs'} \qquad (6.19)$$

Next remembering that the interaction constants $\Phi_{\mu\nu}^{ll',ss'}$ are real, we note that their Fourier transforms (6.17) obey

$$\Psi_{\mu\nu}^{qss'*} = \sum_{l'}^{N} \Phi_{\mu\nu}^{0l',ss'} \left(e^{-i\mathbf{k}_q\cdot\mathbf{R}_{l'}}\right)^*$$

$$= \sum_{l'}^{N} \Phi_{\mu\nu}^{0l',ss'} e^{i\mathbf{k}_q\cdot\mathbf{R}_{l'}} = \Psi_{\mu\nu}^{-qss'} \qquad (6.20)$$

Moreover eq. (6.9) enables us to write $v_\nu^{-qs} = v_\nu^{qs*}$ and $v_\nu^{qs'} = v_\nu^{-qs'*}$. Hence the pair of terms (6.19) can be rewritten as

$$\frac{1}{2}\Psi_{\mu\nu}^{qss'} v_\mu^{qs} v_\nu^{-qs'} + \frac{1}{2}\Psi_{\mu\nu}^{qss'*} v_\mu^{qs*} v_\nu^{-qs'*}$$

$$= \frac{1}{2}\Psi_{\mu\nu}^{qss'} v_\mu^{qs} v_\nu^{-qs'} + \frac{1}{2}(\Psi_{\mu\nu}^{qss'} v_\mu^{qs} v_\nu^{-qs'})^*$$

$$= \Re(\Psi_{\mu\nu}^{qss'} v_\mu^{qs} v_\nu^{-qs'}) \qquad (6.21)$$

which is real. Hence, the full expression (6.18) is real as it should be.

THE KINETIC ENERGY IN RECIPROCAL SPACE

Next, we consider the kinetic energy term in the lattice Hamiltonian (6.1). We start by rewriting the partial differentials with respect to u_μ^{ls} with the help of a chain rule. Using eq. (6.7) to evaluate the derivatives in this chain rule, we find

$$\frac{\partial}{\partial u_\mu^{ls}} = \sum_q^N \frac{\partial v_\mu^{qs}}{\partial u_\mu^{ls}} \frac{\partial}{\partial v_\mu^{qs}} = \frac{1}{\sqrt{N}}\sum_q^N e^{-i\mathbf{k}_q\cdot\mathbf{R}_l} \frac{\partial}{\partial v_\mu^{qs}} \qquad (6.22)$$

Note that the use of derivatives with respect to v_μ^{qs} requires the eigenstates of the Hamiltonian to be functions of these complex variables rather than of their real and

6.1 THE LATTICE HAMILTONIAN

imaginary parts $\Re\{v_\mu^{qs}\}$ and $\Im\{v_\mu^{qs}\}$ separately. We use this expression to calculate the kinetic energy term,

$$-\sum_l^N \left(\frac{\partial}{\partial u_\mu^{ls}}\right)^2 = -\sum_l^N \left(\frac{1}{\sqrt{N}}\sum_q^N e^{-i\mathbf{k}_q\cdot\mathbf{R}_l}\frac{\partial}{\partial v_\mu^{qs}}\right)\left(\frac{1}{\sqrt{N}}\sum_{q'}^N e^{-i\mathbf{k}_{q'}\cdot\mathbf{R}_l}\frac{\partial}{\partial v_\mu^{q's}}\right)$$

$$= -\sum_{q,q'}^N \frac{\partial}{\partial v_\mu^{qs}}\frac{\partial}{\partial v_\mu^{q's}}\frac{1}{N}\sum_l^N e^{-i(\mathbf{k}_q+\mathbf{k}_{q'})\cdot\mathbf{R}_l}$$

$$= -\sum_q^N \sum_{q'}^N \frac{\partial}{\partial v_\mu^{qs}}\frac{\partial}{\partial v_\mu^{q's}}\delta_{q,-q'}$$

$$= -\sum_q^N \frac{\partial}{\partial v_\mu^{qs}}\frac{\partial}{\partial v_\mu^{-qs}} \tag{6.23}$$

where we again use the Kronecker δ-function (6.8). Note that eq. (6.9) renders this expression real.

Inserting the results (6.18) and (6.23) in eq. (6.1) we obtain the transform of the lattice Hamiltonian to reciprocal space,

$$\mathcal{H}_L = \sum_q^N \mathcal{H}_L^q$$

$$= \sum_q^N \left(-\sum_{s,\mu}\frac{\hbar^2}{2M_s}\frac{\partial}{\partial v_\mu^{qs}}\frac{\partial}{\partial v_\mu^{-qs}} + \frac{1}{2}\sum_{ss',\mu\nu}\Psi_{\mu\nu}^{qss'} v_\mu^{qs} v_\nu^{-qs'}\right)$$

$$= \sum_q^{N/2}(\mathcal{H}_L^q + \mathcal{H}_L^{-q})$$

$$= \sum_q^{N/2}\left(-\sum_{s,\mu}\frac{\hbar^2}{2M_s}\frac{\partial}{\partial v_\mu^{qs}}\frac{\partial}{\partial v_\mu^{-qs}} + \frac{1}{2}\sum_{ss',\mu\nu}\Psi_{\mu\nu}^{qss'} v_\mu^{qs} v_\nu^{-qs'}\right.$$

$$\left. -\sum_{s,\mu}\frac{\hbar^2}{2M_s}\frac{\partial}{\partial v_\mu^{-qs}}\frac{\partial}{\partial v_\mu^{qs}} + \frac{1}{2}\sum_{ss',\mu\nu}\Psi_{\mu\nu}^{-qss'} v_\mu^{-qs} v_\nu^{qs'}\right) \tag{6.24}$$

Thus the lattice Hamiltonian is split into $N/2$ pairs of terms $\mathcal{H}_L^q$ and $\mathcal{H}_L^{-q}$, one for each pair $(\mathbf{k}_q, \mathbf{k}_{-q})$ of wave vectors in the first Brillouin zone. As the terms for different pairs are not coupled, each pair provides a good quantum number. In this respect the lattice Hamiltonian behaves similarly but not exactly like the single electron Hamiltonian, which splits into independent terms for different values of $\mathbf{k}_q$ in the first Brillouin zone, and where we find $\mathbf{k}_q$ to be a good quantum number. For the moment we will ignore the relation between the terms $\mathcal{H}_L^q$ and $\mathcal{H}_L^{-q}$ and we will first have closer look at the individual terms $\mathcal{H}_L^q$.

6.2 The Phonon Spectrum of Semiconductors
6.2.1 Optical and Acoustical Phonon Modes
RELATIVE AND CENTRE OF MASS DISPLACEMENTS

Though we have succeeded in reducing the multi-dimensional harmonic oscillator (6.1) to the sum (6.24) of lower-dimensional harmonic oscillators $\mathcal{H}_L^q$, the latter are still not one-dimensional. In typical semiconductors with two atoms in a primitive cell, they are six-dimensional. In the next step to obtain a set of one-dimensional harmonic oscillators, we rewrite the motion of the two individual nuclei in a primitive cell into the centre of mass motion of these nuclei and their relative motion. We will see that this transformation splits each 6-dimensional harmonic oscillator into two independent three-dimensional harmonic oscillators. The set of harmonic oscillators representing the centre of mass motion is called the acoustical branch of the crystal lattice Hamiltonian, while the remaining set is called the optical branch.

The centre of mass displacement of the nuclei in the lth primitive cell is defined as,

$$u_\mu^{lA} = \frac{M_1 u_\mu^{l1} + M_2 u_\mu^{l2}}{M_1 + M_2} \tag{6.25}$$

and the relative displacement as

$$u_\mu^{lO} = u_\mu^{l1} - u_\mu^{l2} \tag{6.26}$$

To be useful for our purpose the centre of mass displacement and relative displacement must be transformed to reciprocal space. This is easily done using eq. (6.7) and we find

$$v_\mu^{qA} = \frac{1}{\sqrt{N}} \sum_l^N u_\mu^{lA} e^{-i\mathbf{k}_q \cdot \mathbf{R}_l} = \frac{M_1 v_\mu^{q1} + M_2 v_\mu^{q2}}{M_1 + M_2} \tag{6.27}$$

$$v_\mu^{qO} = \frac{1}{\sqrt{N}} \sum_l^N u_\mu^{lO} e^{-i\mathbf{k}_q \cdot \mathbf{R}_l} = v_\mu^{q1} - v_\mu^{q2} \tag{6.28}$$

for the Fourier transforms of the centre of mass displacement and the relative displacement, respectively. The inverse transforms are immediately found from eq. (6.6) and are found to be

$$u_\mu^{lA} = \frac{1}{\sqrt{N}} \sum_q^N v_\mu^{qA} e^{i\mathbf{k}_q \cdot \mathbf{R}_l} \tag{6.29}$$

$$u_\mu^{lO} = \frac{1}{\sqrt{N}} \sum_q^N v_\mu^{qO} e^{i\mathbf{k}_q \cdot \mathbf{R}_l} \tag{6.30}$$

Note that the nuclear masses are real, so according to eq. (6.9),

$$v_\mu^{qA*} = \frac{M_1 v_\mu^{q1*} + M_2 v_\mu^{q2*}}{M_1 + M_2} = \frac{M_1 v_\mu^{-q1} + M_2 v_\mu^{-q2}}{M_1 + M_2} = v_\mu^{-qA} \tag{6.31}$$

$$v_\mu^{qO*} = v_\mu^{q1*} - v_\mu^{q2*} = v_\mu^{-q1} - v_\mu^{-q2} = v_\mu^{-qO} \tag{6.32}$$

ACOUSTICAL AND OPTICAL PLANE STANDING WAVES

To investigate the physical significance of v_μ^{qA} and v_μ^{qO} we rewrite the transforms (6.29) and (6.30) following exactly the same procedure as the one leading from eq. (6.6) to eq. (6.14). Because the only difference consists of writing indices A or O instead of the index s, we will omit the derivation and give the result immediately. We find,

$$u_\mu^{lA} = \frac{1}{\sqrt{N}} \sum_q^{N/2} 2|v_\mu^{qA}| \cos(\mathbf{k}_q \cdot \mathbf{R}_l + \phi_\mu^{qA}) \qquad (6.33)$$

$$u_\mu^{lO} = \frac{1}{\sqrt{N}} \sum_q^{N/2} 2|v_\mu^{qO}| \cos(\mathbf{k}_q \cdot \mathbf{R}_l + \phi_\mu^{qO}) \qquad (6.34)$$

where ϕ_μ^{qA} and ϕ_μ^{qO} are defined by

$$v_\mu^{qA} = |v_\mu^{qA}| e^{i\phi_\mu^{qA}} \qquad (6.35)$$

$$v_\mu^{qO} = |v_\mu^{qO}| e^{i\phi_\mu^{qO}} \qquad (6.36)$$

We see that we have again resolved the nuclear displacements in standing plane waves with wave vector $\mathbf{k}_q$, but now have found two different types. One with amplitude $2|v_\mu^{qA}|$ and phase ϕ_μ^{qA} corresponds to motion of the two nuclei in the primitive cell in the same direction. The other with amplitude $2|v_\mu^{qO}|$ and phase ϕ_μ^{qO} corresponds to motion of the two nuclei in the primitive cell in opposite direction.

The two types of motion are illustrated in Figs 6.2 and 6.3 showing the nuclear displacements u_μ^{ls} in the μ-direction for the case that only one standing plane wave of the type (6.33) or one of the type (6.34) are present respectively. As in Fig. 6.1 the horizontal axis corresponds to a one-dimensional row of atoms in the $\mathbf{k}_q$ direction. Again one should be aware that in real space the μ-axis need not be perpendicular to the wave vector $\mathbf{k}_q$, though they are represented as such in Figs 6.2 and 6.3. The type of nuclear displacement shown in Fig. 6.2 is the same as the one occurring when a standing sound wave is excited in the crystal. Precisely this equivalence led to the index A denoting 'acoustical'. On the other hand the nuclear displacements shown in Fig. 6.3 are of a different nature and carry the index O denoting 'optical'.

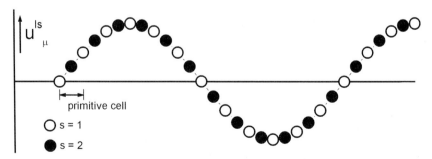

Figure 6.2 The nuclear displacements u_μ^{ls} due to a single plane wave of the type (6.33) with a wave vector $\mathbf{k}_q$, an amplitude $|v_\mu^{qA}|$ and a phase $\phi_\mu^{qA} = 0$.

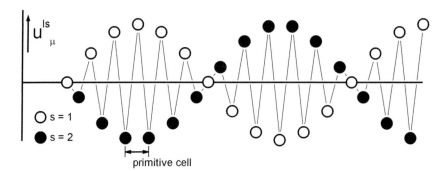

Figure 6.3 The nuclear displacements u_μ^{ls} due to a single plane wave of the type (6.34) with a wave vector $\mathbf{k}_q$, an amplitude $|v_\mu^{qO}|$ and a phase $\phi_\mu^{qO} = 0$.

THE LATTICE HAMILTONIAN AT $\mathbf{k}_q = 0$

Using these definitions we now continue by considering the component $\mathcal{H}_L^0$ of the lattice Hamiltonian corresponding to $\mathbf{k}_q = \mathbf{k}_{-q} = 0$. According to eq. (6.24) it is given by

$$\mathcal{H}_L^0 = -\sum_{s,\mu} \frac{\hbar^2}{2M_s} \frac{\partial}{\partial v_\mu^{0s}} \frac{\partial}{\partial v_\mu^{0s}} + \frac{1}{2} \sum_{ss',\mu\nu} \Psi_{\mu\nu}^{0ss'} v_\mu^{0s} v_\nu^{0s'} \tag{6.37}$$

We remember that the indices s and s' indicate the nucleus in the primitive cell and the indices μ and ν the components of the displacements of these nuclei in a Cartesian frame of coordinates. Thus in semiconductors like Si, Ge and GaAs, s and s' take two values, 1 and 2 only. Moreover, eqs (6.9) and (6.20) require v_μ^{0s} and $\Psi_{\mu\nu}^{0ss'}$ to be real, so all quantities in this Hamiltonian are real.

Elementary arguments show several relations between the various components $\Psi_{\mu\nu}^{0ss'}$. First, the matrix consisting of the components $\Psi_{\mu\nu}^{0ss'}$ has real eigenvalues, so it must be Hermitian[1]. Hence,

$$\Psi_{\mu\nu}^{012} = \Psi_{\mu\nu}^{021*} = \Psi_{\mu\nu}^{021} \tag{6.38}$$

Next, we consider the case that all nuclei with index 1 are displaced by a constant amount ε in the μ as well as in the ν-direction, while all nuclei with index 2 are left in place. Then, using eq. (6.7)

$$\frac{1}{2} \sum_{ss'} \Psi_{\mu\nu}^{0ss'} v_\mu^{0s} v_\nu^{0s'} = \frac{N}{2} \Psi_{\mu\nu}^{011} \varepsilon^2 \tag{6.39}$$

---

[1] This follows immediately from the lemma that a Hermitian matrix remains Hermitian under a unitary transformation. See ref. [26], section 10.19. On the basis of its own eigenstates an operator is represented by a diagonal matrix, while the diagonal elements are its eigenvalues. If the latter are real, this specific matrix representation is Hermitian. Now, the operator's representation on any other orthonormal basis is obtained via a unitary transformation. Hence, these matrix representations are also Hermitian.

This situation is however completely equivalent to displacing all nuclei with index 2 by a constant amount $-\varepsilon$ in the μ as well as in the ν direction, while leaving all nuclei with index 1 in place. In this latter case, we have

$$\frac{1}{2}\sum_{ss'} \Psi_{\mu\nu}^{0ss'} v_\mu^{0s} v_\nu^{0s'} = \frac{N}{2}\Psi_{\mu\nu}^{022}\varepsilon^2 \tag{6.40}$$

so we conclude that

$$\Psi_{\mu\nu}^{011} = \Psi_{\mu\nu}^{022} \equiv \Psi_{\mu\nu}^{0O} \tag{6.41}$$

defining $\Psi_{\mu\nu}^{0O}$. Here the superscript $0O$ stands for $\mathbf{k}_q = 0$ and 'optical', because the displacements ε are relative displacements of the two nuclei in a primitive cell.

Finally, we consider the case that all nuclei are displaced by the same amount ε in the μ as well as in the ν-direction. Of course this case corresponds to a displacement of the crystal as a whole, so the resulting potential energy should be equal to zero. Then,

$$\frac{1}{2}\sum_{ss'} \Psi_{\mu\nu}^{0ss'} v_\mu^{0s} v_\nu^{0s'} = \frac{N}{2}\left(\Psi_{\mu\nu}^{011} + \Psi_{\mu\nu}^{012} + \Psi_{\mu\nu}^{021} + \Psi_{\mu\nu}^{022}\right)\varepsilon^2$$

$$= N\left(\Psi_{\mu\nu}^{0O} + \Psi_{\mu\nu}^{012}\right)\varepsilon^2 = 0 \tag{6.42}$$

where we inserted eqs (6.38) and (6.41). Hence,

$$\Psi_{\mu\nu}^{012} = -\Psi_{\mu\nu}^{0O} \tag{6.43}$$

Combining all these results the potential term in $\mathcal{H}_L^0$ reduces to

$$\frac{1}{2}\sum_{ss',\mu\nu} \Psi_{\mu\nu}^{0ss'} v_\mu^{0s} v_\nu^{0s'} = \frac{1}{2}\sum_{\mu\nu} \Psi_{\mu\nu}^{0O}(v_\mu^{01} v_\nu^{01} - v_\mu^{01} v_\nu^{02} - v_\mu^{02} v_\nu^{01} + v_\mu^{02} v_\nu^{02})$$

$$= \frac{1}{2}\sum_{\mu\nu} \Psi_{\mu\nu}^{0O}(v_\mu^{01} - v_\mu^{02})(v_\nu^{01} - v_\nu^{02})$$

$$= \frac{1}{2}\sum_{\mu\nu} \Psi_{\mu\nu}^{0O} v_\mu^{0O} v_\nu^{0O} \tag{6.44}$$

where because of eq. (6.32) v_μ^{0O} is real, just as v_μ^{01} and v_μ^{02}.

THE STRAIN TENSOR

Thus we have found that for $\mathbf{k}_q = 0$ the centre of mass displacement of the two nuclei in the primitive cell does not entail potential energy. For finite values of $\mathbf{k}_q$ the situation is different because then the centre of mass displacement varies with position and the crystal is deformed. So the difference between centre of mass displacements in subsequent primitive cells determines the magnitude of the potential rather than the amplitude of the centre of mass displacement itself. Mathematically we denote this difference between centre of mass displacements by the strain tensor, which has components defined by

$$S_{\mu\nu}^l = \frac{1}{2}\left(\frac{\partial u_\mu^{lA}}{\partial R_\nu^l} + \frac{\partial u_\nu^{lA}}{\partial R_\mu^l}\right) \tag{6.45}$$

Here, u_μ^{lA} is the centre of mass displacement (6.25) of the nuclei in the lth primitive cell, while we assume that the derivatives with respect to the components R_μ^l of $\mathbf{R}_l$ may be taken because u_μ^l is a smooth function of $\mathbf{R}_l$. This assumption is reasonable as long as we restrict ourselves to the case that the phonon wavelength is long, so the corresponding wave vector $\mathbf{k}_q$ is small. Also, we have symmetrized the strain tensor, because the antisymmetric tensor

$$\frac{1}{2}\left(\frac{\partial u_\mu^{lA}}{\partial R_\nu^l} - \frac{\partial u_\nu^{lA}}{\partial R_\mu^l}\right) \tag{6.46}$$

corresponds to a mere rotation of the crystal as a whole, so the corresponding potential will be equal to zero.

We transform the strain tensor to reciprocal space using eq. (6.29). Then

$$\begin{aligned}
S_{\mu\nu}^l &= \frac{1}{2}\left(\frac{\partial}{\partial R_\nu^l}\frac{1}{\sqrt{N}}\sum_q^N v_\mu^{qA} e^{i\mathbf{k}_q\cdot\mathbf{R}_l} + \frac{\partial}{\partial R_\mu^l}\frac{1}{\sqrt{N}}\sum_q^N v_\nu^{qA} e^{i\mathbf{k}_q\cdot\mathbf{R}_l}\right) \\
&= \frac{1}{2}\frac{1}{\sqrt{N}}\sum_q^N \left(v_\mu^{qA}\frac{\partial}{\partial R_\nu^l}e^{i\mathbf{k}_q\cdot\mathbf{R}_l} + v_\nu^{qA}\frac{\partial}{\partial R_\mu^l}e^{i\mathbf{k}_q\cdot\mathbf{R}_l}\right) \\
&= \frac{i}{2}\frac{1}{\sqrt{N}}\sum_q^N (v_\mu^{qA} k_{q\nu} + v_\nu^{qA} k_{q\mu})e^{i\mathbf{k}_q\cdot\mathbf{R}_l} \tag{6.47}
\end{aligned}$$

HOOKE'S LAW

For small values of $\mathbf{k}_q$, i.e. for long wavelengths, the potential energy should correspond to the macroscopic potential upon deformation of the crystal. Thus, we expect the potential energy to obey Hooke's law and for a finite crystal to be given by

$$\frac{1}{2}\sum_l^N \sum_{\mu\rho\nu\sigma} c_{\mu\rho\nu\sigma} S_{\mu\rho}^l S_{\nu\sigma}^l \tag{6.48}$$

where $c_{\mu\rho\nu\sigma}$ is the elastic tensor. It has 81 real components, but not all of them are independent. For example, according to its definition (6.45) the strain tensor $S_{\mu\rho}^l$ is symmetric, so we also expect that

$$c_{\mu\rho\nu\sigma} = c_{\rho\mu\nu\sigma} = c_{\mu\rho\sigma\nu} = c_{\rho\mu\sigma\nu} \tag{6.49}$$

We transform the potential (6.48) to reciprocal space by means of eq. (6.47). Then, using the symmetry relations (6.49), we find

6.2 THE PHONON SPECTRUM OF SEMICONDUCTORS

$$\frac{1}{2}\sum_{l}^{N}\sum_{\mu\rho\nu\sigma}c_{\mu\rho\nu\sigma}S^l_{\mu\rho}S^l_{\nu\sigma} = \frac{1}{2}\sum_{l}^{N}\sum_{\mu\rho\nu\sigma}c_{\mu\rho\nu\sigma}\frac{1}{\sqrt{N}}\sum_{q}^{N}\frac{i}{2}(v_\mu^{qA}k_{q\rho}+v_\rho^{qA}k_{q\mu})e^{i\mathbf{k}_q\cdot\mathbf{R}_l}$$

$$\times\frac{1}{\sqrt{N}}\sum_{q'}^{N}\frac{i}{2}(v_\nu^{q'A}k_{q'\sigma}+v_\sigma^{q'A}k_{q'\gamma})e^{i\mathbf{k}_{q'}\cdot\mathbf{R}_l}$$

$$=-\frac{1}{2}\sum_{q}^{N}\sum_{q'}^{N}\sum_{\mu\nu}\left(\sum_{\rho\sigma}c_{\mu\rho\nu\sigma}k_{q\rho}k_{q'\sigma}\right)v_\mu^{qA}v_\nu^{q'A}\frac{1}{N}\sum_{l}^{N}e^{i(\mathbf{k}_q+\mathbf{k}_{q'})\cdot\mathbf{R}_l}$$

$$=-\frac{1}{2}\sum_{q}^{N}\sum_{q'}^{N}\sum_{\mu\nu}\left(\sum_{\rho\sigma}c_{\mu\rho\nu\sigma}k_{q\rho}k_{q'\sigma}\right)v_\mu^{qA}v_\nu^{q'A}\delta_{q,-q'}$$

$$=\frac{1}{2}\sum_{q}^{N}\sum_{\mu\nu}\left(\sum_{\rho\sigma}c_{\mu\rho\nu\sigma}k_{q\rho}k_{-q\sigma}\right)v_\mu^{qA}v_\nu^{-qA} \qquad (6.50)$$

where we inserted the Kronecker δ-function (6.8), so $\mathbf{k}_{q'}=-\mathbf{k}_q$. We define

$$\Psi_{\mu\nu}^{qA}=\sum_{\rho\sigma}c_{\mu\rho\nu\sigma}k_{q\rho}k_{-q\sigma} \qquad (6.51)$$

reducing the potential energy to the shape

$$\frac{1}{2}\sum_{l}^{N}\sum_{\mu\rho\nu\sigma}c_{\mu\rho\nu\sigma}S^l_{\mu\rho}S^l_{\nu\sigma}=\frac{1}{2}\sum_{q}^{N}\sum_{\mu\nu}\Psi_{\mu\nu}^{qA}v_\mu^{qA}v_\nu^{-qA} \qquad (6.52)$$

For finite but small values of $\mathbf{k}_q$, this potential energy must be added to the potential term (6.44) for $\mathbf{k}_q=0$. Moreover, because relative motion and centre of mass motion have been separated in the previous treatment, the potential term in the Hamiltonian representing the former motion is expected to keep the shape (6.44), as long as $\mathbf{k}_q$ is small. Therefore, for small values of $\mathbf{k}_q$ the potential in each term $\mathcal{H}_L^q$ in the lattice Hamiltonian (6.24) can be written as

$$\frac{1}{2}\sum_{\mu\nu}\Psi_{\mu\nu}^{qA}v_\mu^{qA}v_\nu^{-qA}+\frac{1}{2}\sum_{\mu\nu}\Psi_{\mu\nu}^{qO}v_\mu^{qO}v_\nu^{-qO} \qquad (6.53)$$

where $\Psi_{\mu\nu}^{qO}\approx\Psi_{\mu\nu}^{00}$. Note that by their definitions (6.41) and (6.51) $\Psi_{\mu\nu}^{qO}$ and $\Psi_{\mu\nu}^{qA}$ are real. Then, following the same argument as in eq. (6.21) and using eqs (6.31) and (6.32), the sum of two terms corresponding to $\mathbf{k}_q$ and $-\mathbf{k}_q$ is always real, thus assuring that the total potential is real as it should be.

REDUCED DISPLACEMENTS

Next we introduce reduced displacements in reciprocal space. For the individual nuclear displacements they are given by eq. (2.48),

$$w_\mu^{qs}=\sqrt{\frac{M_s}{\hbar}}v_\mu^{qs} \qquad (6.54)$$

having dimension $[t]^{1/2}$. However, we need to express them as relative and centre of mass displacements. Moreover, such reduced displacements must be obtained from those given by eq. (6.54) by a unitary transformation. The transformation fulfilling this purpose is given by

$$\begin{pmatrix} w_\mu^{qO} \\ w_\mu^{qA} \end{pmatrix} = \begin{pmatrix} \sqrt{\dfrac{M_2}{M_1+M_2}} & -\sqrt{\dfrac{M_1}{M_1+M_2}} \\ \sqrt{\dfrac{M_1}{M_1+M_2}} & \sqrt{\dfrac{M_2}{M_1+M_2}} \end{pmatrix} \begin{pmatrix} w_\mu^{q1} \\ w_\mu^{q2} \end{pmatrix} \qquad (6.55)$$

This is directly obvious by writing it in the shape

$$\sqrt{\dfrac{M_1+M_2}{M_1 M_2}}\, w_\mu^{qO} = \dfrac{w_\mu^{q1}}{\sqrt{M_1}} - \dfrac{w_\mu^{q2}}{\sqrt{M_2}} = \dfrac{1}{\sqrt{\hbar}}\left(v_\mu^{q1} - v_\mu^{q2}\right) = \dfrac{1}{\sqrt{\hbar}} v_\mu^{qO} \qquad (6.56)$$

$$\dfrac{1}{\sqrt{M_1+M_2}}\, w_\mu^{qA} = \dfrac{M_1}{M_1+M_2}\dfrac{w_\mu^{q1}}{\sqrt{M_1}} + \dfrac{M_2}{M_1+M_2}\dfrac{w_\mu^{q2}}{\sqrt{M_2}}$$

$$= \dfrac{1}{\sqrt{\hbar}} \dfrac{M_1 v_\mu^{q1} + M_2 v_\mu^{q2}}{M_1+M_2} = \dfrac{1}{\sqrt{\hbar}} v_\mu^{qA} \qquad (6.57)$$

Moreover, it is easily checked that the transposed matrix

$$\begin{pmatrix} \sqrt{\dfrac{M_2}{M_1+M_2}} & \sqrt{\dfrac{M_1}{M_1+M_2}} \\ -\sqrt{\dfrac{M_1}{M_1+M_2}} & \sqrt{\dfrac{M_2}{M_1+M_2}} \end{pmatrix} \qquad (6.58)$$

is also the inverse matrix, so it fulfils the requirement of being unitary. Hence for the transformation of the kinetic energy term of $\mathcal{H}_L^0$ eq. (2.55) applies. So, first using eq. (6.54) and next transforming by means of eq. (6.55), we find

$$-\sum_{s,\mu} \dfrac{\hbar^2}{2M_s} \dfrac{\partial}{\partial v_\mu^{qs}} \dfrac{\partial}{\partial v_\mu^{-qs}} = -\dfrac{\hbar}{2}\sum_{s,\mu} \dfrac{\partial}{\partial w_\mu^{qs}} \dfrac{\partial}{\partial w_\mu^{-qs}}$$

$$= -\dfrac{\hbar}{2}\sum_\mu \dfrac{\partial}{\partial w_\mu^{qA}} \dfrac{\partial}{\partial w_\mu^{-qA}} - \dfrac{\hbar}{2}\sum_\mu \dfrac{\partial}{\partial w_\mu^{qO}} \dfrac{\partial}{\partial w_\mu^{-qO}} \qquad (6.59)$$

Here we also took into account that the transformation matrix is real. For the transformation of the potential energy terms we simply use eqs (6.56) and (6.57), while we define

$$D_{\mu\nu}^{qO} = \dfrac{M_1+M_2}{M_1 M_2}\, \Psi_{\mu\nu}^{qO} \qquad (6.60)$$

$$D_{\mu\nu}^{qA} = \dfrac{1}{M_1+M_2}\, \Psi_{\mu\nu}^{qA} \qquad (6.61)$$

6.2 THE PHONON SPECTRUM OF SEMICONDUCTORS

which are real because $\Psi_{\mu\nu}^{qO}$ and $\Psi_{\mu\nu}^{qA}$ are real. Moreover, just as the interaction coefficients $D_{\lambda\lambda'}$ defined by eq. (2.49), $D_{\mu\nu}^{qO}$ and $D_{\mu\nu}^{qA}$ have dimension $[t]^{-2}$. As a result, for small values of $\mathbf{k}_q$ we find

$$\mathcal{H}_L^q = -\frac{\hbar}{2}\sum_\mu \frac{\partial}{\partial w_\mu^{qA}}\frac{\partial}{\partial w_\mu^{-qA}} + \frac{\hbar}{2}\sum_{\mu\nu} D_{\mu\nu}^{qA} w_\mu^{qA} w_\nu^{-qA}$$

$$-\frac{\hbar}{2}\sum_\mu \frac{\partial}{\partial w_\mu^{qO}}\frac{\partial}{\partial w_\mu^{-qO}} + \frac{\hbar}{2}\sum_{\mu\nu} D_{\mu\nu}^{qO} w_\mu^{qO} w_\nu^{-qO} \quad (6.62)$$

While the two latter terms represent the optical branch of the crystal lattice Hamiltonian, the former two terms represent the acoustical branch. Recall that in the superscripts of $D_{\mu\nu}^{qO}$ and $D_{\mu\nu}^{qA}$ the indices O and A denote 'optical and 'acoustical', respectively. Note that we have derived the acoustical branch from Hooke's law representing macroscopic deformation. Thus, this branch represents macroscopic vibrations of the crystal, i.e. genuine acoustical waves.

6.2.2 Transverse and Longitudinal Modes

ROTATION OF THE FRAME OF REFERENCE

In the third step of the transformation to one-dimensional harmonic oscillators we diagonalize the two 3×3 submatrices $D_{\mu\nu}^{qA}$ and $D_{\mu\nu}^{qO}$. For this purpose we need two 3×3 unitary transformation matrices with elements $U_{\mu j}^{qA}$ and $U_{\mu j}^{qO}$. Before going ahead, we first consider their direct physical meaning which can be inferred as follows.

Let us consider the acoustical branch and the corresponding 3×3 unitary transformation matrix. Because all submatrix elements $D_{\mu\nu}^{qA}$ are real, the elements $U_{\mu j}^{qA}$ of the transformation matrix can also be chosen to be real. Hence, this transformation matrix is not just unitary but orthogonal. Now its elements $U_{\mu j}^{qA}$ are numbered by the index j taking three values denoting the three resulting one-dimensional harmonic oscillators and the index $\mu = x, y, z$, denoting the axes of the Cartesian frame of reference used to describe the nuclear displacement. As the transformation matrix is orthogonal, it therefore represents a rotation from the latter frame of reference to a new Cartesian frame of reference representing mutually orthogonal nuclear displacements. Generally one of them proves to be longitudinal, so the nuclei move parallel to the direction of the $\mathbf{k}_q$-vector. Then the other two are necessarily transverse, so the nuclei move perpendicular to this vector. Moreover, the latter two types of motion are again mutually orthogonal. Following the terminology introduced in section 2.3.2, we denote these three types of motion as the *longitudinal acoustical (LA) phonon mode* and the two *transverse acoustical (TA) phonon modes*. Furthermore their mutually orthogonal directions are called the direction of the polarization of these three phonon modes. As a result the vector

$$\begin{pmatrix} U_{x,j}^{qA} \\ U_{y,j}^{qA} \\ U_{z,j}^{qA} \end{pmatrix} = \begin{pmatrix} \hat{p}_{j,x}^{qA} \\ \hat{p}_{j,y}^{qA} \\ \hat{p}_{j,z}^{qA} \end{pmatrix} \quad (6.63)$$

represents the unit vector $\hat{\mathbf{p}}_j^{qA}$ in the polarization direction of the jth acoustical phonon mode.

For the optical branch similar arguments apply. Here the transformation can be interpreted as a rotation from the (x, y, z)-frame to a frame based on three mutually perpendicular polarization directions corresponding to three phonon modes, now called the *longitudinal (LO)* and *transverse optical (TO)* phonon modes. So the vector

$$\begin{pmatrix} U_{x,j}^{qO} \\ U_{y,j}^{qO} \\ U_{z,j}^{qO} \end{pmatrix} = \begin{pmatrix} \hat{p}_{j,x}^{qO} \\ \hat{p}_{j,y}^{qO} \\ \hat{p}_{j,z}^{qO} \end{pmatrix} \tag{6.64}$$

represents the unit vector $\hat{\mathbf{p}}_j^{qO}$ in the polarization direction of the jth optical phonon mode.

COMPLEX NORMAL COORDINATES

Now we have a better understanding of the physical meaning of these transformation matrices, we apply them. Equivalently to eq. (2.51) and using the notation introduced in eqs (6.63) and (6.64), we write

$$D_{\mu\nu}^{qA} = \sum_j \hat{p}_{j,\mu}^{qA} (\omega_j^{qA})^2 \hat{p}_{j,\nu}^{qA} \tag{6.65}$$

$$D_{\mu\nu}^{qO} = \sum_j \hat{p}_{j,\mu}^{qO} (\omega_j^{qO})^2 \hat{p}_{j,\nu}^{qO} \tag{6.66}$$

Clearly $(\omega_j^{qA})^2$ and $(\omega_j^{qO})^2$ are real because the components of polarization vectors as well as $D_{\mu\nu}^{qA}$ and $D_{\mu\nu}^{qO}$ are real. Moreover, as discussed in section 2.3.2, they need to be positive for the crystal to be stable. Hence, also the quantities ω_j^{qA} and ω_j^{qO} themselves are real. Finally, $D_{\mu\nu}^{qA}$ and $D_{\mu\nu}^{qO}$ having dimension $[t]^{-2}$ and the polarization vector being dimensionless, ω_j^{qA} and ω_j^{qO} have dimension $[t]^{-1}$ corresponding to frequency.

Following eq. (2.54) we define normal coordinates Q_j^{qA} and Q_j^{qA}, such that

$$w_\mu^{qA} = \sum_j \hat{p}_{j,\mu}^{qA} Q_j^{qA} \tag{6.67}$$

$$w_\mu^{qO} = \sum_j \hat{p}_{j,\mu}^{qO} Q_j^{qO} \tag{6.68}$$

Remembering eqs (6.31) and (6.32), and observing that the transformations (6.56) and (6.57) to reduced displacements concern real coefficients only, we see immediately from eqs. (6.67) and (6.68) that

$$Q_j^{qA*} = Q_j^{-qA} \tag{6.69}$$

$$Q_j^{qO*} = Q_j^{-qO} \tag{6.70}$$

As a result, equivalently to eq. (2.57), we may write for the total lattice Hamiltonian,

$$\mathcal{H}_L = \sum_q^N \mathcal{H}_L^q = \sum_q^N (\mathcal{H}_L^{qA} + \mathcal{H}_L^{qO}) \tag{6.71}$$

6.2 THE PHONON SPECTRUM OF SEMICONDUCTORS

where we introduce

$$\mathcal{H}_L^{qA} = \frac{\hbar}{2} \sum_j \left[\frac{\partial}{\partial Q_j^{qA}} \frac{\partial}{\partial Q_j^{-qA}} + (\omega_j^{qA})^2 Q_j^{qA} Q_j^{-qA} \right] \qquad (6.72)$$

$$\mathcal{H}_L^{qO} = \frac{\hbar}{2} \sum_j \left[\frac{\partial}{\partial Q_j^{qO}} \frac{\partial}{\partial Q_j^{-qO}} + (\omega_j^{qO})^2 Q_j^{qO} Q_j^{-qO} \right] \qquad (6.73)$$

REAL NORMAL COORDINATES

We have split the Hamiltonian into one-dimensional harmonic oscillators, precisely as we proposed to do. An obvious problem remains however. The coordinates Q_j^{qA} and Q_j^{qO} in these oscillators are complex, so we cannot immediately apply the treatment given in appendix B. In fact we are confronted with an even more fundamental difficulty. According to eqs. (6.69) and (6.70) the complex normal coordinates Q_j^{-qA} and Q_j^{-qO} at position $-\mathbf{k}_q$ are *not* independent of the complex normal coordinates Q_j^{qA} and Q_j^{qO} at position $\mathbf{k}_q$. Hence, the one-dimensional harmonic oscillators $\mathcal{H}_L^{-qA}$ and $\mathcal{H}_L^{-qO}$ corresponding to the former wave vector are not independent of the one-dimensional harmonic oscillators $\mathcal{H}_L^{qA}$ and $\mathcal{H}_L^{qO}$ corresponding to the latter wave vector. Clearly, a further transformation of coordinates is necessary to remove this final dependence between the one-dimensional harmonic oscillators.

To solve these problems we start by arranging the sum over q in the lattice Hamiltonian in pairs of opposite $\mathbf{k}_q$,

$$\mathcal{H}_L = \sum_q^N (\mathcal{H}_L^{qA} + \mathcal{H}_L^{qO})$$

$$= \sum_q^{N/2} [(\mathcal{H}_L^{qA} + \mathcal{H}_L^{-qA}) + (\mathcal{H}_L^{qO} + \mathcal{H}_L^{-qO})] \qquad (6.74)$$

Here, as earlier, $\sum_q^{N/2}$ implies that we sum over half the first Brillouin zone while, if $\mathbf{k}_q$ is included in the sum, then $-\mathbf{k}_q$ is not included. Thus, the one-dimensional harmonic oscillators within each pair depend on each other, while those in different pairs are completely independent.

As a result, we may transform each pair of terms individually. For this purpose we introduce real normal coordinates,

$$Q_j^{qA1} = \sqrt{2} \, \Re\{Q_j^{qA}\} = \frac{1}{\sqrt{2}} (Q_j^{qA} + Q_j^{qA*}) \qquad (6.75)$$

$$Q_j^{qA2} = \sqrt{2} \, \Im\{Q_j^{qA}\} = \frac{1}{i\sqrt{2}} (Q_j^{qA} - Q_j^{qA*}) \qquad (6.76)$$

$$Q_j^{qO1} = \sqrt{2} \, \Re\{Q_j^{qO}\} = \frac{1}{\sqrt{2}} (Q_j^{qO} + Q_j^{qO*}) \qquad (6.77)$$

$$Q_j^{qO2} = \sqrt{2} \, \Im\{Q_j^{qO}\} = \frac{1}{i\sqrt{2}} (Q_j^{qO} - Q_j^{qO*}) \qquad (6.78)$$

The transformation is completely equivalent for the acoustical and the optical phonon branches, so we will just consider the former branch. First, we transform the kinetic energy terms. Now for $x = \sqrt{2}\,\Re\{Q\} = (Q + Q^*)/\sqrt{2}$ and $y = \sqrt{2}\,\Im\{Q\} = (Q - Q^*)/i\sqrt{2}$,

$$\frac{\partial^2}{\partial Q \partial Q^*} = \frac{\partial}{\partial Q}\left(\frac{\partial x}{\partial Q^*}\frac{\partial}{\partial x} + \frac{\partial y}{\partial Q^*}\frac{\partial}{\partial y}\right)$$

$$= \frac{\partial x}{\partial Q}\frac{\partial x}{\partial Q^*}\frac{\partial^2}{\partial x^2} + \frac{\partial x}{\partial Q}\frac{\partial y}{\partial Q^*}\frac{\partial}{\partial x}\frac{\partial}{\partial y} + \frac{\partial y}{\partial Q}\frac{\partial x}{\partial Q^*}\frac{\partial}{\partial y}\frac{\partial}{\partial x} + \frac{\partial y}{\partial Q}\frac{\partial y}{\partial Q^*}\frac{\partial^2}{\partial y^2}$$

$$= \frac{\partial^2}{\partial x^2}\frac{1}{2} + \frac{-1}{2i}\frac{\partial}{\partial x}\frac{\partial}{\partial y} + \frac{1}{2i}\frac{\partial}{\partial y}\frac{\partial}{\partial x} + \frac{1}{2}\frac{\partial^2}{\partial y^2}$$

$$= \frac{1}{2}\frac{\partial^2}{\partial x^2} + \frac{1}{2}\frac{\partial^2}{\partial y^2} \tag{6.79}$$

and similarly for the complex conjugate of this double derivative. So, remembering eq. (6.69), the kinetic energy terms of each pair can be written as

$$\frac{\partial^2}{\partial Q_j^{qA} \partial Q_j^{-qA}} + \frac{\partial^2}{\partial Q_j^{-qA} \partial Q_j^{qA}} = \frac{\partial^2}{\partial Q_j^{qA} \partial Q_j^{qA*}} + \frac{\partial^2}{\partial Q_j^{qA*} \partial Q_j^{qA}}$$

$$= \frac{\partial^2}{\partial (Q_j^{qA1})^2} + \frac{\partial^2}{\partial (Q_j^{qA2})^2} \tag{6.80}$$

Next we consider the potential energy terms. We recall eq. (6.20) requiring the interaction constants for the terms corresponding to $\mathbf{k}_q$ and $-\mathbf{k}_q$ to be each others complex conjugates. Moreover, according to our discussion concerning eq. (6.65), these interaction constants are real after transforming to complex normal coordinates. Hence, they are the same for $\mathbf{k}_q$ and $-\mathbf{k}_q$,

$$(\omega_j^{-qA})^2 = (\omega_j^{qA})^2 \quad \text{and} \quad (\omega_j^{-qO})^2 = (\omega_j^{qO})^2 \tag{6.81}$$

Using also the definitions (7.75) and (7.76), we then see immediately that

$$(\omega_j^{qA})^2 [(Q_j^{qA1})^2 + (Q_j^{qA2})^2] = \frac{(\omega_j^{qA})^2}{2}[(Q_j^{qA} + Q_j^{qA*})^2 - (Q_j^{qA} - Q_j^{qA*})^2]$$

$$= \frac{(\omega_j^{qA})^2}{2}[(Q_j^{qA} + Q_j^{-qA})^2 - (Q_j^{qA} - Q_j^{-qA})^2]$$

$$= (\omega_j^{qA})^2 (Q_j^{qA} Q_j^{-qA} + Q_j^{-qA} Q_j^{qA}) \tag{6.82}$$

As a result each pair of terms $\mathcal{H}_L^{qA} + \mathcal{H}_L^{-qA}$ in the acoustical branch of the Hamiltonian can be rearranged into two new one-dimensional harmonic oscillators,

$$\mathcal{H}_L^{qA} + \mathcal{H}_L^{-qA} = \sum_{\alpha=1}^{2}\left[\frac{\partial^2}{\partial (Q_j^{qA\alpha})^2} + (\omega_j^{qA})^2 (Q_j^{qA\alpha})^2\right] \tag{6.83}$$

Thus, we have reached our two goals. First, the coordinates in these new harmonic oscillators are real allowing us to use the solutions of appendix B. Moreover, these real coordinates are completely independent, so we may obtain their eigenstates and eigenvalues independently. Note however, that the two terms in the sum over α are exactly the same, except for the coordinate Q_j^{qA1} being replaced by Q_j^{qA2}. Hence, each of the eigenstates of the first one-dimensional harmonic oscillator is degenerate with the equivalent eigenstate of the second of these oscillators.

As already mentioned above, the optical phonon branch can be treated in exactly the same manner. So the total lattice Hamiltonian reduces to

$$\mathcal{H}_L = \frac{\hbar}{2} \sum_q \sum_j \sum_{\alpha=1}^{2} \left\{ \left[\frac{\partial^2}{\partial (Q_j^{qA\alpha})^2} + (\omega_j^{qA})^2 (Q_j^{qA\alpha})^2 \right] + \left[\frac{\partial^2}{\partial (Q_j^{qO\alpha})^2} + (\omega_j^{qO})^2 (Q_j^{qO\alpha})^2 \right] \right\} \tag{6.84}$$

6.2.3 Phonon Dispersion Curves

ENERGY OF THE CRYSTAL LATTICE

In the previous sections we performed the laborious task of transforming the lattice Hamiltonian from its shape (6.1) to independent one-dimensional harmonic oscillators (6.84). At first sight we lost some clarity. While originally the coordinates in the lattice Hamiltonian were simply the displacements of the individual nuclei, in eq. (6.84) they are the less transparent real normal coordinates. However, the advantage of the new shape (6.84) is undeniably the possibility to calculate its eigenvalues using the results of appendix B. Following eq. (B.14) they are found to be,

$$E_L = \sum_q^{N/2} \sum_j \sum_{\alpha=1}^{2} \left[\hbar \omega_j^{qA} \left(n_j^{qA\alpha} + \frac{1}{2} \right) + \sum_q^{N/2} \sum_j \hbar \omega_j^{qO} \left(n_j^{qO\alpha} + \frac{1}{2} \right) \right] \tag{6.85}$$

where $n_j^{qA\alpha}$ and $n_j^{qO\alpha}$ are integer quantum numbers ranging from 0 to ∞. Thus, we see twelve types of energy ladders, six for the acoustical branch and six for the optical branch. However, in the acoustical branch as well as in the optical branch the ladders exhibit the same level spacing two by two. This degeneracy is due to the fact that the terms in the Hamiltonian (6.84) are two by two equal.

By plotting ω_j^{qA} and ω_j^{qO} as a function of the position $\mathbf{k}_q$ in the first Brillouin zone, one obtains so-called phonon dispersion curves. Figs 6.4 and 6.5 present such curves for Si and GaAs as collected in ref. [24]. The lower curves represent the three acoustical phonon modes corresponding to the first term in eq. (6.84). This can be inferred from the fact that these curves approach zero for $\mathbf{k}_q \to 0$, in accordance with the fact that $D_{\mu\nu}^{qA}$ is proportional to k^2, so it vanishes when $\mathbf{k}_q \to 0$. Moreover, for small $\mathbf{k}_q$, $D_{\mu\nu}^{qA}$ is proportional to $|\mathbf{k}_q|^2$, so ω_j^{qA} is proportional to $|\mathbf{k}_q|$. This proportionality is also clearly visible in Figs 6.4 and 6.5. One of the lower curves represents the longitudinal acoustical (LA) phonon mode with oscillator frequency

$$\omega_j^{qA} = \omega^{qLA} \xrightarrow{\mathbf{k}_q \to 0} c^{LA} |\mathbf{k}_q| \tag{6.86}$$

Here the proportionality of ω^{qLA} and $|\mathbf{k}_q|$ has been written explicitly. Note that the constant c^{LA} has the dimension of velocity. The remaining two lower curves represent

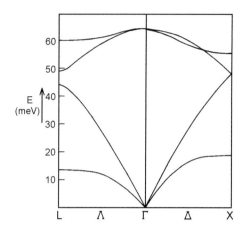

Figure 6.4 Phonon dispersion curves for Si (adapted from Fig. 50, p. 370, ref. [24], © Springer Verlag).

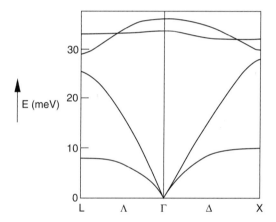

Figure 6.5 Phonon dispersion curves for GaAs (adapted from Fig. 69, p. 528, ref. [24], © Springer Verlag).

the two transverse acoustical (TA) phonon modes. Due to crystal symmetry they are degenerate in the Δ- and Λ-directions. Their oscillator frequencies are denoted by

$$\omega_j^{qA} = \omega^{qTA} \xrightarrow{\mathbf{k}_q \to 0} c^{TA}|\mathbf{k}_q| \tag{6.87}$$

On the other hand the upper curves in Figs 6.4 and 6.5 represent optical phonon modes. One of them represents the longitudinal optical (LO) phonon mode where nuclear motion is parallel to the $\mathbf{k}_q$-vector in reciprocal space. The corresponding oscillator frequency is indicated by

$$\omega_j^{qO} = \omega^{qLO} \tag{6.88}$$

The other upper curves represent the two transverse optical (TO) phonon modes, where nuclear motion is perpendicular to the $\mathbf{k}_q$-vector in reciprocal space. Again

6.2 THE PHONON SPECTRUM OF SEMICONDUCTORS

these modes are degenerate in the Δ- and Λ-directions. Their oscillator frequencies are given by

$$\omega_j^{qO} = \omega^{qTO} \qquad (6.89)$$

ACOUSTICAL PLANE STANDING WAVES

As mentioned above, we have seemingly lost some clarity by transforming to normal coordinates. Still, in the following way we obtain a clear physical interpretation of the real normal coordinates (6.75) and (6.76) corresponding to the acoustical modes. We express the centre of mass displacement (6.25) of the two nuclei in the lth primitive cell in the μ-direction in Q_j^{qA1} and Q_j^{qA2}. Inserting eqs (6.29), (6.57), (6.67), (6.75) and (6.76), we find

$$\begin{aligned}
u_\mu^{lA} &= \frac{1}{\sqrt{N}} \sum_q^N v_\mu^{qA} e^{i\mathbf{k}_q \cdot \mathbf{R}_l} \\
&= \frac{1}{\sqrt{N}} \sum_q^N \sqrt{\frac{\hbar}{(M_1 + M_2)}} w_\mu^{qA} e^{i\mathbf{k}_q \cdot \mathbf{R}_l} \\
&= \frac{1}{\sqrt{N}} \sum_q^N \sqrt{\frac{\hbar}{(M_1 + M_2)}} \sum_j \hat{p}_{j,\mu}^{qA} Q_j^{qA} e^{i\mathbf{k}_q \cdot \mathbf{R}_l} \\
&= \sqrt{\frac{\hbar}{N(M_1 + M_2)}} \sum_q^N \sum_j \hat{p}_{j,\mu}^{qA} Q_j^{qA} e^{i\mathbf{k}_q \cdot \mathbf{R}_l} \\
&= \sqrt{\frac{\hbar}{N(M_1 + M_2)}} \sum_q^{N/2} \sum_j (\hat{p}_{j,\mu}^{qA} Q_j^{qA} e^{i\mathbf{k}_q \cdot \mathbf{R}_l} + \hat{p}_{j,\mu}^{-qA} Q_j^{-qA} e^{-i\mathbf{k}_q \cdot \mathbf{R}_l}) \\
&= \sqrt{\frac{2\hbar}{N(M_1 + M_2)}} \times \sum_q^{N/2} \sum_j \hat{p}_{j,\mu}^{qA} [Q_j^{qA1} \cos(\mathbf{k}_q \cdot \mathbf{R}_l) - Q_j^{qA2} \sin(\mathbf{k}_q \cdot \mathbf{R}_l)] \quad (6.90)
\end{aligned}$$

where the last step follows the procedure of eq. (6.11) while taking into account that $\hat{p}_{j,\mu}^{qA} = |\hat{p}_{j,\mu}^{-qA}|$ and choosing the sign convention $\hat{p}_{j,\mu}^{qA} = +\hat{p}_{j,\mu}^{-qA}$.

Upon comparison with eqs (6.10) and (6.11) we see immediately that eq. (6.90) resolves the displacement in standing waves with a wave vector $\mathbf{k}_q$. As in eq. (6.10), the penultimate line constructs these standing waves from two waves with opposite wave vector $\mathbf{k}_q$ and $-\mathbf{k}_q$ and amplitudes proportional to the corresponding complex normal coordinates. On the other hand, as in eq. (6.11), the last line of eq. (6.90) resolves the displacement in pairs of standing waves with the *same* wave vector $\mathbf{k}_q$, but shifted in phase by 90° and with independent amplitudes proportional to Q_j^{qA1} and Q_j^{qA2}. Again we see that we sum over only half the number of wave vectors in the first Brillouin zone. Once a wave vector $\mathbf{k}_q$ is used, the inverse vector $-\mathbf{k}_q$ is redundant and omitted from the sum. As earlier this is expressed in eq. (6.90) by writing $\sum_q^{N/2}$.

At this point it is interesting to consider the motion of the lattice in the classical limit. In that limit each of the terms in eq. (6.84) yields an independent oscillation. For example, the real normal coordinates $Q_j^{qA\alpha}$ oscillate at a frequency ω_j^{qA}. However, the corresponding terms in eq. (6.90) vary sinusoidally in real space with a wave vector $\mathbf{k}_q$. Hence, the proportionality constants c^{LA} and c^{TA} in eqs (6.86) and (6.87) correspond to the classical phase velocity of these waves, i.e. they represent the velocity of transverse and longitudinal sound waves in the crystal. Moreover, the two terms in the penultimate line of eq. (6.90) can then be interpreted as two of such waves that propagate in opposite direction. Thus, the complex normal coordinates Q_j^{qA} and Q_j^{-qA} acquire the physical meaning of determining the amplitudes of those two propagating waves.

Further insight in the physical meaning of the harmonic oscillators in eq. (6.84) is obtained as follows. The real normal coordinates have dimension $[t]^{1/2}$, which, according to eq. (6.90), is brought back to dimension $[l]$ by multiplication with the factor

$$\sqrt{\frac{\hbar}{m^A}} = \sqrt{\frac{\hbar}{N(M_1 + M_2)}} \tag{6.91}$$

We write

$$x = \sqrt{\frac{\hbar}{m^A}} Q_j^{qA1} \tag{6.92}$$

Then the corresponding term in the Hamiltonian (6.84) reduces to

$$\frac{\hbar}{2}\left[\frac{\partial^2}{\partial (Q_j^{qA1})^2} + (\omega_j^{qA})^2 (Q_j^{qA1})^2\right] = \frac{\hbar}{2}\frac{\hbar}{m^A}\frac{\partial^2}{\partial x^2} + \frac{\hbar}{2}\frac{m^A}{\hbar}(\omega_j^{qA})^2 x^2$$

$$= \frac{\hbar^2}{2m^A}\frac{\partial^2}{\partial x^2} + \frac{1}{2}m^A(\omega_j^{qA})^2 x^2 \tag{6.93}$$

which is exactly the Hamiltonian (B.3) for a harmonic oscillator with mass m and a frequency ω_j^{qA}. Hence, we may associate a mass

$$m^A = N(M_1 + M_2) \tag{6.94}$$

with the harmonic oscillator describing a standing wave in the acoustical phonon branch. It is not accidental that this is exactly the mass of the crystal. Returning to Fig. 6.2, we see that acoustical modes correspond to displacements of nuclei in the same direction, i.e. a macroscopic type of motion involving the total mass of the crystal.

OPTICAL PLANE STANDING WAVES

A similar discussion allows more insight into the nature of the optical phonon modes. Now we express the relative displacement (6.26) of the two nuclei in the lth primitive cell in the μ-direction in Q_j^{qO1} and Q_j^{qO2}. Inserting eqs (6.30), (6.56), (6.68), (6.77) and

6.2 THE PHONON SPECTRUM OF SEMICONDUCTORS

(6.78), we find

$$u_\mu^{lO} = \sqrt{\frac{\hbar(M_1+M_2)}{NM_1M_2}} \sum_q^{N/2} \sum_j (\hat{p}_{j,\mu}^{qO} Q_j^{qO} e^{i\mathbf{k}_q \cdot \mathbf{R}_l} + \hat{p}_{j,\mu}^{-qO} Q_j^{-qO} e^{-i\mathbf{k}_q \cdot \mathbf{R}_l})$$

$$= \sqrt{\frac{\hbar(M_1+M_2)}{NM_1M_2}} \times \sum_q^{N/2} \sum_j \hat{p}_{j,\mu}^{qO} [Q_j^{qO1} \cos(\mathbf{k}_q \cdot \mathbf{R}_l) - Q_j^{qO2} \sin(\mathbf{k}_q \cdot \mathbf{R}_l)] \quad (6.95)$$

where we skipped the intermediate steps as they involve precisely the same procedure as eq. (6.90).

As in eq. (6.90) the displacement of the nuclei is either resolved in two waves with opposite wave vectors $\mathbf{k}_q$ and $-\mathbf{k}_q$ or in two standing waves with the same wave vector $\mathbf{k}_q$ but shifted by 90° in phase with respect to each other. Now, the simple relation with propagating acoustical waves is lost. However, the procedure leading from eq. (6.91) to (6.94) still provides further insight in the nature of optical modes. Now, according to eq. (6.95) the real normal coordinates with dimension $[t]^{1/2}$ are brought back to dimension $[l]$ by multiplication with the factor

$$\sqrt{\frac{\hbar}{m^O}} = \sqrt{\frac{\hbar(M_1+M_2)}{NM_1M_2}} \quad (6.96)$$

We write

$$x = \sqrt{\frac{\hbar}{m^O}} Q_j^{qO1} \quad (6.97)$$

So the corresponding term in the Hamiltonian (6.84) reduces again to the shape (B.3),

$$\frac{\hbar}{2}\left[\frac{\partial^2}{\partial(Q_j^{qO1})^2} + (\omega_j^{qA})^2 (Q_j^{qO1})^2\right] = \frac{\hbar}{2}\frac{\hbar}{m^O}\frac{\partial^2}{\partial x^2} + \frac{\hbar}{2}\frac{m^O}{\hbar}(\omega_j^{qO})^2 x^2$$

$$= \frac{\hbar^2}{2m^O}\frac{\partial^2}{\partial x^2} + \frac{1}{2}m^O(\omega_j^{qO})^2 x^2 \quad (6.98)$$

Hence, we may now associate a mass

$$m^O = \frac{NM_1M_2}{(M_1+M_2)} \quad (6.99)$$

with the harmonic oscillator describing a standing wave in the optical phonon branch. This result is understood by returning to Fig. 6.3. The optical phonon branch corresponds to an opposite motion of the nuclei within a primitive cell. Hence, the reduced mass of these nuclei enters the Hamiltonian rather than the total mass. However, it is the crystal as a whole which oscillates. So, the mass of the oscillator is the reduced mass per primitive cell times the number of these cells.

Fig. 6.3 also suggests how the bonds between neighbouring atoms are stretched and bent by the displacements (6.95) due to optical phonon modes. As Fig. 6.3 provides a one-dimensional schematic picture only, we turn to Fig. 6.6 which provides a more realistic view of the relative nuclear displacements due to an optical phonon mode. In the classical limit one should image the depicted displacement to oscillate at an

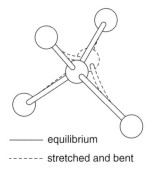

——— equilibrium
------ stretched and bent

Figure 6.6 Stretching and bending of chemical bonds due to optical phonon modes.

optical phonon frequency ω^{qLO} or ω^{qTO}. These displacements correspond to stretching and bending of the chemical bonds between the atoms as also shown in Fig. 6.6, Thus, these frequencies are a direct measure of the strength of these chemical bonds.

To illustrate this, Table 6.1 evaluates the elastic constant $k = m^O(\omega^{qLO})^2$ for crystals of C, Si Ge and Sn containing one primitive cell. In this table the first column gives the atomic weight of these materials in atomic units and the second column the reduced mass m^O according to eq. (6.99) and taking $N = 1$. Then, in the third column m^O is multiplied by Avogadro's number ($N = 6.025 \cdot 10^{26}$ molecules kmol^{-1}) to convert it to SI units. Subsequently, inserting the oscillator frequencies ω^{qLO} in the fourth column as taken from ref. [24], we obtain the elastic constant k in the final column. The resulting values confirm what one suspects about the strength of the chemical bond in these materials. The hardest material, diamond, has the highest value of k, while ongoing to heavier elements, k diminishes together with their hardness.

SUPERLATTICES

To finish this chapter it is instructive to have a brief look at superlattices as we did in section 3.2.6. We recall that superlattices are artificially grown semiconductor crystals consisting of alternating layers of different materials only a few primitive cells thick. An example was shown in Fig. 3.13 and consisted of alternating layers of GaAs and AlAs, each with a thickness of two primitive cells. The procedure to obtain the phonon dispersion curves for this particular superlattice is exactly the same as the method shown in Fig. 3.14 giving its band structure. Again the primitive cell of the superlattice is four times larger as shown in Fig. 3.13, so the first Brillouin zone

Table 6.1 The 'spring constants' of the chemical bonds in C, Si, Ge and Sn as calculated from the optical phonon frequencies.

	$M_1 = M_2$ (a.u.)	m^O (a.u.)	m^O (kg)	ω^{qLO} (s^{-1})	k (J m^{-2})
C	12.011	6.006	$9.97 \cdot 10^{-27}$	$250 \cdot 10^{12}$	$6.23 \cdot 10^3$
Si	23.08	14.05	$23.32 \cdot 10^{-27}$	$97.6 \cdot 10^{12}$	$0.222 \cdot 10^3$
Ge	72.60	36.30	$60.25 \cdot 10^{-27}$	$56.67 \cdot 10^{12}$	$0.194 \cdot 10^3$
Sn	118.70	59.35	$98.51 \cdot 10^{-27}$	$37.05 \cdot 10^{12}$	$0.135 \cdot 10^3$

6.2 THE PHONON SPECTRUM OF SEMICONDUCTORS

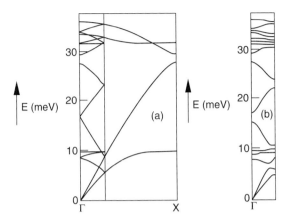

Figure 6.7 (a) Folding of the phonon dispersion curves in a superlattice of GaAs. (The data for a bulk crystal are taken from Fig. 50, p. 370, ref. [24], © Springer Verlag.) (b) A sketch of the effect of the difference between the composing materials.

is four times smaller. As a result the energy levels fold back in the way shown in Fig. 3.14a. The result for the phonon dispersion curves is shown in Fig. 6.7a. One needs to include the fact that the two materials are different and shifting of levels, as in Fig. 3.14b, is also expected to occur in the case of the phonon dispersion curves. Fig. 6.7b shows this effect qualitatively.

There is a serious difference between the present treatment of the crystal lattice and the treatment of the band structure in chapter 3, however. The latter treatment was completely general, so it could be immediately applied to superlattices. Thus, it was possible to identify the Fourier components of the periodic potential in the single electron Hamiltonian causing the change of the band structure in Fig. 3.14. The treatment of the crystal lattice as given in the previous sections made specific use of the fact that there are only two atoms in each primitive cell. Therefore, this treatment cannot be immediately used to describe the phonon spectrum of superlattices.

In particular, the restriction to two atoms per primitive cell allowed us to split the nuclear motion in relative and centre of mass motion in the way presented in section 6.2.1. In the case of primitive cells containing more than two atoms, this splitting must be reconsidered. Still, the centre of mass motion of these more complicated primitive cells can be related to the acoustical phonon branch. Thus, the curves of the phonon spectrum in Fig. 6.7b, where $\omega_j^q \to 0$ at the Γ-point, can still be related to acoustical phonons. We now get more degrees of freedom describing the relative motion between the nuclei and hence a more complicated result for the optical phonon branch. As for normal semiconductors, all these optical phonon levels are characterized by a finite value of ω_j^q at the Γ-point. However, to obtain these finite values one needs to extend the treatment of the previous sections to incorporate more than two atoms per primitive cell. For a more elaborate treatment the reader is referred to ref. [43].

7 Electron–Phonon Coupling

Thus far we have treated the electronic energy levels and the energy levels of the lattice as resulting from completely independent Hamiltonians. This was possible because of the separation of the electron Hamiltonian and the lattice Hamiltonian which was introduced in section 2.3.1. Note however, that various approximations are made before this separation can be achieved. First, the nuclear displacements from their equilibrium positions are assumed to be small, so the Hamiltonian can be expanded as a series (2.35) to (2.37) in these displacements. The zero-order term (2.36) depends on the electron positions only and therefore it may serve as the electron Hamiltonian. Next, to describe the lattice, the electrons are assumed to be in their ground state. The equilibrium positions of the nuclei are defined to be determined by the requirement that the expectation value of the first-order term in the expansion (2.37) vanishes. Finally, the lattice Hamiltonian is taken to be the expectation value (2.45) of the second order term in the expansion.

Already in chapter 2, it was noticed that mixing between electron and lattice states occurs as soon as these approximations are relaxed. In particular, if we introduce excited electron states in the description, then non-zero terms (2.46) appear in first order in the series expansion in the nuclear displacements. These extra terms couple electron and lattice states and, as explained in section 2.3.3, they give rise to transitions where the electrons are promoted to a higher excited state and a phonon is emitted or where the electrons fall down to a lower state under emission of a phonon.

At first sight these electron–lattice couplings seem to represent a refinement of our treatment only, as band structure and phonon spectrum can already be completely determined without taking such couplings into account. However, in practice such a view is short sighted. Many phenomena in semiconductors cannot be understood at all, unless electron–lattice interactions are taken into account. Here we only mention charge transport which will be the subject of the next chapter and indirect optical transitions to be treated in chapter 9.

Unfortunately, electron–lattice coupling is one of the more complex subjects to treat in a study of semiconductors. The reasons for this are manifold. First, it involves both electrons and phonons and hence it combines the complexities of both sides. We cannot discuss its aspects just by electron band structure, the $\mathbf{k}\cdot\mathbf{p}$-approximation and effective mass theory. We need also consider a reciprocal space description of the lattice, the splitting between the optical and acoustical phonons branches and the distinction between longitudinal and transverse modes. A second complication arises from the fact that we have two types of semiconductors, polar and non-polar, leading to a further diversity of interactions. The present chapter is intended to treat these various aspects in a systematic way. However, the reader should not expect to find an in-depth development for all possible situations occurring in all possible semiconductor band structures, as such a treatment would merit a monograph by itself. For more elaborate treatments the reader is referred to ref. [32].

7.1 Electron–Phonon Coupling Mechanisms

7.1.1 Electron–Phonon Coupling in Reciprocal Space

COUPLING BETWEEN BLOCH ELECTRONS AND THE LATTICE

In section 2.3.3 we found that the coupling between electrons and the lattice causes transitions between electron states, while simultaneously a phonon is created or annihilated. Therefore this coupling is generally called electron–phonon coupling. For this coupling to create or annihilate phonons, at least one of the electron states is an excited state. As discussed in section 3.2.5, the concentration of these excited electrons can always be assumed to be extremely small, allowing us to employ an extreme simplification where we assume a single electron in the conduction band or an isolated hole in the valence band, while the crystal is otherwise in the ground state. As also discussed in section 3.2.5, in the presence of donor and acceptor impurities we may retain this picture. Therefore, for the description of the coupling between electrons and the lattice we need to consider two situations only. In the first case the valence band is full and one extra electron occupies an excited Bloch state in the conduction band. In the second case the valence band is also full, except that one electron is missing, the 'hole' in this band. For simplicity we will restrict ourselves to the case of a full valence band and a single electron in the conduction band. Note however, that according to section 3.2.5 our treatment can be directly extended to a single hole in the valence band.

According to eq. (2.46), the electron–phonon coupling can be written as,

$$(\mathcal{H}_{EL})_{ni,mf} = \sum_{l,s,\mu}^{N} (\Phi_\mu^{ls})_{ni,mf}\, u_\mu^{ls} \tag{7.1}$$

where now u_μ^{ls} is the displacement of the sth nucleus in the lth primitive cell in the μ-direction. Furthermore, following eq. (2.39), the deformation potential is given by

$$(\Phi_\mu^{ls})_{ni,mf} = \int_{NV_C} d\mathbf{r}_0 \int_{NV_C} d\mathbf{r}_1 \cdots \int_{NV_C} d\mathbf{r}_j \cdots \Psi_i^{n*}(\nabla_\mu^{ls}\mathcal{V})_0 \Psi_f^m \tag{7.2}$$

Here the notation is also adapted, so ∇_μ^{ls} is the derivative with respect to the coordinate R_μ^{ls} of the sth nucleus in the lth primitive cell in the μ-direction. Moreover, Ψ_i^n and Ψ_f^m denote electronic eigenstates of a crystal containing a single electron in the conduction band. Eventually we will see that the electron–phonon coupling (7.1) causes transitions between these two eigenstates. This observation induces the present notation where the indices (n, i) and (m, f) indicate the initial Bloch state $b_i^n(\mathbf{r})$ and final Bloch state $b_f^m(\mathbf{r})$ occupied by this electron in the conduction band. Here n and m are the band indices and i and f indicate that the initial and final Bloch states are located at the the positions $\mathbf{k}_i$ and $\mathbf{k}_f$ in the first Brillouin zone, respectively. Furthermore, assuming periodic boundary conditions, $\mathbf{k}_i$ and $\mathbf{k}_f$ may take discrete values (6.3) only. Following the notation of eqs (6.3) and (6.4) we may denote these discrete values by integer indices $\mathbf{i} = (i_1, i_2, i_3)$ and $\mathbf{f} = (f_1, f_2, f_3)$, respectively. In this notation the energy of an electron in the states $b_i^n(\mathbf{r})$ and $b_f^m(\mathbf{r})$ is given by E_i^n and E_f^m, respectively.

For the time being we follow the notation of section 3.2.5 and use the index 0 for the position $\mathbf{r}_0$ of the extra electron in the conduction band. Finally, we add the subscript NV_C to the integral to denote that it is performed over the volume of a crystal containing N primitive cells of volume V_C each. We note that restricting the integral to the volume NV_C of the crystal implies that the wave functions Ψ_i^n and Ψ_f^m must be normalized in the same volume. Otherwise eq. (7.1) does not represent a quantum mechanically correct matrix element.

We start by inserting the approximate wave functions (3.93) and (3.94),

$$\Psi_i^n = \frac{1}{\sqrt{NV_C}} \Psi\, b_i^n(\mathbf{r}_0) \tag{7.3}$$

$$\Psi_f^m = \frac{1}{\sqrt{NV_C}} \Psi\, b_f^m(\mathbf{r}_0) \tag{7.4}$$

Here factors $1/\sqrt{NV_C}$ are added because Bloch functions are normalized in a unit volume and as noted above, we require wave functions to be normalized in the volume NV_C of the crystal. As a result the integral (7.2) obtains the shape (3.95),

$$(\Phi_\mu^{ls})_{ni,mf} = \frac{1}{NV_C} \int_{NV_C} d\mathbf{r}_0 \int_{NV_C} d\mathbf{r}_1 \cdots \int_{NV_C} d\mathbf{r}_j \cdots [\Psi^* b_i^{n*}(\mathbf{r}_0)](\nabla_\mu^{ls} \mathcal{V})_0 [\Psi b_f^m(\mathbf{r}_0)]$$

$$= \frac{1}{NV_C} \int_{NV_C} d\mathbf{r}_0\, b_i^{n*}(\mathbf{r}_0)\, \phi_\mu^s(\mathbf{r}_0 - \mathbf{R}_l)\, b_f^m(\mathbf{r}_0) \tag{7.5}$$

where

$$\phi_\mu^s(\mathbf{r}_0 - \mathbf{R}_l) = \int_{NV_C} d\mathbf{r}_1 \cdots \int_{NV_C} d\mathbf{r}_j \cdots \Psi^* (\nabla_\mu^{ls} \mathcal{V})_0 \Psi \tag{7.6}$$

corresponds to eq. (3.97). We recall that $(\Phi_\mu^{ls})_{ni,mf}$ may be called the matrix element of the deformation potential $(\nabla_\mu^{ls} \mathcal{V})_0$ between the eigenstates (7.3) and (7.4) of the full electron Hamiltonian. Then, using the same terminology, $\phi_\mu^s(\mathbf{r}_0 - \mathbf{R}_l)$ is the deformation potential in a formalism where we use the single-electron Hamiltonian, so an electron is described by a Bloch function. Note that $\phi_\mu^s(\mathbf{r}_0 - \mathbf{R}_l)$ is still a function of $\mathbf{r}_0$ because we exclude the integration over $\mathbf{r}_0$. However, because of translational symmetry of the crystal, it is only a function of the difference $(\mathbf{r}_0 - \mathbf{R}_l)$ between the position $\mathbf{r}_0$ of the electron in the conduction band and the position $\mathbf{R}_l$ of the deformed primitive cell, rather than of $\mathbf{r}_0$ and $\mathbf{R}_l$ separately.

7.1 ELECTRON–PHONON COUPLING MECHANISMS

Thus, on the basis of Bloch functions the electron–phonon coupling (7.1) is written as

$$(\mathcal{H}_{EL})_{ni,mf} = \sum_{l,s,\mu}^{N} \left[\frac{1}{NV_C} \int_{NV_C} d\mathbf{r}_0 \, b_i^{n*}(\mathbf{r}_0) \, \phi_\mu^s(\mathbf{r}_0 - \mathbf{R}_l) \, b_f^m(\mathbf{r}_0) \right] u_\mu^{ls} \quad (7.7)$$

One may wonder whether $(\mathcal{H}_{EL})_{ni,mf}$ as represented by eq. (7.7) depends on the size of the crystal. However, it is easy to see that this is not the case. The function $\phi_\mu^s(\mathbf{r}_0 - \mathbf{R}_l)$ represents the potential experienced by a Bloch electron due to a unit displacement in the μ direction of the sth nucleus in the lth primitive cell. Clearly, this is a local potential which vanishes when the distance $(\mathbf{r}_0 - \mathbf{R}_l)$ between the electron and this nucleus becomes very large. Then also the integrand in eq. (7.7) vanishes for larger values of $(\mathbf{r}_0 - \mathbf{R}_l)$ and consequently the integral over $\mathbf{r}_0$ is independent of the integration interval, i.e. independent of the size of the crystal. Moreover, also the sum over N is compensated by the normalization factor $1/NV_C$, so eq. (7.7) is independent of crystal size.

We finally remark that the only electron position figuring in eq. (7.7) is the position $\mathbf{r}_0$ of the conduction electron, whereas all other electron positions $\mathbf{r}_1, \ldots, \mathbf{r}_i, \ldots$ have been integrated out in eq. (7.6). Therefore there is no need to retain the explicit index 0 and we will omit it from now on.

TRANSFORMATION TO RECIPROCAL SPACE

We continue by transforming to reciprocal space. Here we have to take some care, because two different space coordinates are involved. First, we have the space lattice positions $\mathbf{R}_l$ defining the primitive cell containing the nucleus displaced by u_μ^{ls}. Second, there is the position $\mathbf{r}$ of the electron. Eq. (7.7) represents a matrix element which is calculated by integrating over the latter position. Therefore it is not practical to transform the electron position to reciprocal space. Hence, we will only transform space lattice positions $\mathbf{R}_l$, and this will be done by inserting eq. (6.6). Still, though they are functions in normal space, Bloch functions are identified by a reciprocal space lattice coordinate. In our notation the index q is used for reciprocal space positions $\mathbf{k}_q$ representing lattice degrees of freedom, while the indices i and f are used for reciprocal space positions $\mathbf{k}_i$ and $\mathbf{k}_f$ identifying Bloch functions. Then,

$$(\mathcal{H}_{EL})_{ni,mf} = \sum_{l,s,\mu}^{N} \left[\frac{1}{NV_C} \int_{NV_C} d\mathbf{r} \, b_i^{n*}(\mathbf{r}) \, \phi_\mu^s(\mathbf{r} - \mathbf{R}_l) \, b_f^m(\mathbf{r}) \right] \frac{1}{\sqrt{N}} \sum_q^N v_\mu^{qs} e^{i\mathbf{k}_q \cdot \mathbf{R}_l}$$

$$= \frac{1}{\sqrt{N}} \sum_q \sum_{s,\mu}^{N} (\mathcal{D}_\mu^{qs})_{ni,mf} \, v_\mu^{qs} \quad (7.8)$$

where

$$(\mathcal{D}_\mu^{qs})_{ni,mf} = \frac{1}{NV_C} \int_{NV_C} d\mathbf{r} \, b_i^{n*}(\mathbf{r}) \left[\sum_l^N \phi_\mu^s(\mathbf{r} - \mathbf{R}_l) e^{i\mathbf{k}_q \cdot \mathbf{R}_l} \right] b_f^m(\mathbf{r}) \quad (7.9)$$

of the deformation potential in reciprocal space. We introduce the Fourier transform

$$d_\mu^{qs}(\mathbf{r}) = \sum_l^N \phi_\mu^s(\mathbf{r} - \mathbf{R}_l) e^{-i\mathbf{k}_q \cdot (\mathbf{r} - \mathbf{R}_l)} \quad (7.10)$$

of the deformation potential in the single electron Hamiltonian. The latter definition is such that in eq. (7.9) the sum over l is equal to

$$\sum_{l}^{N} \phi_{\mu}^{s}(\mathbf{r} - \mathbf{R}_{l}) e^{i\mathbf{k}_{q} \cdot \mathbf{R}_{l}} = d_{\mu}^{qs}(\mathbf{r}) e^{i\mathbf{k}_{q} \cdot \mathbf{r}} \tag{7.11}$$

Before inserting this sum in eq. (7.9) we consider some of the properties of $d_{\mu}^{qs}(\mathbf{r})$. First, it is still a function of the electron position $\mathbf{r}$. This was to be expected, as we only transformed the space lattice coordinates $\mathbf{R}_l$ to reciprocal space. However, it is a more pleasant function than $\phi_{\mu}^{s}(\mathbf{r} - \mathbf{R}_{l})$, because it obeys the translational symmetry of the crystal. This can be seen by first calculating,

$$d_{\mu}^{qs}(\mathbf{r} - \mathbf{R}_{l'}) = \sum_{l}^{N} \phi_{\mu}^{s}(\mathbf{r} - \mathbf{R}_{l'} - \mathbf{R}_{l}) e^{-i\mathbf{k}_{q} \cdot (\mathbf{r} - \mathbf{R}_{l'} - \mathbf{R}_{l})} \tag{7.12}$$

Subsequently we replace $\mathbf{R}_{l'} + \mathbf{R}_{l}$ by $\mathbf{R}_{l''}$ and the summation over l by a summation over l''. This is allowed because we have assumed periodic boundary conditions. Then, translational symmetry follows immediately,

$$d_{\mu}^{qs}(\mathbf{r} - \mathbf{R}_{l'}) = \sum_{l''}^{N} \phi_{\mu}^{s}(\mathbf{r} - \mathbf{R}_{l''}) e^{-i\mathbf{k}_{q} \cdot (\mathbf{r} - \mathbf{R}_{l''})} = d_{\mu}^{qs}(\mathbf{r}) \tag{7.13}$$

Next we note that the definition of the function $d_{\mu}^{qs}(\mathbf{r})$ does not contain the factor $1/\sqrt{N}$ as e.g. the transform to reciprocal space (6.7) of nuclear displacements. This has the following consequence. As discussed above, the function $\phi_{\mu}^{s}(\mathbf{r} - \mathbf{R}_{l})$ is localized, i.e. it vanishes for large values of $(\mathbf{r} - \mathbf{R}_{l})$. Then, in eq. (7.10), the sum over l is completely determined by a restricted number of terms corresponding to small values of $(\mathbf{r} - \mathbf{R}_{l})$. This implies that the range N and hence the size of the crystal do not influence the result of the summation. In other words, the definition (7.10) renders $d_{\mu}^{qs}(\mathbf{r})$ independent of the size of the crystal.

Now we use this newly defined function $d_{\mu}^{qs}(\mathbf{r})$ to rewrite eq. (7.9) for $(\mathcal{D}_{\mu}^{qs})_{ni,mf}$. Inserting eq. (7.11) we find,

$$(\mathcal{D}_{\mu}^{qs})_{ni,mf} = \frac{1}{NV_C} \int_{NV_C} d\mathbf{r} \, b_i^{n*}(\mathbf{r}) [d_{\mu}^{qs}(\mathbf{r}) e^{i\mathbf{k}_q \cdot \mathbf{r}}] b_f^{m}(\mathbf{r}) \tag{7.14}$$

EXPANSION OF THE BLOCH FUNCTIONS

Next, we insert expression (3.85) for the Bloch functions $b_i^{n}(\mathbf{r})$ and $b_f^{m}(\mathbf{r})$. Moreover we split the integral over $\mathbf{r}$ by a sum of integrals over the individual primitive cells. We find

$$(\mathcal{D}_{\mu}^{qs})_{ni,mf} = \frac{1}{N} \sum_{l'}^{N} \frac{1}{V_C} \int_{V_C^{l'}} d\mathbf{r} \, u_i^{n*}(\mathbf{r}) \, e^{-i\mathbf{k}_i \cdot \mathbf{r}} [d_{\mu}^{qs}(\mathbf{r}) e^{i\mathbf{k}_q \cdot \mathbf{r}}] u_f^{m}(\mathbf{r}) e^{i\mathbf{k}_f \cdot \mathbf{r}}$$

$$= \frac{1}{N} \sum_{l'}^{N} \frac{1}{V_C} \int_{V_C^{l'}} d\mathbf{r} \, u_i^{n*}(\mathbf{r}) \, d_{\mu}^{qs}(\mathbf{r}) \, u_f^{m}(\mathbf{r}) \, e^{i(\mathbf{k}_q - \mathbf{k}_i + \mathbf{k}_f) \cdot \mathbf{r}} \tag{7.15}$$

7.1 ELECTRON–PHONON COUPLING MECHANISMS

Here $\int_{V_C^{l'}}$ indicates integration over the primitive cell with index l'. We continue by replacing $\mathbf{r}$ by $(\mathbf{r} - \mathbf{R}_{l'}) + \mathbf{R}_{l'}$ in the exponent. Then we obtain

$$(\mathcal{D}_\mu^{qs})_{ni,mf} = \frac{1}{N} \sum_{l'}^N e^{i(\mathbf{k}_q - \mathbf{k}_i + \mathbf{k}_f) \cdot \mathbf{R}_{l'}}$$

$$\times \frac{1}{V_C} \int_{V_C^{l'}} d\mathbf{r}\, u_i^{n*}(\mathbf{r})\, d_\mu^{qs}(\mathbf{r})\, u_f^m(\mathbf{r})\, e^{i(\mathbf{k}_q - \mathbf{k}_i + \mathbf{k}_f) \cdot (\mathbf{r} - \mathbf{R}_{l'})} \quad (7.16)$$

Because the functions $u_i^n(\mathbf{r})$, $u_f^m(\mathbf{r})$ and $d_\mu^{qs}(\mathbf{r})$ obey the periodic symmetry of the crystal lattice, the integrals over $\mathbf{r}$ are now clearly independent of the index l' of the primitive cell. This allows us to replace all integrals by the integral over the primitive cell at the origin of our frame of reference where $\mathbf{R}_l = 0$. As a result,

$$(\mathcal{D}_\mu^{qs})_{ni,mf} = \frac{1}{N} \sum_{l'}^N e^{i(\mathbf{k}_q - \mathbf{k}_i + \mathbf{k}_f) \cdot \mathbf{R}_{l'}} \times \frac{1}{V_C} \int_{V_C^0} d\mathbf{r}\, u_i^{n*}(\mathbf{r})\, d_\mu^{qs}(\mathbf{r})\, u_f^m(\mathbf{r})\, e^{i(\mathbf{k}_q - \mathbf{k}_i + \mathbf{k}_f) \cdot \mathbf{r}}$$

$$= \frac{1}{N} \sum_{l'}^N e^{i(\mathbf{k}_q - \mathbf{k}_i + \mathbf{k}_f) \cdot \mathbf{R}_{l'}} \times \langle nk_i | d_\mu^{qs}(\mathbf{r}) e^{i(\mathbf{k}_q - \mathbf{k}_i + \mathbf{k}_f) \cdot \mathbf{r}} | mk_f \rangle \quad (7.17)$$

where we write the integral following the Dirac bracket notation of eq. (4.24). Now the sum over l' can be evaluated separately. It corresponds to the Kronecker δ-function (6.8), so we find

$$(\mathcal{D}_\mu^{qs})_{ni,mf} = \delta_{q,q_{if}} \langle nk_i | d_\mu^{qs}(\mathbf{r}) | mk_f \rangle \quad (7.18)$$

where $\delta_{q,q_{if}}$ requires that

$$\mathbf{k}_q = \mathbf{k}_{q_{if}} = \mathbf{k}_i - \mathbf{k}_f \quad (7.19)$$

for the matrix element (7.18) of the deformation potential in reciprocal space to be non-zero. Here we also introduce the index q_{if} denoting the value of q obeying this δ-function.

Now we have obtained an explicit expression for $(\mathcal{D}_\mu^{qs})_{ni,mf}$, we insert it in eq. (7.8) to evaluate $(\mathcal{H}_{EL})_{ni,mf}$. We find,

$$(\mathcal{H}_{EL})_{ni,mf} = \frac{1}{\sqrt{N}} \sum_q^N \sum_{s,\mu} (\mathcal{D}_\mu^{qs})_{ni,mf}\, v_\mu^{qs}$$

$$= \frac{1}{\sqrt{N}} \sum_q^N \delta_{q,q_{if}} \sum_{s,\mu} \langle nk_i | d_\mu^{qs}(\mathbf{r}) | mk_f \rangle\, v_\mu^{qs}$$

$$= \frac{1}{\sqrt{N}} \sum_{s,\mu} \langle nk_i | d_\mu^{q_{if}s}(\mathbf{r}) | mk_f \rangle\, v_\mu^{q_{if}s}$$

$$= \frac{1}{\sqrt{N}} \sum_{s,\mu} (\mathcal{D}_\mu^{q_{if}s})_{ni,mf}\, v_\mu^{q_{if}s} \quad (7.20)$$

where

$$(\mathcal{D}_\mu^{q_{if}s})_{ni,mf} = \langle nk_i | d_\mu^{q_{if}s}(\mathbf{r}) | mk_f \rangle \quad (7.21)$$

is the only component of $(\mathcal{D}_\mu^{qs})_{ni,mf}$ obeying eq. (7.19) and hence contributing to the electron–phonon coupling.

RELATIVE AND CENTRE OF MASS MOTION

In a more detailed treatment of electron–phonon coupling we have to distinguish between acoustical and optical phonons. We start considering eq. (7.20) for electron–phonon coupling in semiconductors with two nuclei in the primitive cell. Writing the sum over the nuclear index s explicitly, this equation reads as

$$(\mathcal{H}_{EL})_{ni,mf} = \frac{1}{\sqrt{N}} \sum_\mu [\langle nk_i | d_\mu^{q_{if}1}(\mathbf{r}) | mk_f \rangle v_\mu^{q_{if}1} + \langle nk_i | d_\mu^{q_{if}2}(\mathbf{r}) | mk_f \rangle v_\mu^{q_{if}2}] \quad (7.22)$$

Subsequently we transform to centre of mass and relative motion. In reciprocal space the centre of mass and relative displacements are given by eqs (6.27) and (6.28). Solving these equations yields,

$$v_\mu^{q1} = v_\mu^{qA} + \frac{M_2}{M_1 + M_2} v_\mu^{qO} \quad (7.23)$$

$$v_\mu^{q2} = v_\mu^{qA} - \frac{M_1}{M_1 + M_2} v_\mu^{qO} \quad (7.24)$$

So,

$$(\mathcal{H}_{EL})_{ni,mf} = \frac{1}{\sqrt{N}} \sum_\mu \left[\langle nk_i | d_\mu^{q_{if}1}(\mathbf{r}) | mk_f \rangle \left(v_\mu^{q_{if}A} + \frac{M_2}{M_1 + M_2} v_\mu^{q_{if}O} \right) \right.$$
$$\left. + \langle nk_i | d_\mu^{q_{if}2}(\mathbf{r}) | mk_f \rangle \left(v_\mu^{q_{if}A} - \frac{M_1}{M_1 + M_2} v_\mu^{q_{if}O} \right) \right]$$
$$= \frac{1}{\sqrt{N}} \sum_\mu \left[\langle nk_i | d_\mu^{q_{if}1}(\mathbf{r}) + d_\mu^{q_{if}2}(\mathbf{r}) | mk_f \rangle v_\mu^{q_{if}A} \right.$$
$$\left. + \langle nk_i | \frac{M_2 d_\mu^{q_{if}1}(\mathbf{r}) - M_1 d_\mu^{q_{if}2}(\mathbf{r})}{M_1 + M_2} | mk_f \rangle v_\mu^{q_{if}O} \right]$$
$$= \frac{1}{\sqrt{N}} \sum_\mu [\langle nk_i | d_\mu^{q_{if}A}(\mathbf{r}) | mk_f \rangle v_\mu^{q_{if}A} + \langle nk_i | d_\mu^{q_{if}O}(\mathbf{r}) | mk_f \rangle v_\mu^{q_{if}O}]$$
$$= \frac{1}{\sqrt{N}} \sum_\mu [(\mathcal{D}_\mu^{q_{if}A})_{ni,mf} v_\mu^{q_{if}A} + (\mathcal{D}_\mu^{q_{if}O})_{ni,mf} v_\mu^{q_{if}O}] \quad (7.25)$$

The first term in eq. (7.25) is called the electron–acoustical phonon coupling, as it represents the interaction between electrons and the centre of mass displacement of the two nuclei in the primitive cell. The second term representing the interaction with the relative motion, is called the electron–optical phonon coupling. Furthermore, we have introduced the acoustical and optical deformation potentials in reciprocal space

$$d_\mu^{qA}(\mathbf{r}) = d_\mu^{q1}(\mathbf{r}) + d_\mu^{q2}(\mathbf{r}) \quad (7.26)$$

$$d_\mu^{qO}(\mathbf{r}) = \frac{M_2 d_\mu^{q1}(\mathbf{r}) - M_1 d_\mu^{q2}(\mathbf{r})}{M_1 + M_2} \quad (7.27)$$

7.1 ELECTRON–PHONON COUPLING MECHANISMS

and their matrix elements

$$(\mathcal{D}_\mu^{qA})_{ni,mf} = \delta_{q,q_{if}} \langle nk_i | d_\mu^{qA}(\mathbf{r}) | mk_f \rangle \tag{7.28}$$

$$(\mathcal{D}_\mu^{qO})_{ni,mf} = \delta_{q,q_{if}} \langle nk_i | d_\mu^{qO}(\mathbf{r}) | mk_f \rangle \tag{7.29}$$

Because of the δ-functions, these matrix elements vanish unless eq. (7.19) holds.

7.1.2 k·p-Treatment of Electron–Phonon Coupling
SERIES EXPANSIONS

In many cases we are only interested in the behaviour of electrons in electron states near band extrema, i.e. electron states where $\mathbf{k}_i$ and $\mathbf{k}_f$ are small. In these situations matrix elements may be calculated using the lowest order $\mathbf{k}\cdot\mathbf{p}$-approximation and eqs (7.28) and (7.29) simplify to

$$(\mathcal{D}_\mu^{qA})_{ni,mf} \approx \delta_{q,q_{if}} \langle n0 | d_\mu^{qA}(\mathbf{r}) | m0 \rangle \tag{7.30}$$

$$(\mathcal{D}_\mu^{qO})_{ni,mf} \approx \delta_{q,q_{if}} \langle n0 | d_\mu^{qO}(\mathbf{r}) | m0 \rangle \tag{7.31}$$

where only the basis states of the $\mathbf{k}\cdot\mathbf{p}$-approximation appear.

Moreover, because $\mathbf{k}_i$, $\mathbf{k}_f$ are small, also $\mathbf{k}_q = \mathbf{k}_i - \mathbf{k}_f$ is small and we may venture to expand the deformation potentials as a function of $\mathbf{k}_q$. We write up to first order,

$$d_\mu^{qA}(\mathbf{r}) = d_\mu^{0A}(\mathbf{r}) + i \sum_\nu e_{\mu\nu}^A(\mathbf{r}) k_{q\nu} \tag{7.32}$$

$$d_\mu^{qO}(\mathbf{r}) = d_\mu^{0O}(\mathbf{r}) + i \sum_\nu e_{\mu\nu}^O(\mathbf{r}) k_{q\nu} \tag{7.33}$$

and

$$(\mathcal{D}_\mu^{qA})_{ni,mf} = \delta_{q,q_{if}} \left((D_\mu^A)_{nm} + i \sum_\nu (E_{\mu\nu}^A)_{nm} k_{q\nu} \right) \tag{7.34}$$

$$(\mathcal{D}_\mu^{qO})_{ni,mf} = \delta_{q,q_{if}} \left((D_\mu^O)_{nm} + i \sum_\nu (E_{\mu\nu}^O)_{nm} k_{q\nu} \right) \tag{7.35}$$

As we will see below, the factors i are inserted in order to render the coefficients

$$(D_\mu^A)_{nm} = \langle n0 | d_\mu^{0A}(\mathbf{r}) | m0 \rangle \tag{7.36}$$

$$(E_{\mu\nu}^A)_{nm} = \langle n0 | e_{\mu\nu}^A(\mathbf{r}) | m0 \rangle \tag{7.37}$$

$$(D_\mu^O)_{nm} = \langle n0 | d_\mu^{0O}(\mathbf{r}) | m0 \rangle \tag{7.38}$$

$$(E_{\mu\nu}^O)_{nm} = \langle n0 | e_{\mu\nu}^O(\mathbf{r}) | m0 \rangle \tag{7.39}$$

real.

THE OPTICAL DEFORMATION POTENTIAL NEAR k = 0

To evaluate the zero-order terms in these expansions, we continue by considering the limit of eq. (7.22) for $\mathbf{k}_i - \mathbf{k}_f = \mathbf{k}_{q_{if}} = \mathbf{k}_q \to 0$,

$$(\mathcal{H}_{EL})_{ni,mf} = \frac{1}{\sqrt{N}} \sum_\mu [\langle n0 | d_\mu^{01}(\mathbf{r}) | m0 \rangle v_\mu^{01} + \langle n0 | d_\mu^{02}(\mathbf{r}) | m0 \rangle v_\mu^{02}] \tag{7.40}$$

As in section 6.2.1, an elementary argument yields a simple relation between the two effective deformation potentials. Displacing both nuclei in the primitive cell over the same distance ε in the μ-direction corresponds to a displacement of the crystal as a whole. Hence, such a displacement should not lead to a coupling term in the Hamiltonian. The expression between brackets in eq. (7.40) is then given by

$$\langle n0|d_\mu^{01}(\mathbf{r})|m0\rangle v_\mu^{01} + \langle n0|d_\mu^{02}(\mathbf{r})|m0\rangle v_\mu^{02} = [\langle n0|d_\mu^{01}(\mathbf{r})|m0\rangle + \langle n0|d_\mu^{02}(\mathbf{r})|m0\rangle]\varepsilon$$

$$= [\langle n0|d_\mu^{01}(\mathbf{r}) + d_\mu^{02}(\mathbf{r})|m0\rangle]\varepsilon \qquad (7.41)$$

Requiring it to be equal to zero, implies

$$d_\mu^{01}(\mathbf{r}) = -d_\mu^{02}(\mathbf{r}) \qquad (7.42)$$

We insert this result in eqs. (7.26) and (7.27). This immediately leads to the observation that for the limit $\mathbf{k}_i - \mathbf{k}_f = \mathbf{k}_{q_{if}} = \mathbf{k}_q \to 0$,

$$d_\mu^{0O}(\mathbf{r}) = d_\mu^{01}(\mathbf{r}) \qquad (7.43)$$

$$d_\mu^{0A}(\mathbf{r}) = 0 \qquad (7.44)$$

Then, inserting this result in eqs (7.38) and (7.39), in lowest order the electron–optical phonon coupling (7.35) may be approximated by

$$(\mathcal{D}_\mu^{qO})_{ni,mf} \approx \delta_{q,q_{if}}(D_\mu^O)_{nm} = \delta_{q,q_{if}}\langle n0|d_\mu^{0O}(\mathbf{r})|m0\rangle = \delta_{q,q_{if}}\langle n0|d_\mu^{01}(\mathbf{r})|m0\rangle \qquad (7.45)$$

On the other hand, in this lowest order the electron–acoustical phonon coupling vanishes

$$(\mathcal{D}_\mu^{qA})_{ni,mf} \approx \delta_{q,q_{if}}(D_\mu^A)_{nm} = 0 \qquad (7.46)$$

Note that in literature $(D_\mu^O)_{nm}$ is generally called the *optical deformation potential*.

THE ACOUSTICAL DEFORMATION POTENTIAL NEAR $\mathbf{k} = 0$

Because $d_\mu^{0A}(\mathbf{r}) = 0$, we have to go to first order in $\mathbf{k}_q = \mathbf{k}_{q_{if}} = \mathbf{k}_i - \mathbf{k}_f$, to obtain the electron–acoustical phonon coupling. This can also be understood as follows. We start by taking into account that in acoustical phonon modes both atoms in a primitive cell move in the same direction. Hence, it is rather the difference between nuclear displacements in subsequent primitive cells which determines the strength of the coupling than the amplitude of the nuclear displacement itself. In section 6.2.1 we denoted this difference between nuclear displacements mathematically by the strain tensor, which has components defined by

$$S_{\mu\nu}^l = \frac{1}{2}\left(\frac{\partial u_\mu^{lA}}{\partial R_\nu^l} + \frac{\partial u_\nu^{lA}}{\partial R_\mu^l}\right) \qquad (7.47)$$

where u_μ^{lA} is the centre of mass displacement of the lth primitive cell in the μ-direction. Furthermore, we assume again that the derivatives with respect to the components R_μ^l

7.1 ELECTRON–PHONON COUPLING MECHANISMS

of $\mathbf{R}_l$ may be taken because these centre of mass displacements are on a smooth function of $\mathbf{R}_l$. When transformed to reciprocal space, the strain tensor is given by eq. (6.47),

$$S^l_{\mu\nu} = \frac{i}{2} \frac{1}{\sqrt{N}} \sum_q^N (v^{qA}_\mu k_{q\nu} + v^{qA}_\nu k_{q\mu}) e^{i\mathbf{k}_q \cdot \mathbf{R}_l} \quad (7.48)$$

Hence, if in normal space the electron–phonon coupling is expected to be proportional to the strain tensor with components $S^l_{\mu\nu}$, in reciprocal space this coupling should be proportional to its Fourier transform

$$\frac{i}{2}(v^{qA}_\mu k_{q\nu} + v^{qA}_\nu k_{q\mu}) \quad (7.49)$$

We see that it is not only proportional to the displacements v^{qA}_μ in reciprocal space, but also to the length $k_{q\mu}$ of the components of the phonon wave vectors. Hence, it corresponds to the first-order term in eq. (7.34). Therefore, up to first order,

$$(D^{qA}_\mu)_{ni,mf} = i\delta_{q,q_{if}} \sum_\nu (E^A_{\mu\nu})_{nm} k_{q\nu} = i\delta_{q,q_{if}} \sum_\nu \langle n0 | e^A_{\mu\nu}(\mathbf{r}) | m0 \rangle k_{q\nu} \quad (7.50)$$

Note that in literature the tensor $(E^A_{\mu\nu})_{nm}$ is generally called the *acoustical deformation potential*. This name is not very consistent, as this tensor must be multiplied by $\mathbf{k}_q$ in order to obtain the dimension of the optical deformation potential.

Combining our results (7.45), (7.46) and (7.50), for small $\mathbf{k}_q = \mathbf{k}_{q_{if}} = \mathbf{k}_f - \mathbf{k}_i$, the electron–phonon coupling (7.25) is written as

$$(\mathcal{H}_{EL})_{ni,mf} = \frac{1}{\sqrt{N}} \sum_\mu (D^O_\mu)_{nm} v^{q_{if}O}_\mu + i\frac{1}{\sqrt{N}} \sum_{\mu\nu} (E^A_{\mu\nu})_{nm} k_{q_{if}\nu} v^{q_{if}A}_\mu \quad (7.51)$$

7.1.3 Normal Coordinates

COMPLEX NORMAL COORDINATES

Thus, we have succeeded in transforming the nuclear displacements to reciprocal space and to separate relative and centre of mass motion. We now continue the procedure introduced in the previous chapter and transform to normal coordinates. We start by introducing reduced displacements defined by eqs (6.56) and (6.57). Then

$$(\mathcal{H}_{EL})_{ni,mf} = \sqrt{\frac{\hbar(M_1 + M_2)}{NM_1 M_2}} \sum_\mu (D^O_\mu)_{nm} w^{q_{if}O}_\mu$$

$$+ i\sqrt{\frac{\hbar}{N(M_1 + M_2)}} \sum_{\mu\nu} (E^A_{\mu\nu})_{nm} k_{q_{if}\nu} w^{q_{if}A}_\mu \quad (7.52)$$

Next, we transform to complex normal coordinates $Q^{q_{if}O}_j$ and $Q^{q_{if}A}_j$, using eqs (6.67) and (6.68). Then,

$$(\mathcal{H}_{EL})_{ni,mf} = \sqrt{\frac{\hbar(M_1+M_2)}{NM_1M_2}} \sum_\mu (D_\mu^O)_{nm} \sum_j p_{j,\mu}^{q_{if}O} Q_j^{q_{if}O}$$

$$+ i\sqrt{\frac{\hbar}{N(M_1+M_2)}} \sum_{\mu\nu} (E_{\mu\nu}^A)_{nm} k_{q_{if}\nu} \sum_j p_{j,\mu}^{q_{if}A} Q_j^{q_{if}A}$$

$$= \sqrt{\frac{\hbar(M_1+M_2)}{NM_1M_2}} \sum_j Q_j^{q_{if}O} \sum_\mu p_{j,\mu}^{q_{if}O} (D_\mu^O)_{nm}$$

$$+ i\sqrt{\frac{\hbar}{N(M_1+M_2)}} \sum_j Q_j^{q_{if}A} \sum_{\mu\nu} p_{j,\mu}^{q_{if}A} (E_{\mu\nu}^A)_{nm} k_{q_{if}\nu}$$

$$= \frac{\hbar}{\sqrt{N}} \sum_j \sqrt{\frac{(M_1+M_2)}{\hbar\omega_j^{q_{if}O} M_1 M_2}} \sqrt{\omega_j^{q_{if}O}} Q_j^{q_{if}O} \sum_\mu p_{j,\mu}^{q_{if}O} (D_\mu^O)_{nm}$$

$$+ i\frac{\hbar}{\sqrt{N}} \sum_j \frac{1}{\sqrt{\hbar\omega_j^{q_{if}A} (M_1+M_2)}} \sqrt{\omega_j^{q_{if}A}} Q_j^{q_{if}A} \sum_{\mu\nu} p_{j,\mu}^{q_{if}A} (E_{\mu\nu}^A)_{nm} k_{q_{if}\nu}$$

$$= \frac{\hbar}{\sqrt{N}} \sum_j [(\Xi_j^{DO})_{ni,mf} \sqrt{\omega_j^{q_{if}O}} Q_j^{q_{if}O} + i(\Xi_j^{DA})_{ni,mf} \sqrt{\omega_j^{q_{if}A}} Q_j^{q_{if}A}] \quad (7.53)$$

where we define two abbreviations. First,

$$(\Xi_j^{DO})_{ni,mf} = \sqrt{\frac{M_1+M_2}{\hbar\omega_j^{q_{if}O} M_1 M_2}} \sum_\mu \hat{p}_{j,\mu}^{q_{if}O} (D_\mu^O)_{nm} \quad (7.54)$$

Here $\hat{p}_{j,\mu}^{q_{if}O}$ is the μ-component of the unit vector $\hat{\mathbf{p}}_j^{q_{if}O}$ in the polarization direction of the jth phonon mode. Note also that j takes three values only, denoting the longitudinal optical (LO) and the two transverse optical (TO) phonon modes, respectively. So the sum over μ can be interpreted as the expectation value of the optical deformation potential transformed from the original (x,y,z)-coordinates to a frame of reference spanned by the three polarization directions of the three optical phonon modes. Furthermore, $\omega_j^{q_{if}O}$ is the frequency of an optical phonon mode as defined by eq. (6.66). We define

$$(\Xi_j^{DA})_{ni,mf} = \frac{1}{\sqrt{\hbar\omega_j^{q_{if}A}(M_1+M_2)}} \sum_{\mu,\nu} \hat{p}_{j,\mu}^{q_{if}A} (E_{\mu\nu}^A)_{nm} k_{q_{if}\nu} \quad (7.55)$$

As above $\hat{p}_{j,\mu}^{q_{if}A}$ is the μ-component of the unit vector $\hat{\mathbf{p}}_j^{q_{if}A}$ in the polarization direction of the jth phonon mode. Note that also here j takes three values. Now they denote the longitudinal acoustical (LA) and the two transverse acoustical (TA) phonon modes. Then the sums over ν and μ are to be interpreted as follows. The sum over ν corresponds to the direct product of the wave vector $\mathbf{k}_{q_{if}}$ and the 3×3 matrix with elements $(E_{\mu\nu})_{nm}$. The following sum over μ transforms the resulting vector from the original (x,y,z)-coordinates to a frame of reference spanned by the three polarization

7.1 ELECTRON–PHONON COUPLING MECHANISMS

directions of the three acoustical phonon modes. Furthermore, $\omega_j^{q_{if}A}$ is the frequency of an acoustical phonon mode as defined by eq. (6.65). We finally recall that the index q_{if} denotes the wave vector obeying $\mathbf{k}_q = \mathbf{k}_{q_{if}} = \mathbf{k}_i - \mathbf{k}_f$. Note that the definitions (7.54) and (7.55) were chosen to agree with the earlier definition (2.68).

TRANSPOSED MATRIX ELEMENTS

At this point it is fruitful to interrupt the development of the individual matrix elements $(\mathcal{H}_{EL})_{ni,mf}$ and to consider the general shape of the matrix consisting of these elements. Now, the electron–phonon interaction is $\mathcal{H}_{EL}$ is a physical observable and hence Hermitian, so interchanging the two Bloch functions $b_i^n(\mathbf{r})$ and $b_f^m(\mathbf{r})$ in eq. (7.53) yields

$$(\mathcal{H}_{EL})_{mf,ni} \equiv (\mathcal{H}_{EL})_{ni,mf}^T = (\mathcal{H}_{EL})_{ni,mf}^*$$

$$= \frac{\hbar}{\sqrt{N}} \sum_j [(\Xi_j^{DO})^*_{ni,mf} \sqrt{\omega_j^{q_{if}O}}\, Q_j^{-q_{if}O} - i(\Xi_j^{DA})^*_{ni,mf} \sqrt{\omega_j^{q_{if}A}}\, Q_j^{-q_{if}A}] \quad (7.56)$$

where we remember that, according to the treatment in section 6.2.2, $\omega_j^{q_{if}O}$ and $\omega_j^{q_{if}A}$ are real and positive frequencies.

Next we recall that $\hat{p}_{j,\mu}^{q_{if}O}$ and $\hat{p}_{j,\mu}^{q_{if}A}$ represent the components of unit vectors in normal space, so they are also real. Moreover, as mentioned before, we will see that the deformation potentials $(D_\mu^O)_{nm}$ and $(E_{\mu\nu}^A)_{nm}$ are generally real. As a result, all quantities entering the definitions (7.54) and (7.55) of $(\Xi_j^{DO})_{ni,mf}$ and $(\Xi_j^{DA})_{ni,mf}$ are real and hence they are real themselves. So we may write,

$$(\mathcal{H}_{EL})_{mf,ni} = \frac{\hbar}{\sqrt{N}} \sum_j [(\Xi_j^{DO})_{ni,mf} \sqrt{\omega_j^{q_{if}O}}\, Q_j^{-q_{if}O} - i(\Xi_j^{DA})_{ni,mf} \sqrt{\omega_j^{q_{if}A}}\, Q_j^{-q_{if}A}] \quad (7.57)$$

Upon comparison with eq. (7.53), we see that transposing the matrix element $(\mathcal{H}_{EL})_{ni,mf}$ corresponds precisely to replacing the phonon wave vector $\mathbf{k}_{q_{if}}$ by its opposite value $-\mathbf{k}_{q_{if}}$. The effect of this replacement is seen in two places. First, $Q_j^{q_{if}O}$ and $Q_j^{q_{if}A}$ are replaced by $Q_j^{-q_{if}O}$ and $Q_j^{-q_{if}A}$. Second, the sign of the term corresponding to the acoustical phonon branch changes because $\mathbf{k}_{q_{if}}$ enters linearly in $(\Xi_j^{DA})_{ni,mf}$.

Note that the result (7.57) follows from the fact that the electron–phonon interaction is a physical observable, so it is represented by a Hermitian operator. Hence, $\mathcal{H}_{EL}$ always yields *both* matrix elements (7.53) and its complex conjugate (7.57), i.e. the electron–phonon interaction always couples two Bloch states $b_i^n(\mathbf{r})$ and $b_f^m(\mathbf{r})$ simultaneously with two phonon wave vectors, $\mathbf{k}_{q_{if}}$ and $-\mathbf{k}_{q_{if}}$. This observation may be formalized by rewriting the selection rule (7.19) as

$$\mathbf{k}_q = \pm \mathbf{k}_{q_{if}} = \pm(\mathbf{k}_i - \mathbf{k}_f) \quad (7.58)$$

REAL NORMAL COORDINATES

There are further reasons to keep track of both phonon wave vectors $\mathbf{k}_{q_{if}}$ and $-\mathbf{k}_{q_{if}}$. We recall that according to eqs (6.69) and (6.70) the complex normal coordinates

$Q_j^{q_{if}O}$ and $Q_j^{q_{if}A}$ at position $\mathbf{k}_{if}$ are *not* independent of the complex normal coordinates $Q_j^{-q_{if}O}$ and $Q_j^{-q_{if}A}$ at position $-\mathbf{k}_{if}$ in reciprocal space. In section 6.2.2 independent normal coordinates were obtained by arranging all phonon wave vectors in pairs $(\mathbf{k}_q, -\mathbf{k}_q)$ and introducing new, real normal coordinates (6.75) to (6.78) for each pair. This also solved the problem that the normal coordinates $Q_j^{q_{if}O}$ and $Q_j^{q_{if}A}$ are complex, so they cannot serve as coordinates of the real harmonic oscillators treated in appendix B.

In order to keep our present treatment consistent with the derivation in chapter 6, we now introduce real normal coordinates. Then we are able to describe transitions of an electron from one Bloch state $b_i^n(\mathbf{r})$ to another Bloch state $b_f^m(\mathbf{r})$, while simultaneously one of the independent one-dimensional real harmonic oscillators is promoted to a higher eigenstate or degraded to a lower one, i.e. while a phonon is created or annihilated.

To introduce real normal coordinates in eq. (7.53) we solve eqs (6.75)–(6.78),

$$Q_j^{\pm qO} = \frac{1}{\sqrt{2}}(Q_j^{qO1} \pm iQ_j^{qO2}) \qquad (7.59)$$

$$Q_j^{\pm qA} = \frac{1}{\sqrt{2}}(Q_j^{qA1} \pm iQ_j^{qA2}) \qquad (7.60)$$

where the index q denotes a pair $(\mathbf{k}_q, -\mathbf{k}_q)$, so any summation over q is taken over half the first Brillouin zone. Obviously, in eqs (7.59) and (7.60) the plus sign applies in the case that $\mathbf{k}_q$ is chosen from the pair and the minus sign in case $-\mathbf{k}_q$ is chosen. Then, $(\mathcal{H}_{EL})_{ni,mf}$ splits into four terms,

$$(\mathcal{H}_{EL})_{ni,mf} = \frac{\hbar}{\sqrt{N}} \sum_j \left[(\Xi_j^{DO})_{ni,mf} \sqrt{\omega_j^{q_{if}O}} \frac{1}{\sqrt{2}} (Q_j^{q_{if}O1} + iQ_j^{q_{if}O2}) \right.$$
$$\left. + i(\Xi_j^{DA})_{ni,mf} \sqrt{\omega_j^{q_{if}A}} \frac{1}{\sqrt{2}} (Q_j^{q_{if}A1} + iQ_j^{q_{if}A2}) \right] \qquad (7.61)$$

while $(\mathcal{H}_{EL})_{mf,ni}$ is found to be given by the complex conjugate of this result. So,

$$(\mathcal{H}_{EL})_{mf,ni} = \frac{\hbar}{\sqrt{N}} \sum_j \left[(\Xi_j^{DO})_{ni,mf} \sqrt{\omega_j^{q_{if}O}} \frac{1}{\sqrt{2}} (Q_j^{q_{if}O1} - iQ_j^{q_{if}O2}) \right.$$
$$\left. - i(\Xi_j^{DA})_{ni,mf} \sqrt{\omega_j^{q_{if}A}} \frac{1}{\sqrt{2}} (Q_j^{q_{if}A1} - iQ_j^{q_{if}A2}) \right] \qquad (7.62)$$

7.1.4 Matrix Elements between Phonon States

PHONON CREATION AND ANNIHILATION OPERATORS

Thus far we have called $(\mathcal{H}_{EL})_{ni,mf}$ a matrix element, but this denomination is only partly correct. $(\mathcal{H}_{EL})_{ni,mf}$ is indeed a matrix element with respect to the electronic degrees of freedom, as it is obtained by integrating over all electron positions. However, $(\mathcal{H}_{EL})_{ni,mf}$ is still an operator with respect to the nuclear degrees of freedom. All we have done in the past sections, is to transform these nuclear degrees of freedom from the displacements u_μ^{ls} of the individual nuclei to real normal co-

7.1 ELECTRON–PHONON COUPLING MECHANISMS

ordinates. However, we still need to write eqs (7.61) and (7.62) as matrix elements between the eigenstates of the lattice.

The procedure of the past sections has rendered this last step almost trivial. The reason is that each of the real normal coordinates is the coordinate of an independent harmonic oscillator. Hence, the eigenstates of the lattice can be written as the product of the eigenstates of these independent harmonic oscillators, i.e.

$$\prod_{q,j} |n_j^{qA1}\rangle |n_j^{qA2}\rangle |n_j^{qO1}\rangle |n_j^{qO2}\rangle \tag{7.63}$$

To determine the matrix elements of $(\mathcal{H}_{EL})_{ni,mf}$ between such eigenstates we rewrite the real normal coordinates in phonon annihilation and creation operators. Defining these operators in the same way as in eqs (2.61) and (2.62), we find

$$Q_j^{q_{if} O\alpha} = \frac{1}{\sqrt{2\omega_j^{q_{if} O}}} (a_j^{q_{if} O\alpha} + a_j^{q_{if} O\alpha\dagger}) \qquad \alpha = 1,2 \tag{7.64}$$

$$Q_j^{q_{if} A\alpha} = \frac{1}{\sqrt{2\omega_j^{q_{if} A}}} (a_j^{q_{if} A\alpha} + a_j^{q_{if} A\alpha\dagger}) \qquad \alpha = 1,2 \tag{7.65}$$

Upon insertion of these expressions in eq. (7.61), $(\mathcal{H}_{EL})_{ni,mf}$ becomes

$$(\mathcal{H}_{EL})_{ni,mf} = \frac{\hbar}{\sqrt{2N}} \sum_j \left[(\Xi_j^{DO})_{ni,mf} \frac{1}{\sqrt{2}} (a_j^{q_{if} O1} + a_j^{q_{if} O1\dagger}) \right.$$
$$+ i(\Xi_j^{DO})_{ni,mf} \frac{1}{\sqrt{2}} (a_j^{q_{if} O2} + a_j^{q_{if} O2\dagger}) + i(\Xi_j^{DA})_{ni,mf} \frac{1}{\sqrt{2}} (a_j^{q_{if} A1} + a_j^{q_{if} A1\dagger})$$
$$\left. - (\Xi_j^{DA})_{ni,mf} \frac{1}{\sqrt{2}} (a_j^{q_{if} A2} + a_j^{q_{if} A2\dagger}) \right] \tag{7.66}$$

Note that the phonon annihilation and creation operators are real. Hence, $(\mathcal{H}_{EL})_{mf,ni}$ is simply obtained by inverting the signs of the second and third terms.

TRANSITION MATRIX ELEMENTS

In eq. (7.66) the index j may take three values, one representing a longitudinal mode and two representing transverse modes. Thus, expression (7.66) consists of 12 terms, each representing coupling with a different phonon mode. However, all these phonon modes are completely independent. As a result, the 12 different terms in the electron–phonon coupling (7.66) are completely independent as well. This is fortunate, because it allows us to consider each of these terms separately and to investigate which transitions are induced by each of these individual terms.

Moreover, all terms in eq. (7.66) have a very similar shape. So, once the effects due to one of these 12 terms has been investigated, the reader may easily extend the treatment to the other terms. Here we choose arbitrarily the terms corresponding to one of the longitudinal optical phonon modes. For this term,

$$(\mathcal{H}_{EL})_{ni,mf}^{LO1} = (\mathcal{H}_{EL})_{mf,ni}^{LO1} = \frac{\hbar}{2\sqrt{N}} (\Xi^{DLO})_{ni,mf} (a^{q_{if} LO1} + a^{q_{if} LO1\dagger}) \tag{7.67}$$

where the subscript j is replaced by a superscript L denoting 'longitudinal'. Appendix B allows the immediate calculation of the matrix element of $(\mathcal{H}_{EL})^{LO1}_{ni,mf} = (\mathcal{H}_{EL})^{LO1}_{mf,ni}$ between two eigenstates $|n^{q_{ij}LO1}\rangle$ and $|m^{q_{ij}LO1}\rangle$ of this particular phonon mode. Equivalently to eq. (2.70) the matrix elements of $a^{q_{ij}LO1}$ and $a^{q_{ij}LO1\dagger}$ between these two eigenstates are given by

$$\langle n^{q_{ij}LO1}|a^{q_{ij}LO1}|m^{q_{ij}LO1}\rangle = \sqrt{n^{q_{ij}LO1}+1}\,\delta_{m^{q_{ij}LO1},n^{q_{ij}LO1}+1}$$

$$\langle n^{q_{ij}LO1}|a^{q_{ij}LO1\dagger}|m^{q_{ij}LO1}\rangle = \sqrt{n^{q_{ij}LO1}}\,\delta_{m^{q_{ij}LO1},n^{q_{ij}LO1}-1} \tag{7.68}$$

Hence,

$$\langle n^{q_{ij}LO1}|(\mathcal{H}_{EL})^{LO1}_{ni,mf}|m^{q_{ij}LO1}\rangle = \langle n^{q_{ij}LO1}|(\mathcal{H}_{EL})^{LO1}_{mf,ni}|m^{q_{ij}LO1}\rangle$$

$$= \frac{\hbar}{2\sqrt{N}}(\Xi^{DLO})_{ni,mf}\left(\sqrt{n^{q_{ij}LO1}+1}\,\delta_{m^{q_{ij}LO1},n^{q_{ij}LO1}+1}\right.$$

$$\left. + \sqrt{n^{q_{ij}LO1}}\,\delta_{m^{q_{ij}LO1},n^{q_{ij}LO1}-1}\right) \tag{7.69}$$

Clearly the phonon quantum number $n^{q_{ij}LO1}$ may change by ± 1 only.

ALLOWED AND FORBIDDEN TRANSITIONS

At this point it is instructive to construct a section of the matrix consisting of the elements (7.69). We choose the eigenstates defined in Table 7.1, which also gives the corresponding energies. The first column presents the quantum numbers ni and nf corresponding to the initial and final Bloch states $b_i^n(\mathbf{r})$ and $b_f^m(\mathbf{r})$, while the second column gives the corresponding band energies. The third column gives the quantum numbers $n^{q_{ij}LO1} - 1$, $n^{q_{ij}LO1}$ and $n^{q_{ij}LO1} + 1$ of three successive phonon states, while the fourth column presents the corresponding phonon energies. Table 7.2 presents the matrix elements (7.71) on the basis of the states given in Table 7.1, though divided by $(\hbar/2\sqrt{N})(\Xi^{DLO})_{ni,mf}$ to keep the expressions short. We encounter eight non-zero matrix elements providing four types of coupling between the six selected eigenstates. These couplings are represented systematically in Table 7.3. Here the first two columns present the quantum numbers of the states that are coupled, the next column gives their energy difference, while the last column provides the magnitude of the matrix elements involved in their coupling. Again this matrix element is divided by $(\hbar/2\sqrt{N})(\Xi^{DLO})_{ni,mf}$ to keep the expressions short.

Table 7.1 Quantum numbers and energies corresponding to a small set of eigenstates of the crystal.

nf	E_f^n	$n^{q_{ij}LO1}+1$	$\hbar\omega^{q_{ij}LO1}(n^{q_{ij}LO1}+1)$
nf	E_f^n	$n^{q_{ij}LO1}$	$\hbar\omega^{q_{ij}LO1}n^{q_{ij}LO1}$
nf	E_f^n	$n^{q_{ij}LO1}-1$	$\hbar\omega^{q_{ij}LO1}(n^{q_{ij}LO1}-1)$
ni	E_i^n	$n^{q_{ij}LO1}+1$	$\hbar\omega^{q_{ij}LO1}(n^{q_{ij}LO1}+1)$
ni	E_i^n	$n^{q_{ij}LO1}$	$\hbar\omega^{q_{ij}LO1}n^{q_{ij}LO1}$
ni	E_i^n	$n^{q_{ij}LO1}-1$	$\hbar\omega^{q_{ij}LO1}(n^{q_{ij}LO1}-1)$

7.1 ELECTRON–PHONON COUPLING MECHANISMS

Table 7.2 Matrix elements of the electron–phonon coupling on the basis of the states given in Table 7.1.

0	0	0	0	$\sqrt{n^{q_{ij}LO1}+1}$	0
0	0	0	$\sqrt{n^{q_{ij}LO1}+1}$	0	$\sqrt{n^{q_{ij}LO1}}$
0	0	0	0	$\sqrt{n^{q_{ij}LO1}}$	0
0	$\sqrt{n^{q_{ij}LO1}+1}$	0	0	0	0
$\sqrt{n^{q_{ij}LO1}+1}$	0	$\sqrt{n^{q_{ij}LO1}}$	0	0	0
0	$\sqrt{n^{q_{ij}LO1}}$	0	0	0	0

Table 7.3 States coupled by the electron–phonon coupling, their energy differences and the matrix elements coupling these states.

$\|nf\rangle\|n^{q_{ij}LO1}+1\rangle$	$\|ni\rangle\|n^{q_{ij}LO1}\rangle$	$E_f^n - E_i^n + \hbar\omega^{q_{ij}LO1}$	$\sqrt{n^{q_{ij}LO1}+1}$
$\|ni\rangle\|n^{q_{ij}LO1}+1\rangle$	$\|nf\rangle\|n^{q_{ij}LO1}\rangle$	$E_i^n - E_f^n + \hbar\omega^{q_{ij}LO1}$	$\sqrt{n^{q_{ij}LO1}+1}$
$\|nf\rangle\|n^{q_{ij}LO1}\rangle$	$\|ni\rangle\|n^{q_{ij}LO1}-1\rangle$	$E_f^n - E_i^n + \hbar\omega^{q_{ij}LO1}$	$\sqrt{n^{q_{ij}LO1}}$
$\|ni\rangle\|n^{q_{ij}LO1}\rangle$	$\|nf\rangle\|n^{q_{ij}LO1}-1\rangle$	$E_i^n - E_f^n + \hbar\omega^{q_{ij}LO1}$	$\sqrt{n^{q_{ij}LO1}}$

The couplings presented in Table 7.3 are responsible for a variety of effects that will be treated in chapters 8 and 9. The most basic effect consists of transitions where an electron is transferred from one Bloch state to another while a phonon is emitted or absorbed. In chapter 8 we will calculate the rate of this type of transition using first order time dependent perturbation theory. Here we will not go into the details of such a calculation. However, we may now note that first-order perturbation theory requires the energy to be strictly conserved for any transition to be allowed. Here this implies that only those transitions are possible where the energy difference given in the third column is exactly equal to zero. Hence, never more than two of the four indicated transitions are allowed under the same circumstances, i.e. the first and the third if $E_i^n > E_f^n$ and the second and the fourth in the opposite case that $E_i^n < E_f^n$. Moreover, we require

$$E_i^n - E_f^n = \hbar\omega^{q_{ij}LO1} \qquad (7.70)$$

in the former and

$$E_i^n - E_f^n = -\hbar\omega^{q_{ij}LO1} \qquad (7.71)$$

in the latter case.

The results presented in Tables 7.1–7.3 are easily extended to the other phonon modes. For example, for the other longitudinal optical phonon mode where $\alpha = 2$ instead of 1, one obtains exactly the same results, except that all matrix elements in Tables 7.2 and 7.3 need to be multiplied by i.

SOME STATISTICAL ARGUMENTS

In many cases one does not need to calculate the exact quantum numbers for each phonon mode. For example, in order to obtain the rate of electron transitions from one Bloch state $b_i^n(\mathbf{r})$ to another Bloch state $b_i^m(\mathbf{r})$, it is sufficient to know the statistical average of the matrix elements (7.69). Consider the electron transition rate due the longitudinal optical phonon mode with $\alpha = 1$. To calculate this rate one may simply replace the phonon quantum numbers $n^{q_{if}LO1}$ by their statistical average $\overline{n^{q_{if}LO1}}$.

In such cases it is interesting to observe that transition probabilities are always proportional to the square of the absolute value of any matrix element involved, an observation which holds up to any order of perturbation theory. Hence, as we can see from eq. (7.66), transition probabilities involving a phonon mode with $\alpha = 1$ are given by exactly the same expressions as those involving the corresponding mode for which $\alpha = 2$. Moreover, as both longitudinal optical phonon modes are completely equivalent, one also expects that

$$\overline{n^{q_{if}LO1}} = \overline{n^{q_{if}LO2}} \tag{7.72}$$

Hence, one expects the same statistical probability to find a phonon mode in a given quantum state, whether $\alpha = 1$ or $\alpha = 2$. As a result, not only the expressions for any transition probability are equal whether $\alpha = 1$ or $\alpha = 2$, but also the numerical values for these transition rates.

One may take advantage of such circumstances to introduce a simplified expression for eq. (7.66). For this purpose we take the square of the absolute values of each of the 12 terms. Next, we add two by two the terms with $\alpha = 1$ and $\alpha = 2$. Finally, we may take the square root of each of the six resulting terms. So we write,

$$(\mathcal{H}_{EL})_{ni,mf} = (\mathcal{H}_{EL})_{mf,ni}$$

$$= \frac{\hbar}{\sqrt{2N}} \sum_j [(\Xi_j^{DO})_{ni,mf} (a_j^{q_{if}O} + a_j^{q_{if}O\dagger}) + (\Xi_j^{DA})_{ni,mf} (a_j^{q_{if}A} + a_j^{q_{if}A\dagger})] \tag{7.73}$$

where we skip the now irrelevant index $\alpha = 1, 2$. It is easily checked that this expression yields the same transition rates as eq. (7.66), provided eq. (7.72) holds.

It is interesting to see that this latter expression could also have been obtained by inserting directly

$$Q_j^{q_{if}O} = \frac{1}{\sqrt{2\omega_j^{q_{if}LO}}} (a_j^{q_{if}O} + a_j^{q_{if}O\dagger})$$

$$Q_j^{q_{if}A} = \frac{1}{\sqrt{2\omega_j^{q_{if}LO}}} (a_j^{q_{if}A} + a_j^{q_{if}A\dagger})$$

in eq. (7.53). However, there is a subtlety inhibiting us to do this replacement immediately. Eq. (7.53) conforms to the selection rule (7.19) while eq. (7.73) obeys the selection rule (7.58). Thus, the latter expression represents the coupling to twice as many phonon modes as the former equation.

7.1.5 Electron–Phonon Coupling in Effective Mass Theory
COUPLING BETWEEN PLANE WAVES AND THE LATTICE

We continue by investigating electron–phonon coupling as it would appear in the effective mass formalism. In this formalism as introduced in chapter 5, we consider

7.1 ELECTRON–PHONON COUPLING MECHANISMS

electron states within a single band only. So one could try to apply it to describe transitions between an initial and a final Bloch state $b_i^n(\mathbf{r})$ and $b_f^n(\mathbf{r})$. The essence of effective mass theory is to replace these Bloch functions by plane waves

$$e^{i\mathbf{k}_i \cdot \mathbf{r}} \quad \text{and} \quad e^{i\mathbf{k}_f \cdot \mathbf{r}} \tag{7.74}$$

Therefore one could imagine that the effective mass formalism replaces the electron–lattice coupling (7.7) by

$$(\mathcal{H}_{EL})_{ni,nf} = \sum_{l,s,\mu}^{N} \left[\frac{1}{NV_C} \int_{NV_C} d\mathbf{r}\, e^{-i\mathbf{k}_i \cdot \mathbf{r}} \phi_\mu^{s,eff}(\mathbf{r} - \mathbf{R}_l) e^{i\mathbf{k}_f \cdot \mathbf{r}} \right] u_\mu^{ls} \tag{7.75}$$

Here the function $\phi_\mu^{s,eff}(\mathbf{r} - \mathbf{R}_l)$ represents the *effective* potential experienced by an electron at position $\mathbf{r}$ due to a unit displacement in the μ-direction of the sth nucleus in the lth primitive cell. Note that we also implemented the normalization convention of eq. (7.7), allowing us to compare the results of effective mass theory directly with those obtained in the previous section.

TRANSFORMATION TO RECIPROCAL SPACE

To investigate whether eq. (7.75) makes sense and whether the function $\phi_\mu^{s,eff}(\mathbf{r} - \mathbf{R}_l)$ has a physical meaning, we follow the same treatment as in section 7.1.1. We transform the space lattice positions $\mathbf{R}_l$ to reciprocal space by inserting eq. (6.6). Again the index q is used for reciprocal space positions $\mathbf{k}_q$ representing lattice degrees of freedom, while the indices i and f are used for reciprocal space positions $\mathbf{k}_i$ and $\mathbf{k}_f$ representing electron degrees of freedom. Then, eq. (7.75) can again be written in the shape (7.8),

$$(\mathcal{H}_{EL})_{ni,nf} = \sum_{l,s,\mu}^{N} \left[\frac{1}{NV_C} \int_{NV_C} d\mathbf{r}\, e^{-i\mathbf{k}_i \cdot \mathbf{r}} \phi_\mu^{s,eff}(\mathbf{r} - \mathbf{R}_l) e^{i\mathbf{k}_f \cdot \mathbf{r}} \right] \times \frac{1}{\sqrt{N}} \sum_q^{N} v_\mu^{qs} e^{i\mathbf{k}_q \cdot \mathbf{R}_l}$$

$$= \frac{1}{\sqrt{N}} \sum_q \sum_{s,\mu}^{N} (\mathcal{D}_\mu^{qs})_{ni,nf}\, v_\mu^{qs} \tag{7.76}$$

Now, however, we have a different expression for the matrix element of the deformation potential. It is defined on the basis of plane waves instead of Bloch functions and is written as

$$(\mathcal{D}_\mu^{qs})_{ni,nf} = \frac{1}{NV_C} \int_{NV_C} d\mathbf{r}\, e^{-i\mathbf{k}_i \cdot \mathbf{r}} \left[\sum_l^{N} \phi_\mu^{s,eff}(\mathbf{r} - \mathbf{R}_l) e^{i\mathbf{k}_q \cdot \mathbf{R}_l} \right] e^{i\mathbf{k}_f \cdot \mathbf{r}} \tag{7.77}$$

i.e. as a matrix element of an effective deformation potential

$$d_\mu^{qs,eff} = \sum_l^{N} \phi_\mu^{s,eff}(\mathbf{r} - \mathbf{R}_l) e^{-i\mathbf{k}_q \cdot (\mathbf{r} - \mathbf{R}_l)} \tag{7.78}$$

in reciprocal space. The latter definition is such that in eq. (7.77) the sum over l is equal to

$$\sum_l^{N} \phi_\mu^{s,eff}(\mathbf{r} - \mathbf{R}_l) e^{i\mathbf{k}_q \cdot \mathbf{R}_l} = d_\mu^{qs,eff} e^{i\mathbf{k}_q \cdot \mathbf{r}} \tag{7.79}$$

At this point we remark that $d_\mu^{qs,eff}$ is independent of the electron position $\mathbf{r}$. This can be inferred as follows. In the full formalism given in section 7.1.1, we showed its equivalent $d_\mu^{qs}(\mathbf{r})$ to obey the translational symmetry of the crystal lattice. These arguments are still valid, so $d_\mu^{qs,eff}$ also obeys this translational symmetry. However, $d_\mu^{qs,eff}$ is defined in the effective mass formalism where all potentials are averaged over primitive cells. Functions that obey the translational symmetry of the crystal lattice *and* that are constant in such a primitive cell are necessarily constant everywhere. Hence, $d_\mu^{qs,eff}$ is independent of the electron position $\mathbf{r}$. As a result, eq. (7.77) can be rewritten as,

$$(\mathcal{D}_\mu^{qs})_{ni,nf} = \frac{1}{NV_C} \int_{NV_C} d\mathbf{r}\, e^{-i\mathbf{k}_i \cdot \mathbf{r}} (d_\mu^{qs,eff} e^{i\mathbf{k}_q \cdot \mathbf{r}}) e^{i\mathbf{k}_f \cdot \mathbf{r}}$$

$$= d_\mu^{qs,eff} \frac{1}{NV_C} \int_{NV_C} d\mathbf{r}\, e^{i(\mathbf{k}_q - \mathbf{k}_i + \mathbf{k}_f) \cdot \mathbf{r}} \qquad (7.80)$$

The latter integral is completely equivalent to the Kronecker δ-function (3.35), provided that we replace the volume of a primitive cell by the volume of the whole crystal. Hence,

$$(\mathcal{D}_\mu^{qs})_{ni,nf} = d_\mu^{qs,eff} \delta_{q,q_{if}} \qquad (7.81)$$

where the Kronecker δ-function requires that $\mathbf{k}_q = \mathbf{k}_{q_{if}} = \mathbf{k}_i - \mathbf{k}_f$.

EFFECTIVE DEFORMATION POTENTIALS

Upon comparison with eq. (7.18) we see that the effective mass formalism yields a result which is consistent with the more general treatment given in section 7.1.1. Moreover, this comparison yields an expression for the effective deformation potential:

$$d_\mu^{qs,eff} = \langle n\mathbf{k}_i | d_\mu^{qs}(\mathbf{r}) | n\mathbf{k}_f \rangle \qquad (7.82)$$

Actually, if we look more carefully, we find that $(\mathcal{D}_\mu^{q_{if}s})_{ni,nf}$ has the physical meaning of being an effective deformation potential, because it is non-zero only for $\mathbf{k}_q = \mathbf{k}_{q_{if}} = \mathbf{k}_i - \mathbf{k}_f$. Hence, the quantities $(D_\mu^O)_{nn}$ and $(E_{\mu\nu}^A)_{nn}$—that are generally called deformation potentials in literature—are actually *effective* deformation potentials!

7.1.6 Polar Semiconductors

A special situation arises in so-called polar semiconductors like GaAs where the cores of the Ga atoms have charge +3 which differs from the charge +5 of the As cores. In such polar semiconductors the displacement of the atoms due to the excitation of a phonon mode yields an electrical polarization. An electron in the conduction band experiences the long range electric field caused by this polarization and the electron–lattice coupling can be interpreted as the potential of the electron in this electric field. Interestingly, because the electric field is long range, the resulting coupling can be treated using an effective mass formalism. As we will see below, this allows us to obtain explicit expressions for its strength.

POLAR ELECTRON-OPTICAL PHONON COUPLING

When an optical phonon mode is excited, the two cores move in opposite direction and the dipole moment of the lth primitive cell is given by

$$\mathbf{p}_l = e^*(\mathbf{u}^{l1} - \mathbf{u}^{l2}) \tag{7.83}$$

where $\mathbf{u}^{ls}$ is the nuclear displacement vector with components u^{ls}_μ. Here e^* is a proportionality constant which can be interpreted as the effective charge of the dipole moment. This effective charge is approximately the charge of the displaced cores. The dipole moment (7.83) causes a long range electric field which is described by a scalar potential

$$U_l(\mathbf{r}) = \frac{1}{4\pi\varepsilon_0}\mathbf{p}_l \cdot \nabla \frac{1}{|\mathbf{r} - \mathbf{R}_l|} = \frac{e^*}{4\pi\varepsilon_0}(\mathbf{u}^{l1} - \mathbf{u}^{l2}) \cdot \nabla \frac{1}{|\mathbf{r} - \mathbf{R}_l|} \tag{7.84}$$

The total scalar potential in the crystal is the sum of the scalar potentials caused by the dipole moments in each individual primitive cell. Hence, an electron in the crystal experiences a total potential energy given by

$$-e\sum_l^N U_l(\mathbf{r}) = -\frac{ee^*}{4\pi\varepsilon_0}\sum_l^N (\mathbf{u}^{l1} - \mathbf{u}^{l2}) \cdot \nabla \frac{1}{|\mathbf{r} - \mathbf{R}_l|}$$

$$= -\frac{ee^*}{4\pi\varepsilon_0}\sum_{l\mu}^N (u^{l1}_\mu - u^{l2}_\mu) \nabla_\mu \frac{1}{|\mathbf{r} - \mathbf{R}_l|}$$

$$= -\frac{ee^*}{4\pi\varepsilon_0}\sum_{l\mu}^N u^{lO}_\mu \nabla_\mu \frac{1}{|\mathbf{r} - \mathbf{R}_l|} \tag{7.85}$$

So

$$-\frac{ee^*}{4\pi\varepsilon_0}\nabla_\mu \frac{1}{|\mathbf{r} - \mathbf{R}_l|}$$

plays the role of $\phi^{s,eff}_\mu(\mathbf{r} - \mathbf{R}_l)$ in eq. (7.75). The potential energy (7.85) represents electron–optical phonon coupling, because it depends on the relative displacement u^{lO}_μ of the nuclei within a primitive cell only. However, it does not originate in the change of the periodic potential and therefore it cannot be said to be due to a deformation potential. Instead it represents the contribution due to the long range electric field caused by the dipole moments induced by the optical phonons. Generally one denotes the coupling defined in 7.1.1 as *deformation potential optical phonon coupling*, while the present contribution is called *polar electron–optical phonon coupling*.

We transform the relative displacement of the nuclei to reciprocal space using eq. (6.30). Then,

$$-e\sum_l^N U_l(\mathbf{r}) = -\frac{ee^*}{4\pi\varepsilon_0}\sum_{l\mu}^N \frac{1}{\sqrt{N}}\sum_q^N v^{qO}_\mu e^{i\mathbf{k}_q \cdot \mathbf{R}_l} \nabla_\mu \frac{1}{|\mathbf{r} - \mathbf{R}_l|}$$

$$= -\frac{ee^*}{4\pi\varepsilon_0}\frac{1}{\sqrt{N}}\sum_{q\mu}^N v^{qO}_\mu \sum_l^N e^{i\mathbf{k}_q \cdot \mathbf{R}_l} \nabla_\mu \frac{1}{|\mathbf{r} - \mathbf{R}_l|} \tag{7.86}$$

As we are using effective mass theory electron wave vectors $\mathbf{k}_i$ and $\mathbf{k}_f$ and hence phonon wave vectors $\mathbf{k}_q = \mathbf{k}_{q_{if}} = \mathbf{k}_i - \mathbf{k}_f$ are small. This induces us to approximate the sum over l by an integral which can be evaluated (see appendix F). We calculate

$$\sum_l^N e^{-i\mathbf{k}_q \cdot (\mathbf{r}-\mathbf{R}_l)} \nabla_\mu \frac{1}{|\mathbf{r}-\mathbf{R}_l|} \approx \frac{1}{V_C} \int_{NV_C} d\mathbf{R}\, \nabla_\mu \frac{1}{|\mathbf{r}-\mathbf{R}|} e^{-i\mathbf{k}_q \cdot (\mathbf{r}-\mathbf{R})}$$

$$\approx \frac{1}{V_C} \int_\infty d\mathbf{R}'\, \nabla_\mu \frac{1}{|\mathbf{R}'|} e^{i\mathbf{k}_q \cdot \mathbf{R}'} = -\frac{1}{V_C} 4\pi i \frac{k_{q\mu}}{|\mathbf{k}_q|^2} \quad (7.87)$$

where we replaced $(\mathbf{R} - \mathbf{r})$ by $\mathbf{R}'$. Note that the result is independent of the electron position $\mathbf{r}$. Note also that the factor V_C^{-1} needs to be added because each term in the sum over l represents a volume V_C. Hence

$$-e \sum_l^N U_l(\mathbf{r}) = \frac{ee^*}{4\pi\varepsilon_0} \frac{1}{\sqrt{N}} \sum_{q\mu}^N v_\mu^{qO} \frac{1}{V_C} 4\pi i \frac{k_{q\mu}}{|\mathbf{k}_q|^2} e^{i\mathbf{k}_q \cdot \mathbf{r}}$$

$$= i \frac{ee^*}{\varepsilon_0 V_C} \frac{1}{\sqrt{N}} \sum_{q\mu}^N v_\mu^{qO} \frac{k_{q\mu}}{|\mathbf{k}_q|^2} e^{i\mathbf{k}_q \cdot \mathbf{r}}$$

$$= \frac{1}{\sqrt{N}} \sum_{q\mu}^N v_\mu^{qO} d_\mu^{qPO,\text{eff}} e^{i\mathbf{k}_q \cdot \mathbf{r}} \quad (7.88)$$

where

$$d_\mu^{qPO,\text{eff}} = i \frac{ee^*}{\varepsilon_0 V_C} \frac{k_{q\mu}}{|\mathbf{k}_q|^2} \quad (7.89)$$

The coefficient (7.89) is the equivalent of the effective deformation potential (7.78). The latter's matrix element between two plane waves is evaluated in eq. (7.80) and the result given by eq. (7.81). Equivalently we write,

$$(\mathcal{D}_\mu^{qPO})_{ni,nf} = \delta_{q,q_{if}} d_\mu^{qPO,\text{eff}} = i \frac{ee^*}{\varepsilon_0 V_C} \frac{k_{q\mu}}{|\mathbf{k}_q|^2} \delta_{q,q_{if}} \quad (7.90)$$

MATRIX ELEMENTS FOR POLAR ELECTRON–OPTICAL PHONON COUPLING

We now have enough information to evaluate the matrix element $(\mathcal{H}_{EL}^{PO})_{ni,nf}$ for polar electron–optical phonon coupling. We start by considering that this matrix element must have the same shape as the electron–optical phonon coupling term in eq. (7.25). Hence,

$$(\mathcal{H}_{EL}^{PO})_{ni,nf} = \frac{1}{\sqrt{N}} \sum_q \sum_\mu^N (\mathcal{D}_\mu^{qPO})_{ni,nf}\, v_\mu^{qO} \quad (7.91)$$

Next we follow the procedure leading from eq. (7.52) to eq. (7.54). First we insert reduced displacements defined by eq. (6.56) and subsequently we transform to

7.1 ELECTRON–PHONON COUPLING MECHANISMS

complex normal coordinates using eq. (6.68). Then we find

$$(\mathcal{H}_{EL}^{PO})_{ni,nf} = \frac{1}{\sqrt{N}} \sum_q \sum_\mu (D_\mu^{qPO})_{ni,nf} \, v_\mu^{qO}$$

$$= \frac{1}{\sqrt{N}} \sum_q \sum_\mu \delta_{q,q_{if}} \, d_\mu^{qPO,eff} \, v_\mu^{qO}$$

$$= \frac{1}{\sqrt{N}} \sum_\mu d_\mu^{q_{if}PO,eff} \, v_\mu^{q_{if}O}$$

$$= \sqrt{\frac{\hbar(M_1 + M_2)}{NM_1 M_2}} \sum_\mu d_\mu^{q_{if}PO,eff} \, w_\mu^{q_{if}O}$$

$$= \frac{\hbar}{\sqrt{N}} \sum_j \sqrt{\frac{M_1 + M_2}{\hbar \omega_j^{q_{if}O} M_1 M_2}} \sum_\mu d_\mu^{q_{if}PO,eff} \, \hat{p}_{\mu,j}^{q_{if}O} \sqrt{\omega_j^{q_{if}O}} Q_j^{q_{if}O}$$

$$= \frac{\hbar}{\sqrt{N}} \sum_j (\Xi_j^{PO})_{ni,nf} \sqrt{\omega_j^{q_{if}O}} Q_j^{q_{if}O} \qquad (7.92)$$

where

$$(\Xi_j^{PO})_{ni,nf} = \sqrt{\frac{M_1 + M_2}{\hbar \omega_j^{q_{if}O} M_1 M_2}} \sum_\mu d_\mu^{q_{if}PO,eff} \, \hat{p}_{\mu,j}^{q_{if}O}$$

$$= i \sqrt{\frac{M_1 + M_2}{\hbar \omega_j^{q_{if}O} M_1 M_2}} \sum_\mu \frac{ee^*}{\varepsilon_0 V_C} \frac{k_{q_{if}\mu}}{|\mathbf{k}_{q_{if}}|^2} \hat{p}_{\mu,j}^{q_{if}O} \qquad (7.93)$$

and where we insert eq. (7.89) for $d_\mu^{q_{if}PO,eff}$.

We now recall that, according to eq. (6.64), $\hat{p}_{j,\mu}^{q_{if}O}$ is a component of the unit polarization vector $\hat{\mathbf{p}}_j^{q_{if}O}$ of the jth optical phonon mode. Hence the sum over μ in eq. (7.93) represents the projection of this unit polarization vector on the phonon wave vector. Clearly, as only this projection enters in $(\mathcal{H}_{EL}^{PO})_{ni,nf}$, only *longitudinal* phonon modes contribute to this type of electron–lattice coupling. Thus, in the end j takes one value only: the one corresponding to the longitudinal optical (LO) phonon mode. We write $j = LO$, then

$$\hat{p}_\mu^{qLO} = \frac{k_{q\mu}}{|\mathbf{k}_q|} \quad \text{so} \quad \sum_\mu \hat{p}_\mu^{qLO} \frac{k_{q\mu}}{|\mathbf{k}_q|^2} = \sum_\mu \frac{k_{q\mu}^2}{|\mathbf{k}_q|^3} = \frac{1}{|\mathbf{k}_q|} \qquad (7.94)$$

and

$$(\mathcal{H}_{EL}^{PLO})_{ni,nf} = \frac{\hbar}{\sqrt{N}} (\Xi^{PLO})_{ni,nf} \sqrt{\omega^{q_{if}LO}} Q^{q_{if}LO} \qquad (7.95)$$

where

$$(\Xi^{PLO})_{ni,nf} = i \frac{ee^*}{\varepsilon_0 V_C} \sqrt{\frac{M_1 + M_2}{\hbar \omega^{q_{if}LO} M_1 M_2}} \frac{1}{|\mathbf{k}_{q_{if}}|} \qquad (7.96)$$

This final result has an unpleasant feature which cannot be left without comment. The resulting matrix element is proportional to $1/|\mathbf{k}_q|$, so it diverges for $\mathbf{k}_q \to 0$. In practice the resulting divergence is thought to be blocked by a process which is called shielding. It is argued that the dipole fields giving rise to the electron–optical phonon coupling in polar semiconductors, are shielded by free charge carriers in the crystal, so this field has a limited range. Furthermore it can be shown that such a limitation of the range of the dipole field always leads to a finite value for the transition matrix element. We will come back to this issue in the description of electron-impurity scattering in chapter 8.

PIEZOELECTRIC ELECTRON-PHONON COUPLING

Not only the excitation of optical phonon modes but also that of acoustical phonon modes causes the creation of an electrical polarization in polar semiconductors. One should be aware however, that the same arguments apply as previously and that the resulting polarization will be proportional to the strain tensor rather than the nuclear displacements themselves. Hence, the μ-component of the electrical dipole moment of the lth primitive cell will be given by

$$p_\mu^l = V_C \sum_{\nu,\rho} \frac{\varepsilon_{\mu\nu\rho}^{pe}}{\varepsilon_r} S_{\nu\rho}^l, \qquad (7.97)$$

where the proportionality constants $\varepsilon_{\mu\nu\rho}^{pe}$ are the components of the so-called piezoelectric tensor. The factor $(1/\varepsilon_r)$ makes it the same piezoelectric tensor as used to calculate the piezoelectric effect. Here ε_r is the macroscopic relative permittivity of the crystal. Therefore the electron–phonon coupling arising from the polarization (7.97) is called *piezoelectric electron–phonon coupling* distinguishing it from *deformation potential acoustical phonon coupling* as treated in section 7.1.1. Note that the electronic polarization is included in the piezoelectric effect, so we need not add a relative permittivity ε_r in the resulting potential energy. Furthermore, the dimension of this piezoelectric tensor is chosen to yield polarization per unit volume, so a factor V_C has to be added to obtain the electric dipole moment of a primitive cell. In general this tensor has 27 components, but, depending on crystal symmetry, many may be equal to each other, while many others may be equal to zero. We are concerned with crystals like GaAs with the tetragonal zinc blende structure. Then, as discussed in appendix D, only one type of non-zero component exists, which we may denote by

$$\varepsilon^{pe} = \varepsilon_{xyz}^{pe} = \varepsilon_{yzx}^{pe} = \varepsilon_{zxy}^{pe} = \varepsilon_{zyx}^{pe} = \varepsilon_{yxz}^{pe} = \varepsilon_{xzy}^{pe} \qquad (7.98)$$

Then, in eq. (7.97) in the sum over ν and ρ two equal terms remain and

$$p_\mu^l = V_C \frac{2\varepsilon^{pe}}{\varepsilon_r} S_{(\mu+1)(\mu+2)}^l \qquad (7.99)$$

Clearly, the only difference between the present expression and eq. (7.83) for the dipole moment of a primitive cell due to the excitation of an optical phonon mode lies in the replacement of $e^* u_\mu^l$ by $V_C(2\varepsilon^{pe}/\varepsilon_r) S_{(\mu+1)(\mu+2)}^l$. Hence, again we invoke the

7.1 ELECTRON–PHONON COUPLING MECHANISMS

effective mass approximation and we follow the arguments leading to eq. (7.85). Then an electron in the crystal experiences a total potential energy given by

$$-e \sum_{l}^{N} U_l(\mathbf{r}) = -V_C \frac{2e\varepsilon^{pe}}{4\pi\varepsilon_0\varepsilon_r} \sum_{l\mu} S^l_{(\mu+1)(\mu+2)} \nabla_\mu \frac{1}{|\mathbf{r} - \mathbf{R}_l|} \quad (7.100)$$

Here we introduce a cyclic notation, where the components of a vector bear indices μ, $(\mu+1)$, $(\mu+2)$, $(\mu+3) = \mu$, etc. Next, equivalently to eqs (7.86) to (7.88), we transform to reciprocal space. Inserting eqs (6.47) and (7.87) we find

$$\sum_{l\mu}^{N} S^l_{(\mu+1)(\mu+2)} \nabla_\mu \frac{1}{|\mathbf{r} - \mathbf{R}_l|}$$

$$= \sum_{l\mu}^{N} \frac{i}{2} \frac{1}{\sqrt{N}} \sum_{q} (v^{qa}_{(\mu+1)} k_{q(\mu+2)} + v^{qa}_{(\mu+2)} k_{q(\mu+1)}) e^{i\mathbf{k}_q \cdot \mathbf{R}_l} \nabla_\mu \frac{1}{|\mathbf{r} - \mathbf{R}_l|}$$

$$= \frac{i}{2} \frac{1}{\sqrt{N}} \sum_{q\mu} (v^{qa}_{(\mu+1)} k_{q(\mu+2)} + v^{qa}_{(\mu+2)} k_{q(\mu+1)}) \sum_{l}^{N} e^{i\mathbf{k}_q \cdot \mathbf{R}_l} \nabla_\mu \frac{1}{|\mathbf{r} - \mathbf{R}_l|}$$

$$\approx -\frac{i}{2} \frac{1}{\sqrt{N}} \sum_{q\mu}^{N} (v^{qa}_{(\mu+1)} k_{q(\mu+2)} + v^{qa}_{(\mu+2)} k_{q(\mu+1)}) \frac{1}{V_C} 4\pi i \frac{k_{q\mu}}{|\mathbf{k}_q|^2} e^{i\mathbf{k}_q \cdot \mathbf{r}}$$

$$= \frac{1}{V_C} \frac{2\pi}{\sqrt{N}} \sum_{q\mu}^{N} (v^{qa}_{(\mu+1)} k_{q(\mu+2)} + v^{qa}_{(\mu+2)} k_{q(\mu+1)}) \frac{k_{q\mu}}{|\mathbf{k}_q|^2} e^{i\mathbf{k}_q \cdot \mathbf{r}}$$

$$= \frac{1}{V_C} \frac{4\pi}{\sqrt{N}} \sum_{q\mu}^{N} v^{qa}_{\mu} \frac{k_{q(\mu+1)} k_{q(\mu+2)}}{|\mathbf{k}_q|^2} e^{i\mathbf{k}_q \cdot \mathbf{r}} \quad (7.101)$$

Here we did use our cyclic notation to our advantage by renumbering the indices in the first term from μ to $(\mu+2)$ and in the second term from μ to $(\mu+1)$. As a result,

$$-e \sum_{l}^{N} U_l(\mathbf{r}) = -V_C \frac{2e\varepsilon^{pe}}{4\pi\varepsilon_0\varepsilon_r} \frac{1}{V_C} \frac{4\pi}{\sqrt{N}} \sum_{q\mu}^{N} v^{qA}_\mu \frac{k_{q(\mu+1)} k_{q(\mu+2)}}{|\mathbf{k}_q|^2} e^{i\mathbf{k}_q \cdot \mathbf{r}}$$

$$= -\frac{2e\varepsilon^{pe}}{\varepsilon_0\varepsilon_r} \frac{1}{\sqrt{N}} \sum_{q\mu}^{N} v^{qA}_\mu \frac{k_{q(\mu+1)} k_{q(\mu+2)}}{|\mathbf{k}_q|^2} e^{i\mathbf{k}_q \cdot \mathbf{r}}$$

$$= \frac{1}{\sqrt{N}} \sum_{q\mu}^{N} v^{qA}_\mu d^{qPA,eff}_\mu e^{i\mathbf{k}_q \cdot \mathbf{r}} \quad (7.102)$$

where

$$d^{qPA,eff}_\mu = -\frac{2e\varepsilon^{pe}}{\varepsilon_0\varepsilon_r} \frac{k_{q(\mu+1)} k_{q(\mu+2)}}{|\mathbf{k}_q|^2} \quad (7.103)$$

Then, equivalently to eq. (7.90)

$$(\mathcal{D}^{qPA}_\mu)_{ni,nf} = d^{qPA,eff}_\mu \delta_{q,q_{if}} = -\frac{2e\varepsilon^{pe}}{\varepsilon_0\varepsilon_r} \frac{k_{q(\mu+1)} k_{q(\mu+2)}}{|\mathbf{k}_q|^2} \delta_{q,q_{if}} \quad (7.104)$$

MATRIX ELEMENTS FOR PIEZOELECTRIC ELECTRON–PHONON COUPLING

We finish the treatment of piezoelectric electron–phonon coupling following the same procedure as the one leading from eqs (7.91) to (7.93). The matrix element for piezoelectric electron–phonon coupling must have the same shape as the electron–acoustical phonon coupling term in eq. (7.25),

$$(\mathcal{H}_{EL}^{PA})_{ni,nf} = \frac{1}{\sqrt{N}} \sum_q^N \sum_\mu (\mathcal{D}_\mu^{qPA})_{ni,nf} \, v_\mu^{qA} \tag{7.105}$$

Next we transform to reduced coordinates using eq. (6.57) and subsequently to complex normal coordinates by eq. (6.68). As a result we find a contribution to the electron–acoustical phonon coupling given by

$$(\mathcal{H}_{EL}^{PA})_{ni,nf} = \frac{1}{\sqrt{N}} \sum_q^N \sum_\mu (\mathcal{D}_\mu^{qPA})_{ni,nf} \, v_\mu^{qA}$$

$$= \frac{1}{\sqrt{N}} \sum_q^N \sum_\mu d_\mu^{qPA,eff} \, \delta_{q,q_{if}} \, v_\mu^{qA}$$

$$= \frac{1}{\sqrt{N}} \sum_\mu d_\mu^{q_{if}PA,eff} \, v_\mu^{q_{if}A}$$

$$= \sqrt{\frac{\hbar}{N(M_1 + M_2)}} \sum_\mu d_\mu^{q_{if}PA,eff} \, w_\mu^{q_{if}A}$$

$$= \frac{\hbar}{\sqrt{N}} \sum_j \frac{1}{\sqrt{\hbar \omega_j^{q_{if}A}(M_1 + M_2)}} \sum_\mu d_\mu^{q_{if}PA,eff} \, \hat{p}_{\mu,j}^{q_{if}A} \sqrt{\omega_j^{q_{if}A}} \, Q_j^{q_{if}A}$$

$$= i \frac{\hbar}{\sqrt{N}} \sum_j (\Xi_j^{PA})_{ni,nf} \sqrt{\omega_j^{q_{if}A}} \, Q_j^{q_{if}A} \tag{7.106}$$

where,

$$(\Xi_j^{PA})_{ni,nf} = \frac{-i}{\sqrt{\hbar \omega_j^{q_{if}A}(M_1 + M_2)}} \sum_\mu d_\mu^{q_{if}PA,eff} \, \hat{p}_{\mu,j}^{q_{if}A}$$

$$= i \frac{2e\varepsilon^{pe}}{\varepsilon_0 \varepsilon_r} \frac{1}{\sqrt{\hbar \omega_j^{q_{if}A}(M_1 + M_2)}} \sum_\mu \left(\hat{p}_{\mu,j}^{q_{if}A} \frac{k_{q_{if}(\mu+1)} k_{q_{if}(\mu+2)}}{|\mathbf{k}_{q_{if}}|^2} \right) \tag{7.107}$$

Again $\hat{p}_{j,\mu}^{q_{if}A}$ is a component of the unit polarization vector of the jth phonon mode and again $\mathbf{k}_{q_{if}}$ is the wave vector of this phonon mode, so the factors $k_{q_{if}\mu}/|\mathbf{k}_{q_{if}}|$ represent the direction cosines of this wave vector. Unfortunately we thus have obtained a quite complicated angular dependence that cannot be simplified in the same manner as in the case of polar electron–optical phonon coupling.

7.2 Electron–Phonon Coupling near the Γ-point

In chapter 4 we saw the power of the **k·p**-approximation when we need to evaluate the band structure near the Γ-point. It does more however, it also provides a valuable tool to obtain expressions for the matrix elements for electron–phonon coupling in this region of the first Brillouin zone. To show this, we finish the present chapter evaluating the electron–phonon coupling at the Γ-point. In particular we will consider the quantities $(\Xi_j^{DO})_{ni,mf}$, $(\Xi_j^{DA})_{ni,mf}$, $(\Xi_j^{PO})_{ni,mf}$ and $(\Xi_j^{PA})_{ni,mf}$, as given by eqs (7.54), (7.55), (7.96) and (7.107).

Basically, the approximations that are made in the **k·p**-approximation, were already made from section 7.1.2 onwards. Hence, eqs (7.54), (7.55), (7.96) and (7.107) are effectively expressions for $(\Xi_j^{DO})_{ni,mf}$, $(\Xi_j^{DA})_{ni,mf}$, $(\Xi_j^{PO})_{ni,mf}$ and $(\Xi_j^{PA})_{ni,mf}$, as obtained by the **k·p** approximation. However, these are general expressions, comparable to those obtained in sections 4.1.3 and 4.1.4 for the band structure. Here we intend to consider the electron–phonon coupling in the case of the specific model treated in section 4.2. Thus, we want to obtain expressions equivalent to those obtained in sections 4.2.2, 4.2.5 and 4.2.6. In the following sections we will first consider the more simple case of electrons near the Γ-point minimum of the conduction band. Next we extend our treatment to holes at the top of the valence band.

7.2.1 Conduction Band Electron–Phonon Coupling

NON-POLAR ELECTRON–OPTICAL PHONON COUPLING

We start our discussion considering electron–optical phonon coupling in non-polar semiconductors. We are concentrating on electron–phonon coupling at the Γ-point minimum of the conduction band, so always the band indices are $n = m = c$. Then, according to eq. (7.54), $(\Xi_j^{DO})_{ci,cf}$ is given by,

$$(\Xi_j^{DO})_{ci,cf} = \sqrt{\frac{M_1 + M_2}{\hbar \omega_j^{q_{if} O} M_1 M_2}} \sum_\mu \hat{p}_{j,\mu}^{q_{if} O} (D_\mu^O)_{cc} \qquad (7.108)$$

Note that j takes three values, denoting the longitudinal optical (LO) and the two transverse optical (TO) phonon modes, respectively. In eq. (7.108) M_1 and M_2 are the masses of the two nuclei in the primitive cell. Furthermore, according to eq. (6.64), $\hat{p}_{j,\mu}^{q_{if} O}$ is the μ-component of the unit polarization vector of the jth optical phonon mode with wave vector $\mathbf{k}_q = \mathbf{k}_{q_{if}} = \mathbf{k}_f - \mathbf{k}_i$ and frequency $\omega_j^{q_{if} O}$. Finally, as defined in eq. (7.38),

$$(D_\mu^O)_{cc} = \langle c0 | d_\mu^{00}(\mathbf{r}) | c0 \rangle \qquad (7.109)$$

Note that the pre-factor in eq. (7.108) has dimension [energy × mass]$^{-1/2}$ = [momentum] = $[\hbar k]^{-1}$ while the components $\hat{p}_{j,\mu}^{q_{if} O}$ of the unit polarization vector are dimensionless, so $(D_\mu^O)_{nm}$ has the dimension of energy per unit of length.

In this special case $n = m = c$ it is easily argued that expression (7.109) for the expectation value of the optical deformation potential vanishes. For this purpose we consider an optical phonon mode at $\mathbf{k}_q = 0$. Excitation of this phonon mode implies a relative displacement of the two nuclei in the primitive cell and hence a deformation of this primitive cell. Because $\mathbf{k}_q = 0$, this deformation is the same for all primitive cells, so the crystal deforms as a whole. The phonon mode creating this deformation corresponds furthermore to a harmonic oscillator, so the potential energy of the crystal lattice increases *quadratically* with this deformation. We now consider the energy of an electron in the conduction band. This energy changes also when the phonon mode is excited and the amount is *linearly* proportional to $(D_\mu^O)_{cc}$. Hence, the electron energy changes *linearly* with the deformation. Now, for a small enough deformation with a correctly chosen sign, the linear *decrease* of the electron energy will always be larger than the quadratic *increase* of the lattice energy. So, deformation lets the crystal as a whole *gain* energy and we must conclude that the crystal will be *permanently deformed*, unless

$$(D_\mu^O)_{cc} = 0 \tag{7.110}$$

From observation we know that cubic semiconductor crystals are not deformed by electrons in the conduction band, so eq. (7.110) is expected to hold. However, it must be emphasized that deformations are observed in *other* types of crystals, referred to as the Jahn–Teller effect after their discoverers. For a detailed discussion of this effect, see ref. [44].

For non-polar semiconductors the absence of the Jahn–Teller effect has the direct consequence that the electron–optical phonon coupling at the conduction band minimum at the Γ-point vanishes in lowest order $\mathbf{k} \cdot \mathbf{p}$-approximation. This result is less important than expected at first sight, however. The reason is that the only non-polar semiconductors, silicon and germanium, are indirect semiconductors. In their case the lowest conduction band minima lie at other locations in the Brillouin zone where the above argument cannot be applied. As a result, conduction band electrons in silicon and germanium do still experience electron–optical phonon coupling.

NON-POLAR ELECTRON–ACOUSTICAL PHONON COUPLING

Next we switch to electron–acoustical phonon coupling in non-polar semiconductors. Again we are concentrating on electron–phonon coupling at the Γ-point minimum of the conduction band, so always the band indices are $n = m = c$. Following eq. (7.55), $(\Xi_j^{DA})_{ci,cf}$ is now given by

$$(\Xi_j^{DA})_{ci,cf} = \frac{1}{\sqrt{\hbar \omega_j^{q_{if}A}(M_1 + M_2)}} \sum_{\mu,\nu} \hat{p}_{j,\mu}^{q_{if}A} (E_{\mu\nu}^A)_{cc} k_{q_{if}\nu} \tag{7.111}$$

where we note that j takes three values, corresponding to the longitudinal acoustical (LA) and the two transverse acoustical (TA) phonon modes, respectively. We also recall that according to eq. (6.63), $\hat{p}_{j,\mu}^{q_{if}A}$ is the μ-component of the unit polarization vector of the jth phonon mode with wave vector $\mathbf{k}_q = \mathbf{k}_{q_{if}} = \mathbf{k}_f - \mathbf{k}_i$ and frequency $\omega_j^{q_{if}A}$. As furthermore defined in eq. (7.37),

$$(E_{\mu\nu}^A)_{cc} = \langle c0 | e_{\mu\nu}^A(\mathbf{r}) | c0 \rangle \tag{7.112}$$

7.2 ELECTRON–PHONON COUPLING NEAR THE Γ-POINT

which is independent of any wave vector, including those of the initial and final Bloch functions. Hence, $(\Xi_j^{DA})_{ci,cf}$ depends linearly on the phonon wave vector $\mathbf{k}_q$ and vanishes when this wave vector approaches zero. Now, precisely because $(\Xi_j^{DA})_{ci,cf}$ vanishes in the limit $\mathbf{k}_q = \mathbf{k}_{q_{if}} = \mathbf{k}_f - \mathbf{k}_i \to 0$, the absence of a Jahn–Teller effect does not exclude finite values for the diagonal elements. Hence,

$$(E^A_{\mu\nu})_{cc} \neq 0 \tag{7.113}$$

Note furthermore that the pre-factor in eq. (7.111) has dimension $[\hbar k]^{-1}$, while the components of the unit polarization vector are dimensionless, so $(E^A_{\mu\nu})_{cc}$ has the dimension of energy.

For the conduction band minimum at the Γ-point we need to determine nine quantities $(E^A_{\mu\nu})_{cc}$. Fortunately, the cubic symmetry of the crystal strongly reduces the number of them which is non-zero. Group theory would provide the most elegant method to obtain this reduction [4], but as noted earlier, such a treatment is beyond the scope of this book. Here we simply evoke that the quantities $(E^A_{\mu\nu})_{cc}$ constitute a tensor of rank 2. We refer to the discussion in section 4.2.6 showing that such a tensor should obey tetrahedral symmetry. According to appendix D, it then has one independent non-zero component only,

$$(E^A_{\mu\nu})_{cc} = E\delta_{\mu,\nu} \tag{7.114}$$

So, just one constant E determines the electron–acoustical phonon coupling at the conduction band minimum at the Γ-point.

Expression (7.114) allows us to evaluate eq. (7.111) for electron–acoustical phonon coupling at the conduction band minimum at the Γ-point. We find

$$(\Xi_j^{DA})_{ci,cf} = \frac{1}{\sqrt{\hbar\omega_j^{q_{if} A}(M_1 + M_2)}} \sum_{\mu,\nu} \hat{p}_{j,\mu}^{q_{if} A} k_{q_{if},\nu} E \delta_{\mu,\nu}$$

$$= \frac{1}{\sqrt{\hbar\omega_j^{q_{if} A}(M_1 + M_2)}} \sum_{\mu} \hat{p}_{j,\mu}^{q_{if} A} k_{q_{if},\mu} E \tag{7.115}$$

We now recall that in eq. (7.115) the quantities $\hat{p}_{j,\mu}^{q_{if} A}$ are the components of the unit polarization vector of the jth phonon mode with wave vector $\mathbf{k}_{q_{if}}$. Hence, the sum over μ in eq. (7.115) represents the projection of this polarization vector on this phonon wave vector. Clearly, this projection equals zero for transverse phonon modes, while it is equal to one for the longitudinal phonon mode. Hence, only *longitudinal* phonons contribute to electron–acoustical phonon coupling at the Γ-point minimum of the conduction band. So, using eq. (7.94),

$$(\Xi^{DTA})_{ci,cf} = 0 \tag{7.116}$$

$$(\Xi^{DLA})_{ci,cf} = \frac{1}{\sqrt{\hbar\omega^{q_{if} LA}(M_1 + M_2)}} |\mathbf{k}_{q_{if}}| E \tag{7.117}$$

ELECTRON–PHONON COUPLING IN POLAR SEMICONDUCTORS

Near the conduction band minimum at the Γ-point, polar electron–optical phonon coupling and piezoelectric electron–phonon coupling are simply given by eqs (7.96) and (7.107). So,

$$(\Xi^{PLO})_{ci,cf} = i \frac{ee^*}{\varepsilon_0 V_C} \sqrt{\frac{M_1 + M_2}{\hbar \omega^{q_{if} LO} M_1 M_2}} \frac{1}{|\mathbf{k}_{q_{if}}|} \quad (7.118)$$

and

$$(\Xi_j^{PA})_{ci,cf} = i \frac{2ee^{pe}}{\varepsilon_0 \varepsilon_r} \frac{1}{\sqrt{\hbar \omega_j^{q_{if} A}(M_1 + M_2)}} \sum_\mu \left(\hat{p}_{\mu,j}^{q_{if} A} \frac{k_{q_{if}(\mu+1)} k_{q_{if}(\mu+2)}}{|\mathbf{k}_{q_{if}}|^2} \right) \quad (7.119)$$

7.2.2 HOLE–PHONON COUPLING

Next, we consider the valence band maximum at the Γ-point. Here the situation is considerably more complicated than the one treated above. Now, because of degeneracy at the Γ-point itself, all eigenstates $|nk\rangle$ are linear combinations

$$|v\mu 0\rangle = \sum_\mu |v\mu 0\rangle \langle v\mu 0 | nk \rangle \quad (7.120)$$

of the basis states $|vx0\rangle$, $|vy0\rangle$ and $|vz0\rangle$ at $\mathbf{k} = 0$. The proportionality constants in these linear combinations are the components $\langle v\mu 0 | nk \rangle$ of the eigenvectors of the matrix consisting of the sum of the matrices (4.72) and (4.105). Here $n = hh$, lh or λh, denote the heavy holes, the light holes and the spin-orbit band respectively. We remember that section 4.2.5 presents an analytical method to solve these eigenvectors. We note that even in lowest order these eigenvectors depend on $\mathbf{k}$ and also that each $\langle v\mu 0 | nk \rangle$ is actually a two-dimensional vector itself with components $\langle v\mu 0+ | nk \rangle$ and $\langle v\mu 0- | nk \rangle$, on the basis of spin states.

NON-POLAR HOLE–OPTICAL PHONON COUPLING

We handle the situation for non-polar hole–optical phonon coupling by rewriting eq. (7.54) as

$$(\Xi_j^{DO})_{ni,mf} = \sqrt{\frac{M_1 + M_2}{\hbar \omega_j^{q_{if} O} M_1 M_2}} \sum_\mu \hat{p}_{j,\mu}^{q_{if} O} (D_\mu^O)_{ni,mf} \quad (7.121)$$

where

$$(D_\mu^O)_{ni,mf} = \sum_{\nu,\rho} \langle nk_i | v\nu 0 \rangle (D_\mu^O)_{v\nu,v\rho} \langle v\rho 0 | mk_f \rangle \quad (7.122)$$

The situation is more complicated than in eq. (7.54). In eq. (7.122) the 27 quantities $(D_\mu^O)_{v\nu,v\mu}$ are still independent of $\mathbf{k}_i$ and $\mathbf{k}_f$. However, $(D_\mu^O)_{ni,mf}$ occurring in eq. (7.121) now depends on the initial and final wave vectors $\mathbf{k}_i$ and $\mathbf{k}_f$, because the coefficients $\langle nk_i | v\nu 0 \rangle$ and $\langle v\rho 0 | mk_f \rangle$ depend on these wave vectors

$$(D_\mu^O)_{v\nu,v\rho} = \langle v\nu 0 | d_\mu^{00}(\mathbf{r}) | v\rho 0 \rangle \quad (7.123)$$

where μ, ν and ρ may take the values x, y or z. We recall the above discussion of electron–optical phonon coupling in non-polar semiconductors, where we argued that the absence of a Jahn–Teller effect should render diagonal elements like

7.2 ELECTRON–PHONON COUPLING NEAR THE Γ-POINT

$(D_\mu^O)_{vv,vv} = 0$. However, this does not imply that also hole–optical phonon coupling vanishes in lowest order at the Γ-point. The reason is that the other 18 non-diagonal elements remain finite.

A reduction of the number of independent values of $(D_\mu^O)_{vv,v\rho}$ is obtained by considering the cubic symmetry of the crystal. As in the more refined treatment of the valence band structure presented in section 4.2.6, this is best done using group theory [4], but as remarked there, such an approach is beyond the scope of this book. As in section 4.2.6, we just note that the quantities $(D_\mu^O)_{vv,v\rho}$ form a tensor, this time of rank 3. We refer to the discussion in section 4.2.6 where we argued that such a tensor should obey tetrahedral symmetry. As discussed in appendix D, it then has one independent non-zero component only, which we denote by

$$D = (D_z^O)_{vx,vy} = (D_y^O)_{vz,vx} = (D_x^O)_{vy,vz}$$
$$= (D_z^O)_{vy,vx} = (D_y^O)_{vx,vz} = (D_x^O)_{vz,vy} \qquad (7.124)$$

It is interesting to recall that these are the same relations as those valid for the piezoelectric tensor. Thus, one may write $(D_\mu^O)_{vv,v\rho}$ in matrix form on the basis states $|vx0\rangle$, $|vy0\rangle$ and $|vz0\rangle$ as,

$$D \begin{pmatrix} 0 & \delta_{z\mu} & \delta_{y\mu} \\ \delta_{z\mu} & 0 & \delta_{x\mu} \\ \delta_{y\mu} & \delta_{x\mu} & 0 \end{pmatrix} \qquad (7.125)$$

So only one independent parameter, the optical deformation potential constant D, determines the hole–optical phonon coupling in non-polar semiconductors. Table 7.4 gives the value of

$$d_0 = \sqrt{\frac{2}{3}} \frac{c_0 D}{e}$$

for the two non-polar semiconductors silicon and germanium, as collected in ref. [32]. Here c_0 is the length of the edge of the cubic unit cell as shown in Figs 1.2 and 1.3 and listed in Table 2.2, while e is the elementary charge, so d_0 is given in units of eV.

Now the tensor elements $(D_\mu^O)_{vv,v\rho}$ are known, we insert them in eq. (7.121). Writing the result using a matrix on the basis states $|vx0\rangle$, $|vy0\rangle$ and $|vz0\rangle$, we find,

$$(\Xi_j^{DO})_{ni,mf} = \sqrt{\frac{M_1 + M_2}{\hbar \omega_j^{q_{if}O} M_1 M_2}}$$

$$\times (\langle nk_i|vx0\rangle \ \langle nk_i|vy0\rangle \ \langle nk_i|vz0\rangle) \begin{pmatrix} 0 & \hat{p}_{j,z}^{q_{if}O} & \hat{p}_{j,y}^{q_{if}O} \\ \hat{p}_{j,z}^{q_{if}O} & 0 & \hat{p}_{j,x}^{q_{if}O} \\ \hat{p}_{j,y}^{q_{if}O} & \hat{p}_{j,x}^{q_{if}O} & 0 \end{pmatrix} \begin{pmatrix} \langle vx0|mk_f\rangle \\ \langle vy0|mk_f\rangle \\ \langle vz0|mk_f\rangle \end{pmatrix}$$

$$(7.126)$$

Table 7.4 Deformation potentials for non-polar semiconductors in eV.

	d_0	l/e	m/e	n/e
Si	29.3	−2.3	4.3	9.2
Ge	40.0	−2.2	4.1	12.1

NON-POLAR HOLE–ACOUSTICAL PHONON COUPLING

Next, we consider non-polar hole–acoustical phonon coupling in the valence band maximum at the Γ-point. Equivalently to the case non-polar hole–optical phonon coupling, we write eq. (7.55) as

$$(\Xi_j^{DA})_{ni,mf} = \frac{1}{\sqrt{\hbar\omega_j^{q_{if}}(M_1+M_2)}} \sum_{\mu,\nu} \hat{p}_{j,\mu}^{q_{if}A} (E_{\mu\nu}^A)_{ni,mf} k_{q_{if}\nu} \qquad (7.127)$$

where we introduce 81 quantities,

$$(E_{\mu\nu}^A)_{ni,mf} = \sum_{\sigma,\rho} \langle nk_i | v\sigma 0\rangle (E_{\mu\nu}^A)_{v\sigma,v\rho} \langle v\rho 0 | mk_f\rangle \qquad (7.128)$$

Here μ, ν, σ and ρ may take the values x, y or z. Fortunately, using the cubic symmetry of the crystal, one may again restrict the number of independent parameters that is involved [4]. As earlier we note that the quantities $(E_{\mu\nu}^A)_{v\sigma,v\rho}$ form a tensor, now of rank 4, which should obey tetrahedral symmetry. According to appendix D, such a tensor has three independent non-zero components only, here to be denoted as

$$l = (E_{xx}^A)_{vx,vx} = (E_{yy}^A)_{vy,vy} = (E_{zz}^A)_{vz,vz} \qquad (7.129)$$

$$m = (E_{xx}^A)_{vy,vy} = (E_{yy}^A)_{vz,vz} = (E_{zz}^A)_{vx,vx} \qquad (7.130)$$

$$n = (E_{xy}^A)_{vx,vy} = (E_{yz}^A)_{vy,vz} = (E_{zx}^A)_{vz,vx}$$
$$= (E_{yx}^A)_{vy,vx} = (E_{zy}^A)_{vz,vy} = (E_{xz}^A)_{vx,vz} \qquad (7.131)$$

Hence, on the basis states $|vx0\rangle$, $|vy0\rangle$ and $|vz0\rangle$, one obtains for $(E_{\mu\nu}^A)_{v\sigma,v\rho}$,

$$m(\delta_{x\mu}\delta_{x\nu} + \delta_{y\mu}\delta_{y\nu} + \delta_{z\mu}\delta_{z\nu}) \begin{pmatrix} 1 & 0 & 0 \\ 0 & 1 & 0 \\ 0 & 0 & 1 \end{pmatrix}$$

$$+ \begin{pmatrix} (l-m)\delta_{x\mu}\delta_{x\nu} & \frac{n}{2}(\delta_{x\mu}\delta_{y\nu} + \delta_{y\mu}\delta_{x\nu}) & \frac{n}{2}(\delta_{z\mu}\delta_{x\nu} + \delta_{x\mu}\delta_{z\nu}) \\ \frac{n}{2}(\delta_{x\mu}\delta_{y\nu} + \delta_{y\mu}\delta_{x\nu}) & (l-m)\delta_{y\mu}\delta_{y\nu} & \frac{n}{2}(\delta_{y\mu}\delta_{z\nu} + \delta_{z\mu}\delta_{y\nu}) \\ \frac{n}{2}(\delta_{z\mu}\delta_{x\nu} + \delta_{x\mu}\delta_{z\nu}) & \frac{n}{2}(\delta_{y\mu}\delta_{z\nu} + \delta_{z\mu}\delta_{y\nu}) & (l-m)\delta_{z\mu}\delta_{z\nu} \end{pmatrix} \qquad (7.132)$$

Table 7.4 lists the parameters l/e, m/e and n/e for the two non-polar semiconductors silicon and germanium as collected in ref. [32]. As these parameters are divided by the elementary charge e, they are given in units of eV.

The thus obtained expression (7.132) for the tensor elements $(E_{\mu\nu}^A)_{v\sigma,v\rho}$ allows us to evaluate eq. (7.127) and write it as a matrix on the basis states $|vx0\rangle$, $|vy0\rangle$ and $|vz0\rangle$. Though the exercise is straightforward, the result is dreadful and therefore it is left for the dedicated reader to try.

POLAR HOLE–PHONON COUPLING

Finally it would be interesting to consider hole–optical phonon and hole–acoustical phonon scattering near the Γ-point of polar crystals. Unfortunately, this does not amount to an almost trivial exercise as we found it to be in the case of electron–phonon coupling near the Γ-point in these crystals. The reason is that our results (7.96) and (7.107) were obtained using effective mass theory for non-degenerate bands. Thus, we cannot apply these results immediately to the degenerate bands at the top of the valence band. For further reading the reader is referred to ref. [31].

8 Charge Transport

8.1 Quasi-Classical Motion

8.1.1 Elements of Charge Transport

CARRIER MOTION AND ELECTRICAL CURRENTS

Virtually all applications of semiconductors rely in some way or another on transport of charge carriers through these materials. Thus, when designing semiconductor devices it is of primordial importance to be able to predict the electrical current density in the material as a function of the strength of the electromagnetic field. It is not trivial to do such a prediction theoretically. Microscopically, electrical currents consist of moving electrons in the conduction band and holes in the valence band. This motion is highly complicated, because electrons and holes are not only accelerated by externally applied fields, but also by internal fields due to space charge effects. Moreover, they are slowed down by a variety of processes, among others due to carrier–phonon coupling and to carrier–carrier and carrier–impurity interactions. Clearly, motion differs for each of the carriers and the total current density $\mathbf{J}$ must be obtained by averaging the velocities $\mathbf{v}_e$ and $\mathbf{v}_h$ of the individual electrons and holes,

$$\mathbf{J} = -eN_e\bar{\mathbf{v}}_e + eN_h\bar{\mathbf{v}}_h \tag{8.1}$$

where $e = 1.602 \cdot 10^{-19}$ C is the elementary charge, N_e the density of electrons and N_h the density of holes.

In weak electric fields $\mathbf{E}$ the average velocities $\bar{\mathbf{v}}_e$ and $\bar{\mathbf{v}}_h$ and hence the current density $\mathbf{J}$ depend linearly on the field strength. Then one may define a conductivity σ and electron and hole mobilities μ_e and μ_h such that,

$$\mathbf{J} = \sigma\mathbf{E} \tag{8.2}$$

$$\bar{\mathbf{v}}_e = \mu_e\mathbf{E} \tag{8.3}$$

$$\bar{\mathbf{v}}_h = \mu_h\mathbf{E} \tag{8.4}$$

Thus, to be able to predict the electrical current density, one needs to know the densities N_e and N_h of electrons and holes and their respective mobilities μ_e and μ_h. In practice the work to be done consists of three steps. First, one needs to determine and solve the equations of motion of the charge carriers and the rates of the various transitions due to their interactions with phonons, impurities, etc. Next, one has to implement an averaging procedure in order to obtain the mean carrier velocities $\bar{v}_e$ and $\bar{v}_h$. Finally, one needs to invoke still more statistics to determine the carrier densities N_e and N_h.

A MODEL FOR CHARGE TRANSPORT

The effective mass formalism derived in chapter 5 is particularly powerful when performing such a study. In many common situations externally applied fields, carrier–phonon coupling and carrier–impurity coupling are sufficiently weak, so their effects may be treated in lowest order. This allows the assumption that these interactions do not mix. Then, electrons and holes are considered to move in free flight through the crystal, while this free flight is interrupted at irregular intervals by scattering events, i.e. sudden changes of state due to electron–phonon or electron–impurity coupling. During free flight the effective mass formalism applies. As we will see below, scattering at impurities can also be very well described using an effective mass formalism. Moreover, as described in the previous chapter, deformation potential electron–phonon coupling can be treated within an effective mass formalism as well, provided that we introduce effective deformation potentials. Finally, polar carrier–optical phonon coupling and piezoelectric carrier–acoustical phonon coupling are basically described using an effective mass formalism. Thus, all aspects of charge transport can be treated using this formalism.

When the size of the wave packet is much smaller than the distance covered by the charge carriers during their free flight, further approximations can be made. Then, not only the effective mass formalism applies, but we may take its classical limit as well[1]. The resulting description of charge carrier motion is called quasi-classical: free flight is treated classically, while scattering due to electron–lattice coupling and at impurities is treated quantum mechanically.

SCOPE OF THE PRESENT CHAPTER

In the present chapter we will concentrate on the first step of the procedure leading to predictions of the electrical current density **J** as a function of the electrical field strength **E**. In particular we will refrain from obtaining analytical expressions for the mobility, i.e. the average velocity of charge carriers per unit of electric field. The reason is threefold. First, various textbooks handle this problem extensively (see refs [32] and [35]). Second, present computer simulation techniques allow a much more precise calculation of the average velocity, than can be obtained analytically [36]. Finally, the necessary statistical treatment is beyond the scope of this book. For a treatment of statistics of semiconductors the reader is referred to ref. [5] and for predictions of current densities in actual devices to the book of Sze [38] which is still a leading reference on the subject.

The aim of this chapter is to provide the raw material needed to set up a so-called Monte Carlo simulation of carrier transport in semiconductors. In such Monte Carlo

---

[1] Ignoring interference effects that may occur on a larger space scale, as long as the wave function maintains its phase.

programs one simulates the motion of a single carrier or of an ensemble of carriers. The motion during free flight is calculated classically, while the probability of transitions due to scattering are determined using the relevant transition matrix elements. The initial conditions, the length of free flights, the choice of scattering processes and the state after such a scattering are all chosen stochastically, i.e. at 'random', but weighed with the appropriate transition probabilities. While the simulation continues the requested quantities are sampled. Finally after a large number of scattering events the process is halted and averages of the sampled quantities are calculated. Thus, average carrier velocity, average energy, distribution functions for energy and momentum, etc., can be obtained without the need to solve complicated analytical transport equations. Clearly, the raw material needed to set up such a simulation program consists of the equations of motion and the rates of the various scattering processes. These are the subjects that will be treated in the sections below.

Apart from avoiding the need to solve complicated analytical transport equations, Monte Carlo simulations have a further advantage. In present day microelectronic circuits electric fields may be very high and the average velocities $\bar{v}_e$ and $\bar{v}_h$ and hence the current density $\mathbf{J}$ do not depend linearly on the applied electric field. Now, Monte Carlo simulations are very well suited to handle this problem because they yield results as easily in the non-linear regime as in the linear regime, contrary to analytical methods.

8.1.2 The Classical Limit

CLASSICAL EQUATIONS OF MOTION

We start considering the equations of motion needed to describe charge transport. Therefore, in this section we will derive the classical motion of a band electron during free flight in external electric and magnetic fields that are constant in space. Then, the vector potential can be chosen to obey eq. (5.68),

$$\mathbf{A}(\mathbf{r}, t) = \frac{1}{2} [\mathbf{r} \times \mathbf{B}(t)] \tag{8.5}$$

Moreover, we assume the effective mass tensor to be symmetric, so $m^*_{\mu\nu} = m^*_{\nu\mu}$. Then the effective Hamiltonian (5.64) is given by,

$$\mathcal{H}_{eff} = \frac{1}{2m_e} \sum_{\mu,\nu} \frac{\left[\frac{\hbar}{i}\nabla_\mu + e\bar{A}_\mu(\mathbf{r},t)\right]\left[\frac{\hbar}{i}\nabla_\nu + e\bar{A}_\nu(\mathbf{r},t)\right]}{m^*_{\mu\nu}} - e\bar{U}(\mathbf{r},t) \tag{8.6}$$

where we skip the constant term E_0^n, while according to eq. (5.74) the term (5.65) vanishes. Following appendix E, this free flight is then governed by the classical equations of motion (E.26) and (E.44),

$$v_\mu = \frac{d}{dt}\langle r_\mu\rangle = \frac{1}{m_e}\sum_\nu \frac{1}{m^*_{\mu\nu}}\left[\left\langle\frac{\hbar}{i}\nabla_\nu\right\rangle + e\bar{A}_\nu(\langle\mathbf{r}\rangle,t)\right] \tag{8.7}$$

$$\frac{\partial}{\partial t}\left[\left\langle\frac{\hbar}{i}\nabla_\mu\right\rangle + e\bar{A}_\mu(\langle\mathbf{r}\rangle,t)\right] = -e[\mathbf{v}\times\mathbf{B}(t)]_\mu - e\bar{E}_\mu(\langle\mathbf{r}\rangle,t) \tag{8.8}$$

8.1 QUASI-CLASSICAL MOTION

Here, we inserted the charge $q = -e$ of an electron and eq. (E.3) for $\mathcal{P}_\mu$. Furthermore, as discussed in section 5.2.2, the bars over $\bar{\mathbf{A}}(\langle\mathbf{r}\rangle, t)$, $\bar{U}(\langle\mathbf{r}\rangle, t)$ and $\bar{\mathbf{E}}(\langle\mathbf{r}\rangle, t)$ indicate weighed averages over primitive cells and correspond to the macroscopic values of these potentials and fields. Furthermore, as in appendix E we assume these quantities to vary little over the wave packets. Thus, we write them as functions of the expectation value $\langle\mathbf{r}\rangle$ of the position of the wave packet rather than $\mathbf{r}$. As we use the effective mass formalism, the expectation values in eqs (8.7) and (8.8) have to be calculated using the envelope function $F^n(\mathbf{r}, t)$, so

$$\langle r_\mu \rangle = \int_\infty d\mathbf{r}\, F^{n*}(\mathbf{r}, t) r_\mu F^n(\mathbf{r}, t) \tag{8.9}$$

$$\left\langle \frac{\hbar}{i}\nabla_\nu \right\rangle = \int_\infty d\mathbf{r}\, F^{n*}(\mathbf{r}, t)\left(\frac{\hbar}{i}\nabla_\nu\right) F^n(\mathbf{r}, t) \tag{8.10}$$

Here the integrations need to be taken over infinite normal space, because the envelope functions $F^n(\mathbf{r}, t)$ are normalized in infinite normal space.

EVALUATION OF THE EXPECTATION VALUES

Clearly, $\langle\mathbf{r}\rangle$ represents the classical position of the electron in normal space. To interpret $\langle(\hbar/i)\nabla_\nu\rangle$, we need to be aware that the wave function $F^n(\mathbf{r}, t)$ depends on the gauge chosen for the potentials over $\bar{\mathbf{A}}(\langle\mathbf{r}\rangle, t)$ and $\bar{U}(\langle\mathbf{r}\rangle, t)$. Without taking proper precautions, this may lead to a gauge dependence of measurable quantities like $\langle(\hbar/i)\nabla_\nu\rangle$, and hence to physically incorrect results. It is beyond the scope of this book to discuss methods to avoid such undesired gauge dependencies of measurable quantities, as this subject belongs to the realm of quantum field theory. For our present purposes it is sufficient to follow a pragmatic procedure. We choose the gauge such that the vector potential is zero and remains zero at the centre of the wave packet,

$$\bar{\mathbf{A}}(\langle\mathbf{r}\rangle, t) = 0 \tag{8.11}$$

which can be achieved by replacing eq. (8.5) by

$$\mathbf{A}(\mathbf{r}, t) = \frac{1}{2}[(\mathbf{r} - \langle\mathbf{r}\rangle) \times \mathbf{B}(t)] \tag{8.12}$$

and we propose that with this particular choice, gauge-dependent terms in the calculation of $\langle(\hbar/i)\nabla_\nu\rangle$ vanish. At the end of the calculation we will check whether our assumption did indeed yield a physically correct result, i.e. that the resulting expectation values are gauge independent. For a more extended discussion of this problem the reader is referred to [34]. Thus, we evaluate $\langle(\hbar/i)\nabla_\nu\rangle$ using eq. (5.33) for $F^n(\mathbf{r}, t)$ and eq. (8.11) for $\bar{\mathbf{A}}(\langle\mathbf{r}\rangle, t)$. Then,

$$\begin{aligned}\left\langle \frac{\hbar}{i}\nabla_\nu \right\rangle &= \int_\infty d\mathbf{r}\, F^{n*}(\mathbf{r}, t)\left(\frac{\hbar}{i}\nabla_\nu\right) F^n(\mathbf{r}, t) \\ &= \int_\infty d\mathbf{r}\left[\frac{1}{(2\pi)^{3/2}}\int_{V_B} d\mathbf{k}'\, e^{-i\mathbf{k}'\cdot\mathbf{r}}\widetilde{G}^{n*}(\mathbf{k}', t)\right]\left(\frac{\hbar}{i}\nabla_\nu\right) \\ &\quad \times \left[\frac{1}{(2\pi)^{3/2}}\int_{V_B} d\mathbf{k}\, e^{i\mathbf{k}\cdot\mathbf{r}}\widetilde{G}^n(\mathbf{k}, t)\right]\end{aligned} \tag{8.13}$$

We recall that

$$\left(\frac{\hbar}{i}\nabla_\nu\right)e^{i\mathbf{k}\cdot\mathbf{r}} = e^{i\mathbf{k}\cdot\mathbf{r}}(\hbar k_\nu)$$

so,

$$\left\langle\frac{\hbar}{i}\nabla_\nu\right\rangle = \int_\infty d\mathbf{r}\left[\frac{1}{(2\pi)^{3/2}}\int_{V_B} d\mathbf{k}'\, e^{-i\mathbf{k}'\cdot\mathbf{r}}\tilde{G}^{n*}(\mathbf{k}',t)\right] \times \left[\frac{1}{(2\pi)^{3/2}}\int_{V_B} d\mathbf{k}\, e^{i\mathbf{k}\cdot\mathbf{r}}\hbar k_\nu \tilde{G}^n(\mathbf{k},t)\right]$$

$$= \int_{V_B} d\mathbf{k}' \int_{V_B} d\mathbf{k}\left[\frac{1}{(2\pi)^3}\int_\infty d\mathbf{r}\, e^{-i(\mathbf{k}'-\mathbf{k})\cdot\mathbf{r}}\right]\tilde{G}^{n*}(\mathbf{k}',t)\,\hbar k_\nu\,\tilde{G}^n(\mathbf{k},t)$$

$$= \int_{V_B} d\mathbf{k}' \int_{V_B} d\mathbf{k}\,\delta(\mathbf{k}'-\mathbf{k})\,\tilde{G}^{n*}(\mathbf{k}',t)\,\hbar k_\nu\,\tilde{G}^n(\mathbf{k},t)$$

$$= \int_{V_B} d\mathbf{k}\,\tilde{G}^{n*}(\mathbf{k},t)\,\hbar k_\nu\,\tilde{G}^n(\mathbf{k},t) = \hbar\langle k_\nu\rangle \quad (8.14)$$

where we use the Dirac δ-function (3.22). At this point we recall that in section 5.2.1 we defined $\tilde{G}^n(\mathbf{k},t)$ to be equal to zero outside the first Brillouin zone. Hence, the integrations over $\mathbf{k}$ and $\mathbf{k}'$ in eqs (8.13) and (8.14) could just as well have been performed over infinite space. This is fortunate because, assuming the envelope function $F^n(\mathbf{r},t)$ to be normalized in infinite normal space, the normalization convention of eq. (3.24) yields $\tilde{G}^n(\mathbf{k},t)$ to be normalized in infinite reciprocal space. Therefore, eq. (8.14) represents a correctly calculated expectation value of crystal momentum $\langle\hbar\mathbf{k}\rangle$ in reciprocal space.

Thus, using the envelope function in normal space to calculate the expectation value of the momentum operator $\langle(\hbar/i)\nabla\rangle$, yields the same result as calculating the expectation value of crystal momentum $\langle\hbar\mathbf{k}\rangle$ with the envelope function in reciprocal space. Note that this result confirms the physical meaning of crystal momentum as the momentum of the electron in the effective mass approximation.

As a result, the equations of motion (8.7) and (8.8) can be written as

$$v_\mu = \frac{d}{dt}\langle r_\mu\rangle = \frac{1}{m_e}\sum_\nu\frac{\hbar\langle k_\nu\rangle}{m^*_{\mu\nu}} \quad (8.15)$$

$$\frac{\partial}{\partial t}\hbar\langle k_\mu\rangle = -e[\mathbf{v}\times\mathbf{B}(t)]_\mu - e\bar{E}_\mu[\langle\mathbf{r}\rangle,t] \quad (8.16)$$

We see that we have indeed obtained a physically correct result. Had we taken $\bar{A}(\langle\mathbf{r}\rangle,t)\neq 0$, then the velocity of the wave packet would have been found to depend on $\mathbf{A}(\langle\mathbf{r}\rangle,t)$. This would have been physically unacceptable, because this potential can be changed at will by gauge transformations.

We note that the velocity of a wave packet can be rewritten as

$$v_\mu = \frac{1}{2m_e}\sum_{\nu,\rho}\frac{\partial}{\partial\hbar\langle k_\mu\rangle}\frac{\hbar\langle k_\nu\rangle\hbar\langle k_\rho\rangle}{m^*_{\nu\rho}} = \frac{\partial\langle E^n_k\rangle}{\partial\hbar\langle k_\mu\rangle} \quad (8.17)$$

So the velocity of a wave packet is given by the slope of the band energy in reciprocal space. This latter result corresponds to the classical Hamilton equation

$$v_\mu = \frac{\partial H}{\partial p_\mu} \quad (8.18)$$

provided that we identify the band energy $\langle E_k^n \rangle$ with the classical Hamiltonian H and crystal momentum $\hbar \langle k_\mu \rangle$ with classical momentum p_μ.

Our final result (8.17) has a more general validity than suggested by the derivation given in the present section. Here, this result was obtained from eq. (8.7) which only holds for band extrema where the energy E_k^n can be approximated to depend quadratically on k_μ. Using a different approach from the one given in the present chapter [34], it can be proven that eq. (8.17) also holds when the band energy has a much more complicated dependence on the **k**-vector. Thus, eqs (8.16) and (8.17) may be used to describe the classical limit of the motion of band electrons for arbitrary bands. This observation is particularly useful for holes in the valence band where, according to eqs (4.106)–(4.112), the valence band energy E_k^n depends in a complicated manner on the individual components k_μ of the position **k** in reciprocal space.

FULL BANDS AND TRANSPORT

Eq. (8.17) has a further consequence which is vital for the understanding of charge transport in semiconductors: it can be used to prove that a full band does not contribute to such charge transport. For this purpose we first show that the band energy E_k^n obeys inversion symmetry with respect to $\mathbf{k} = 0$, so the energy E_k^n of an electron in the nth band at position **k** in reciprocal space is equal to the energy E_{-k}^n of an electron in the same band, but at position −**k** in reciprocal space.

To prove inversion symmetry of the band energy we take into account that—neglecting the electron spin—this band energy represents the eigenvalues of the real and Hermitian single electron Hamiltonian $\mathcal{H}'(\mathbf{r})$ given by eq. (3.60). We use the lemma proven in section 4.1.2, eqs (4.12) to (4.15). For our purpose we formulate it as follows: if the Bloch state $b_k^n(\mathbf{r})$ is a complex eigenstate of the real and Hermitian single electron Hamiltonian $\mathcal{H}'(\mathbf{r})$, then its complex conjugate $b_k^{n*}(\mathbf{r})$ is also an eigenstate of $\mathcal{H}'(\mathbf{r})$ and moreover $b_k^n(\mathbf{r})$ and $b_k^{n*}(\mathbf{r})$ are degenerate. Now,

$$b_k^n(\mathbf{r}) = e^{i\mathbf{k}\cdot\mathbf{r}} u_k^n(\mathbf{r}) \tag{8.19}$$

$$b_k^{n*}(\mathbf{r}) = e^{-i\mathbf{k}\cdot\mathbf{r}} u_k^{n*}(\mathbf{r}) \tag{8.20}$$

Now, again neglecting spin, within a single band there is one solution for each value of **k** only, so if $b_k^n(\mathbf{r})$ is the Bloch state at +**k**, $b_k^{n*}(\mathbf{r})$ is necessarily its counterpart at −**k**. Moreover, both states are degenerate, proving inversion symmetry[2].

Now inversion symmetry of the band energy is proven, it is easily seen that the velocity of an electron in the nth band at position **k** in reciprocal space is exactly opposite to the velocity of an electron in the same band, but at position −**k** in reciprocal space. According to eq. (8.17),

$$v_\mu(-\mathbf{k}) = \left(\frac{\partial \langle E_k^n \rangle}{\partial \hbar \langle k_\mu \rangle}\right)_{-\mathbf{k}} = -\left(\frac{\partial \langle E_k^n \rangle}{\partial \hbar \langle k_\mu \rangle}\right)_{\mathbf{k}} = -v_\mu(\mathbf{k}) \tag{8.21}$$

---

[2] When the electron spin is included, but spin-orbit coupling neglected, the total number of states just doubles, without changing the argument. When also spin-orbit coupling is taken into account, the argument becomes more complicated but does not break down [4].

so the centre of mass velocity of the two electrons is equal to zero. Now, in a full band all values of **k** are occupied, so for each electron at position $+\mathbf{k}$, there is also an electron at position $-\mathbf{k}$. Hence, the centre of mass velocity of all electrons in the band is zero, i.e. these electrons are collectively immobilized so they cannot contribute to charge transport. As a result only partly filled bands provide charge transport, i.e. only electrons in the conduction band or holes in the valence band. As either of the two types of charge carriers occurs in a very low concentration only, charge transport is due to a very small proportion of the band electrons in semiconductors.

8.2 Carrier Scattering
8.2.1 Phonon Scattering Processes
INITIAL AND FINAL STATES

In quasi-classical motion a band electron is considered to move through the crystal in a classical orbit which is abruptly changed at irregular intervals due to scattering processes. In the present and following sections we consider such scatterings due to electron–phonon interaction. For simplicity we will only consider processes where the electron remains within the same band. Thus we exclude scattering of an electron from its current band to another band. The extension to include these latter transitions is not very difficult, however, and the reader is invited to try it as an exercise. In any case this extension is required in the case of the valence band where phonons may induce transitions between the heavy holes, light holes and spin-orbit bands.

As we saw above, classical motion needs finite size wave packets in normal space as well as in reciprocal space, otherwise classical equations of motion cannot be derived. When scattering is due to electron–phonon interaction this requirement can be relaxed for the scattering processes themselves. The reason is that these processes depend on the position of the electron in reciprocal space only, so the extension of the wave function in normal space is irrelevant. As a result, for the description of these scattering processes it is sufficient to approximate the envelope function in reciprocal space by a δ-function, so the envelope function in normal space reduces to a plane wave. As a result, in effective mass theory a scattering process is described as a transition from one plane wave,

$$e^{i\mathbf{k}_i \cdot \mathbf{r}} \tag{8.22}$$

to another plane wave

$$e^{i\mathbf{k}_f \cdot \mathbf{r}} \tag{8.23}$$

where the indices i and f denote the initial and final wave vectors $\mathbf{k}_i$ and $\mathbf{k}_f$. Of course, in such an approximation a full wave function is just a single Bloch function and a scattering process corresponds to a transition from one Bloch state,

$$b_i^n(\mathbf{r}) = u_i^n(\mathbf{r})e^{i\mathbf{k}_i \cdot \mathbf{r}} \tag{8.24}$$

with an energy E_i^n to another Bloch state

$$b_f^n(\mathbf{r}) = u_f^n(\mathbf{r})e^{i\mathbf{k}_f \cdot \mathbf{r}} \tag{8.25}$$

with an energy E_f^n, while n indicates the band index.

MATRIX ELEMENTS FOR PHONON SCATTERING

The matrix elements for these transitions were derived in the previous chapter. Four different types of electron–phonon coupling were identified: optical deformation potential and acoustical deformation potential electron–phonon coupling in non-polar semiconductors and in polar semiconductors, polar optical phonon and piezoelectric electron–phonon coupling. Following eq. (7.66), for all these types of electron–phonon coupling the matrix elements can be written in the shape

$$(\mathcal{H}_{EL}^{\beta\gamma})_{ni,nf} = \frac{\hbar}{\sqrt{2N}} \sum_j \left[\alpha_1 (\Xi_j^{\beta\gamma})_{ni,nf} \frac{1}{\sqrt{2}} (a_j^{q_{if}\gamma 1} + a_j^{q_{if}\gamma 1\dagger}) \right.$$
$$\left. + \alpha_2 (\Xi_j^{\beta\gamma})_{ni,nf} \frac{1}{\sqrt{2}} (a_j^{q_{if}\gamma 2} + a_j^{q_{if}\gamma 2\dagger}) \right] \quad (8.26)$$

Here the indices i and f denote the initial and final plane waves (8.22) and (8.23), while we retain the index n indicating the band where the electron resides. Furthermore, N is the number of primitive cells in the crystal. The index $\beta = D$ or P denotes the type of coupling, while $\gamma = O$ or A indicates whether the phonon mode is optical or acoustical. Furthermore the index j denotes the polarization of these modes. Also, $\alpha_1 = 1$ and $\alpha_2 = i$ for optical phonon modes, while $\alpha_1 = i$ and $\alpha_2 = -1$ for acoustical phonon modes.

Eq. (8.26) features two terms for each type of phonon coupling. This is due to the fact that an electron transition from Bloch state (8.24) to Bloch state (8.25) may be due to coupling with two phonon modes with wave vector $\mathbf{k}_q = \pm \mathbf{k}_{q_{if}} = \pm(\mathbf{k}_i - \mathbf{k}_f)$. Thus, $a_j^{q_{if}\gamma 1}$, $a_j^{q_{if}\gamma 1\dagger}$, $a_j^{q_{if}\gamma 2}$ and $a_j^{q_{if}\gamma 2\dagger}$ are annihilation and creation operators for phonons corresponding to these two modes. However, we do not need to worry about this complicating factor. Our present interest is the calculation of the rate of electron transitions from one plane wave state (8.22) to another plane wave state (8.23), i.e. from one Bloch state (8.24) to another Bloch state (8.25). Then the arguments of section 7.1.4 hold and it is sufficient to use the more simple effective expression (7.73) for $(\mathcal{H}_{EL}^{\beta\gamma})_{ni,nf}$ instead of the full expression (7.66),

$$(\mathcal{H}_{EL}^{\beta\gamma})_{ni,nf} = \frac{\hbar}{\sqrt{2N}} \sum_j (\Xi_j^{\beta\gamma})_{ni,nf} (a_j^{q_{if}\gamma} + a_j^{q_{if}\gamma\dagger}) \quad (8.27)$$

The coupling constant $(\Xi_j^{\beta\gamma})_{ni,nf}$ differs for the four types of coupling. First, according to eq. (7.54), for optical deformation potential electron–phonon coupling,

$$(\Xi_j^{DO})_{ni,nf} = \sqrt{\frac{M_1 + M_2}{\hbar \omega_j^{q_{if}O} M_1 M_2}} \sum_\mu \hat{p}_{j,\mu}^{q_{if}O} (D_\mu^O)_{nn} \quad (8.28)$$

Here $\hat{p}_{j,\mu}^{q_{if}O}$ is the μ-component of the unit vector $\hat{\mathbf{p}}_j^{q_{if}O}$ in the polarization direction of the jth phonon mode, while j takes three values denoting the longitudinal optical (LO) and the two transverse optical (TO) phonon modes, respectively. Moreover M_1 and M_2 are the masses of the two atoms in a primitive cell, while $\omega_j^{q_{if}O}$ is the frequency of an optical phonon with wave vector $\mathbf{k}_{q_{if}}$. The quantity $(D_\mu^O)_{nn}$ represents the optical deformation potential and determines the strength of the optical deformation potential coupling.

According to eq. (7.55),

$$(\Xi_j^{DA})_{ni,nf} = \frac{1}{\sqrt{\hbar \omega_j^{q_{if}A}(M_1 + M_2)}} \sum_{\mu,\nu} \hat{p}_{j,\mu}^{q_{if}A} (E_{\mu\nu}^A)_{nn} k_{q_{if}\nu} \qquad (8.29)$$

As above, $\hat{p}_{j,\mu}^{q_{if}A}$ is the μ-component of the unit vector $\hat{\mathbf{p}}_j^{q_{if}A}$ in the polarization direction of the jth phonon mode. Note that also here j takes three values denoting the longitudinal acoustical (LA) and the two transverse acoustical (TA) phonon modes. Now $\omega_j^{q_{if}A}$ is the frequency of an acoustical phonon with wave vector $\mathbf{k}_{q_{if}}$. The quantity $(E_{\mu\nu}^A)_{nn}$ represents the acoustical deformation potential and determines the strength of the acoustical deformation potential coupling.

Furthermore, according to eq. (7.96)

$$(\Xi_j^{PO})_{ni,nf} = (\Xi^{PLO})_{ni,nf} = i \frac{ee^*}{\varepsilon_0 V_C} \sqrt{\frac{(M_1 + M_2)}{\hbar \omega^{q_{if}LO} M_1 M_2}} \frac{1}{|\mathbf{k}_{q_{if}}|} \qquad (8.30)$$

where e is the elementary charge, e^* the effective charge in the dipole moment induced by the optical phonon mode, ε_0 the permittivity of the vacuum and V_C the volume of the primitive cell, which for the usual semiconductors corresponds to the volume per induced dipole moment. Note that only longitudinal optical phonons contribute, so j takes a single value only.

Finally, according to eq. (7.107),

$$(\Xi_j^{PA})_{ni,nf} = i \frac{2e\varepsilon^{pe}}{\varepsilon_0 \varepsilon_r} \frac{1}{\sqrt{\hbar \omega_j^{q_{if}A}(M_1 + M_2)}} \sum_{j,\mu} \left(\hat{p}_{\mu,j}^{q_{if}A} \frac{k_{q_{if}(\mu+1)} k_{q_{if}(\mu+2)}}{|\mathbf{k}_{q_{if}}|^2} \right) \qquad (8.31)$$

where ε^{pe} is the only non-zero component of the piezoelectric tensor in semiconductors with the tetragonal zinc blende structure.

8.2.2 Selection Rules

FIRST-ORDER TRANSITION RATES

We now continue by considering the various transitions that may be induced by the electron–phonon coupling. A discussion of these transitions was given earlier in section 7.1.4, but for convenience we repeat it here. The general expression (8.27) for the matrix element of this coupling between two plane waves allows us to obtain a first insight in the character of these transitions. Eq. (8.27) shows the product of a part operating on the electronic wave functions and a part operating on the lattice. First, we consider the latter part. As already discussed in section 2.3.3 the phonon creation and annihilation operators $a_j^{q_{if}\gamma}$ and $a_j^{q_{if}\gamma\dagger}$ have finite matrix elements between neighbouring phonon states only. Thus, using eq. (2.70) the only non-zero matrix elements of $(\mathcal{H}_{EL}^{\beta\gamma})_{ni,nf}$ are

$$\langle n_j^{q_{if}\gamma} | (\mathcal{H}_{EL}^{\beta\gamma})_{ni,nf} | n_j^{q_{if}\gamma} + 1 \rangle = \hbar \sqrt{\frac{n_j^{q_{if}\gamma} + 1}{2N}} (\Xi_j^{\beta\gamma})_{ni,nf} \qquad (8.32)$$

$$\langle n_j^{q_{if}\gamma} | (\mathcal{H}_{EL}^{\beta\gamma})_{ni,nf} | n_j^{q_{if}\gamma} - 1 \rangle = \hbar \sqrt{\frac{n_j^{q_{if}\gamma}}{2N}} (\Xi_j^{\beta\gamma})_{ni,nf} \qquad (8.33)$$

8.2 CARRIER SCATTERING

Hence, in first-order perturbation theory the only allowed transitions involve emission or absorption of a phonon with frequency $\omega_j^{q_{if}\gamma}$ and a wave vector $\mathbf{k}_q = \pm \mathbf{k}_{q_{if}} = \pm(\mathbf{k}_f - \mathbf{k}_i)$. This implies that the crystal momentum lost or gained by the electron is taken up or provided by the phonon.

A second selection rule is obtained by using Fermi's golden rule (C.67) to calculate the probabilities of such phonon emissions and absorptions according to first-order time-dependent perturbation theory. For all four processes and all three polarizations we can write them as

$$(W_{ni,nf})^{em} = \frac{2\pi}{\hbar} |\langle n_j^{q_{if}\gamma} | (\mathcal{H}_{EL}^{\beta\gamma})_{ni,nf} | n_j^{q_{if}\gamma} + 1 \rangle|^2 \delta(E_f^n - E_i^n + \hbar\omega_j^{q_{if}\gamma})$$

$$= \frac{2\pi}{\hbar} \left| \hbar \sqrt{\frac{n_j^{q_{if}\gamma} + 1}{2N}} (\Xi_j^{\beta\gamma})_{ni,nf} \right|^2 \delta(E_f^n - E_i^n + \hbar\omega_j^{q_{if}\gamma})$$

$$= \frac{2\pi}{\hbar} \frac{(n_j^{q_{if}\gamma} + 1)}{2N} |\hbar(\Xi_j^{\beta\gamma})_{ni,nf}|^2 \delta(E_f^n - E_i^n + \hbar\omega_j^{q_{if}\gamma}) \quad (8.34)$$

$$(W_{ni,nf})^{abs} = \frac{2\pi}{\hbar} |\langle n_j^{q_{if}\gamma} | (\mathcal{H}_{EL}^{\beta\gamma})_{ni,nf} | n_j^{q_{if}\gamma} - 1 \rangle|^2 \delta(E_f^n - E_i^n - \hbar\omega_j^{q_{if}\gamma})$$

$$= \frac{2\pi}{\hbar} \left| \hbar \sqrt{\frac{n_j^{q_{if}\gamma}}{2N}} (\Xi_j^{\beta\gamma})_{ni,nf} \right|^2 \delta(E_f^n - E_i^n - \hbar\omega_j^{q_{if}\gamma})$$

$$= \frac{2\pi}{\hbar} \frac{n_j^{q_{if}\gamma}}{2N} |\hbar(\Xi_j^{\beta\gamma})_{ni,nf}|^2 \delta(E_f^n - E_i^n - \hbar\omega_j^{q_{if}\gamma}) \quad (8.35)$$

Now these transitions scatter the electron from the initial plane wave $\exp(i\mathbf{k}_i \cdot \mathbf{r})$ to the final plane wave $\exp(i\mathbf{k}_f \cdot \mathbf{r})$. Such a transition involves an energy difference $(E_f^n - E_i^n)$, where E_i^n and E_f^n are the energies corresponding to the initial and final states, i.e. the band energies corresponding to the initial and final Bloch states $b_i^n(\mathbf{r})$ and $b_f^n(\mathbf{r})$. As eqs (8.34) and (8.35) are the product of a lattice and an electronic part, they induce these transitions simultaneously with the creation or annihilation of a phonon. Then the δ-function in eqs (8.34) and (8.35) represents conservation of energy and requires the energy of the annihilated or created phonon to be equal to the energy gained or lost by the electron,

$$\hbar\omega_j^{q_{if}\gamma} = \pm(E_f^n - E_i^n) \quad (8.36)$$

CONSEQUENCES OF THE SELECTION RULES

However, eqs (8.34) and (8.35) teach us more. The transition matrix elements given in these equations are proportional to $\sqrt{n_j^{q_{if}\gamma} + 1}$ and $\sqrt{n_j^{q_{if}\gamma}}$, respectively, so, according to eqs (8.34) and (8.35) they induce transitions with a probability proportional to $(n_j^{q_{if}\gamma} + 1)$ and $n_j^{q_{if}\gamma}$ respectively. This implies that these probabilities increase with the number of phonons present and this means in practice that they increase with temperature. We also see that the probabilities for phonon emission and phonon absorption are not equal. Starting at an initial phonon state $|n_j^{q_{if}\gamma}\rangle$, the probability for phonon emission, leading to a final phonon states $|n_j^{q_{if}\gamma} + 1\rangle$, is proportional to

$(n_j^{q_{ij}\gamma} + 1)$. However, the probability for phonon absorption, leading to a final phonon state $|n_j^{q_{ij}\gamma} - 1\rangle$, is proportional to $n_j^{q_{ij}\gamma}$. As mentioned in section 2.3.3, this seemingly small difference is responsible for the important feature that electrons and phonons reach thermal equilibrium through electron–lattice coupling. Moreover, the phonon emission rate (8.34) does not decrease to zero when $n_j^{q_{ij}\gamma}$ approaches zero. Thus, even when the temperature approaches zero, an electron may still emit phonons spontaneously.

As an example of the selection rules, Fig. 8.1a shows the bottom of the conduction band of GaAs and Fig. 8.1c its phonon spectrum on the same horizontal and vertical scale. It is clear that the constraint of simultaneous energy and momentum conservation limits the number of possible transitions severely. Arrow A indicates a transition where an acoustical phonon is emitted by a conduction band electron. We see a considerable change in the **k**-vector of these electrons, but almost no change in its energy. Thus this transition can be considered to be quasi-elastic. However, as indicated by arrows B and C, transitions involving absorption or emission of an

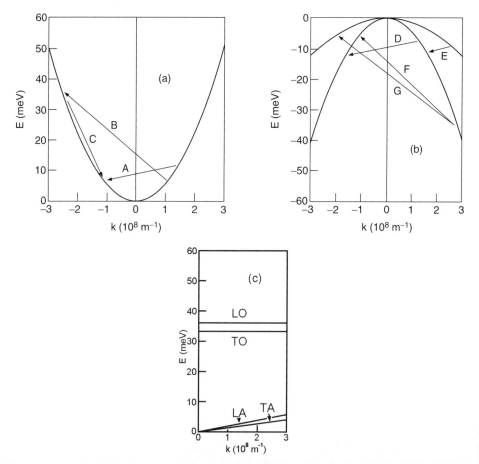

Figure 8.1 (a) The bottom of the conduction band of GaAs. (b) The top of the valence band of GaAs. (c) The central region of the phonon spectrum of GaAs. Arrows show possible transitions due to electron–lattice coupling.

8.2 CARRIER SCATTERING

optical phonon imply an important change in the energy of the electron, so these transitions are truly inelastic.

Our previous treatment was restricted to electrons in the conduction band. But, according to section 3.2.5, it may be directly generalized to holes in the valence band without losing the general features found above. Some features of hole–lattice coupling are shown in Fig. 8.1b showing the top of the valence band of GaAs, again on the same scale as Figs 8.1a and 8.1c. Arrows D and E indicate transitions where an acoustical phonon is emitted by a hole in the valence band. As for electrons we see a considerable change in the **k**-vector of this hole, but almost no change in its energy. Thus, this transition can also be considered to be quasi-elastic. Arrows F and G indicate that transitions involving absorption or emission of an optical phonon change the energy of the hole considerably, rendering such a transition truly inelastic.

8.2.3 Density of States Functions

SELECTION OF POSSIBLE FINAL STATES

When considering the probability of a scattering process, one must realize that the initial state is well defined: the electron is in a state in the nth band described by a plane wave $\exp i\mathbf{k}_i \cdot \mathbf{r}$, while the lattice is in the state

$$\prod_\gamma \prod_j \prod_q |n_j^{q\gamma}\rangle \tag{8.37}$$

corresponding to $n_j^{q\gamma}$ phonons in the γjth phonon mode and having a wave vector $\mathbf{k}_q$ and an energy $\hbar\omega_j^q$. Many final states are possible, though selection rules restrict their number. In principle we need to determine the transition probability to each of these final states. In practice, it is sufficient to consider *differential scattering rates*, that are transition probabilities from a given initial state to a restricted set of final states, and *total scattering rates*, that are the sum of all differential scattering rates, i.e. the total transition probability from a given initial state to all allowed final states. Below we will introduce a specific type of differential scattering rate, where a particular choice of the set of final states has been made.

Which final states are possible, is determined by the selection rules. First, the number of phonons may change by one only, i.e. only one phonon may be emitted or absorbed. Two further selection rules involve conservation of crystal momentum and energy. The first two selection rules are already accounted for when determining the matrix elements (8.32) and (8.33). The last selection rule, conservation of energy, is represented in eqs (8.34) and (8.35) by a δ-function. The calculation of a differential or total scattering rate involves summing over the chosen set of final wave vectors $\mathbf{k}_f$ and hence summing out this δ-function.

These summations are not trivial. First, in order to suppress the δ-functions in eqs (8.34) and (8.35), we require summation over one variable only. Choosing to sum over a single component of the final wave vector $\mathbf{k}_f$, the resulting scattering rate retains the sums over the other two components of $\mathbf{k}_f$. It is practical to choose spherical coordinates (k_f, θ_f, ϕ_f) and to sum over the length $k_f = |\mathbf{k}_f|$ of the final wave vector to suppress the Dirac δ-function $\delta(E_f^n - E_i^n \pm \hbar\omega_j^{q_{if}\gamma})$. Thus, the resulting scattering rate corresponds to the probability that an electron with an initial wave vector $\mathbf{k}_i$ scatters into the direction (θ_f, ϕ_f) in reciprocal space through the action of phonon mode j. We will define this as the differential scattering rate. Then the total scattering rate is a sum of all differential scattering rates over all possible directions (θ_f, ϕ_f) in reciprocal space.

TRANSFORMING THE FINAL WAVE VECTOR TO THE FINAL ENERGY

A second problem occurs because the δ-function concerns energies, while we intend to sum over one of the components of $\mathbf{k}_f$. Moreover, this remaining δ-function is actually a Dirac δ-function which requires *integration* instead of summation. We tackle the problem by approximating the sum over all final wave vectors $\mathbf{k}_f$ by an integral and by writing this integral in spherical coordinates. For simplicity, we exclude so-called 'umklapp' processes where the final wave vector $\mathbf{k}_f$ lies outside the first Brillouin zone. Then, the summation over this final wave vector can be restricted to its N allowed values in the first Brillouin zone, and the total rates for emission or for absorption are given by

$$\sum_f^N (W_{ni,nf})^{em/abs} \approx \frac{N}{V_B} \int_{V_B} d\phi_f d\theta_f \sin\theta_f dk_f \, k_f^2 \, (W_{ni,nf})^{em/abs}$$

$$\approx \frac{N}{V_B} \int_\infty d\phi_f d\theta_f \sin\theta_f dk_f \, k_f^2 \, (W_{ni,nf})^{em/abs}$$

$$= \int_0^{2\pi} d\phi_f \int_0^\pi d\theta_f \sin\theta_f \left[\frac{N}{V_B} \int_0^\infty dk_f \, k_f^2 \, (W_{ni,nf})^{em/abs}\right]$$

$$= \int_0^{2\pi} d\phi_f \int_0^\pi d\theta_f \sin\theta_f (W_i)^{em/abs}(\theta_f, \phi_f) \qquad (8.38)$$

Here, the factor

$$\frac{N}{V_B} = \frac{NV_C}{(2\pi)^3} \qquad (8.39)$$

is included to account for the ratio of the total number N of elements in the sum over f and the volume V_B over which we integrate. Note that V_B is expressed in V_C using eq. (3.13), while NV_C is the volume of the crystal. In the first step the integral is extended to infinite space, because we assume that the transition probabilities $(W_{ni,nf})^{em}$ and $(W_{ni,nf})^{abs}$ vanish for $\mathbf{k}_f$ outside the first Brillouin zone. In eq. (8.38), $d\theta_f \sin\theta_f d\phi_f$ is an element of solid angle and hence

$$(W_i)^{em/abs}(\theta_f, \phi_f) = \frac{N}{V_B} \int_0^\infty dk_f \, k_f^2 \, (W_{ni,nf})^{em/abs} \qquad (8.40)$$

represents the probability per unit of solid angle that an electron is scattered into the (θ_f, ϕ_f) direction. Following our discussion above we call $(W_{ni,nf})^{em/abs}$ the differential phonon emission/absorption rate, which upon integration over the solid angle yields the total phonon emission/absorption rate.

Next, we transform the integral over k_f into an integral over the energy E_f^n of the final plane wave state. The Jacobian of this transformation is the partial derivative of

8.2 CARRIER SCATTERING

E_f^n with respect to k_f, while keeping θ_f and ϕ_f constant. So, the differential phonon emission and absorption rates (8.40) are written as

$$(W_i)^{em/abs}(\theta_f, \phi_f) = \frac{N}{V_B} \int_0^\infty dk_f \, k_f^2 \, (W_{ni,nf})^{em/abs}$$

$$= \frac{N}{V_B} \int_0^\infty dE_f^n \, \frac{k_f^2}{\left(\frac{\partial E_f^n}{\partial k_f}\right)_{\theta_f, \phi_f}} (W_{ni,nf})^{em/abs} \qquad (8.41)$$

where k_f is a function of E_f^n, θ_f and ϕ_f.

THE DIFFERENTIAL DENSITY OF STATES

Introducing the differential density of states,

$$N_d(E_f^n, \theta_f, \phi_f) = \frac{N}{V_B} \frac{k_f^2}{\left(\frac{\partial E_f^n}{\partial k_f}\right)_{\theta_f, \phi_f}} \qquad (8.42)$$

we can shorten the differential scattering rate (8.41) to

$$W_i^{em/abs}(\theta_f, \phi_f) = \int_0^\infty dE_f^n \, N_d(E_f^n, \theta_f, \phi_f)(W_{ni,nf})^{em/abs} \qquad (8.43)$$

The differential density of states (8.42) has a direct physical interpretation. First, we take into account that the final plane wave state corresponds to the final Bloch function. Next, we recall section 3.2.3, where we found that periodic boundary conditions allow N Bloch states per electron band in a Brillouin zone with volume V_B, while these Bloch states are distributed homogeneously over this Brillouin zone. Then, the factor (N/V_B) corresponds to the number of Bloch states per unit volume of reciprocal space, i.e. the density of final states in reciprocal space. Next, we consider a volume element in reciprocal space as shown in Fig. 8.2. It contains all final states positioned in a solid angle $d\phi_f d\theta_f \sin \theta_f$, while the electron energy lies between E_f^n and $E_f^n + dE_f^n$. The volume of this element is given by

$$dk_f k_f^2 d\phi_f d\theta_f \sin \theta_f = \frac{k_f^2}{\left(\frac{\partial E_f^n}{\partial k_f}\right)_{\theta_f, \phi_f}} dE_f^n d\phi_f d\theta_f \sin \theta_f \qquad (8.44)$$

Multiplying with (N/V_B) yields the number of final states in this volume element,

$$\frac{N}{V_B} \frac{k_f^2}{\left(\frac{\partial E_f^n}{\partial k_f}\right)_{\theta_f, \phi_f}} dE_f^n d\phi_f d\theta_f \sin \theta_f = N_d(E_f^n, \theta_f, \phi_f) dE_f^n d\phi_f d\theta_f \sin \theta_f \qquad (8.45)$$

Thus, the differential density of states corresponds to the number of final states per unit of solid angle and per unit of energy.

We note that this definition of the density of states ignores the electron spin. Because of this spin each Bloch state may actually be occupied twice. Hence, the

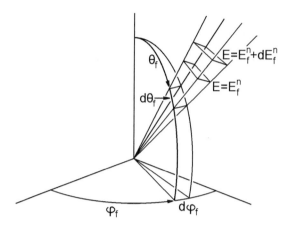

Figure 8.2 Volume element in reciprocal space containing all allowed final states in a solid angle $d\phi_f d\theta_f \sin\theta_f$ and energy interval dE_f^n.

total number of electron states per unit of solid angle and per unit of energy is actually twice as high as the differential density of states defined by eq. (8.42). In the present discussion this factor of two is not relevant, however. The reason is as follows. The electron–phonon coupling as discussed in chapter 7 does not involve interaction between phonons and the electron spin. Hence, electron–phonon coupling does not allow spin flips, i.e. it conserves the spin of a band electron and only those final states are allowed, where the electron spin state is the same as in the initial states. As a result the total number of allowed final electron states per unit of solid angle and per unit of energy is precisely the differential density of states as defined in eq. (8.42).

8.2.4 Density of States in Effective Mass Theory

ANISOTROPIC EFFECTIVE MASS

The differential density of states function (8.42) derived above is generally valid and can be used for any shape of the band energy as a function of the final wave vector $\mathbf{k}_f$. It is instructive to evaluate eq. (8.42) for the special case that effective mass theory applies. Then, the $\mathbf{k}\cdot\mathbf{p}$-approximation may be used to calculate the band energy E_f^n as a function of $\mathbf{k}_f$ and, according to eq. (4.34) for non-degenerate bands,

$$E_f^n = E_{f0}^n + \frac{\hbar^2}{2m_e} \sum_{\mu,\nu} \frac{k_\mu k_\nu}{m^*_{\mu\nu}} \tag{8.46}$$

where E_{f0}^n is the band energy at the Γ-point. As above we choose the zero-point of energy such that $E_{f0}^n = 0$. To be able to perform the differentiation in eq. (8.42), we rewrite $\mathbf{k}_f$ in spherical coordinates,

$$k_{fx} = k_f \sin\theta_f \cos\phi_f = k_f \hat{k}_x(\theta_f, \phi_f) \tag{8.47}$$

$$k_{fy} = k_f \sin\theta_f \sin\phi_f = k_f \hat{k}_y(\theta_f, \phi_f) \tag{8.48}$$

$$k_{fz} = k_f \cos\theta_f = k_f \hat{k}_z(\theta_f, \phi_f) \tag{8.49}$$

8.2 CARRIER SCATTERING

where $\hat{\mathbf{k}}(\theta_f, \phi_f) = [\hat{k}_x(\theta_f, \phi_f), \hat{k}_y(\theta_f, \phi_f), \hat{k}_z(\theta_f, \phi_f)]$ is the unit vector parallel to the final wave vector $\mathbf{k}_f$. Thus, it is a function of θ_f and ϕ_f only. Then,

$$E_f^n = \frac{\hbar^2 k_f^2}{2m_e} \sum_{\mu,\nu} \frac{\hat{k}_\mu(\theta_f, \phi_f)\hat{k}_\nu(\theta_f, \phi_f)}{m^*_{\mu\nu}} \tag{8.50}$$

and

$$\left(\frac{\partial E_f^n}{\partial k_f}\right)_{\theta_f,\phi_f} = \left(\frac{\partial}{\partial k_f}\right)_{\theta_f,\phi_f} \frac{\hbar^2 k_f^2}{2m_e} \sum_{\mu,\nu} \frac{\hat{k}_\mu(\theta_f, \phi_f)\hat{k}_\nu(\theta_f, \phi_f)}{m^*_{\mu\nu}}$$

$$= \frac{\hbar^2 k_f}{m_e} \sum_{\mu,\nu} \frac{\hat{k}_\mu(\theta_f, \phi_f)\hat{k}_\nu(\theta_f, \phi_f)}{m^*_{\mu\nu}} \tag{8.51}$$

Hence, if effective mass theory applies, the differential density of states (8.42) can be written as

$$N_d(E_f^n, \theta_f, \phi_f) = \frac{N}{V_B}\left[k_f^2 \bigg/ \frac{\hbar^2 k_f}{m_e} \sum_{\mu,\nu} \frac{\hat{k}_\mu(\theta_f, \phi_f)\hat{k}_\nu(\theta_f, \phi_f)}{m^*_{\mu\nu}}\right]$$

$$= \frac{N}{V_B}\left[k_f \bigg/ \frac{\hbar^2}{m_e} \sum_{\mu,\nu} \frac{\hat{k}_\mu(\theta_f, \phi_f)\hat{k}_\nu(\theta_f, \phi_f)}{m^*_{\mu\nu}}\right] \tag{8.52}$$

Note that the differential density of states $N_d(E_f^n, \theta_f, \phi_f)$ is a function of the energy E_f^n of the final state and the direction (θ_f, ϕ_f) of the final wave vector. Therefore, in the above result we have to rewrite k_f as a function of these three parameters. This is done by solving eq. (8.50), yielding,

$$k_f(E_f^n, \theta_f, \phi_f) = \left[E_f^n \bigg/ \frac{\hbar^2}{2m_e} \sum_{\mu,\nu} \frac{\hat{k}_\mu(\theta_f, \phi_f)\hat{k}_\nu(\theta_f, \phi_f)}{m^*_{\mu\nu}}\right]^{1/2} \tag{8.53}$$

Then, combining eqs (8.52) and (8.53), we find for the differential density of states function,

$$N_d(E_f^n, \theta_f, \phi_f) = \frac{N}{V_B}(2E_f^n)^{1/2} \bigg/ \left[\frac{\hbar^2}{m_e}\sum_{\mu,\nu}\frac{\hat{k}_\mu(\theta_f, \phi_f)\hat{k}_\nu(\theta_f, \phi_f)}{m^*_{\mu\nu}}\right]^{3/2} \tag{8.54}$$

Thus, the differential density of states grows with the square root of the energy of the final state.

ISOTROPIC EFFECTIVE MASS

An even more special case concerns isotropic effective mass. Then,

$$\frac{1}{m^*_{\mu\nu}} = \frac{1}{m^*}\delta_{\mu,\nu} \tag{8.55}$$

so,

$$\sum_{\mu,\nu} \frac{\hat{k}_\mu(\theta_f,\phi_f)\hat{k}_\nu(\theta_f,\phi_f)}{m^*_{\mu\nu}} = \frac{1}{m^*}\sum_{\mu,\nu} \hat{k}_\mu(\theta_f,\phi_f)\hat{k}_\nu(\theta_f,\phi_f)\delta_{\mu,\nu}$$

$$= \frac{1}{m^*}\sum_\mu \hat{k}^2_\mu(\theta_f,\phi_f) = \frac{1}{m^*} \quad (8.56)$$

where we use $\hat{k}(\theta_f,\phi_f)$ is a unit vector with length 1. As a result, for an isotropic effective mass, eqs (8.52) and (8.54) reduce to

$$N_d(E^n_f,\theta_f,\phi_f) = \frac{N}{V_B}\frac{m_e m^* k_f}{\hbar^2} = \frac{N}{V_B}\frac{(2E^n_f)^{1/2}(m_e m^*)^{3/2}}{\hbar^3} \quad (8.57)$$

Again the resulting expression for the differential density of states shows a square root dependence on the final energy. Now, because of isotropy, the angular dependence has disappeared.

8.2.5 Phonon Scattering Rates

GENERAL BAND STRUCTURE

Now we are in the position to evaluate the rate of a scattering process, i.e. a process where an electron with an initial wave vector $\mathbf{k}_i$ is scattered into the (θ_f,ϕ_f)-direction. As above we restrict this calculation to the case that the electron remains in the same, nth, band. Two processes may take place, phonon emission described by eq. (8.34) and phonon absorption described by eq. (8.35). These equations express the rates of these processes in the initial states of the electron and the lattice. However, eqs (8.34) and (8.35) still lack detail with respect to the final electron state. Therefore they cannot yet be used to evaluate the rates of specific scattering processes. It is precisely the differential density of state function which allows us to integrate the δ-functions out of eqs (8.34) and (8.35) and to introduce the required detail on the final state of scattering process.

Inserting eq. (8.34) in eq. (8.43), the differential scattering rate for phonon emission is found to be given by

$$(W_i)^{em}(\theta_f,\phi_f) = \int_0^\infty dE^n_f \, N_d(E^n_f,\theta_f,\phi_f)\,(W_{ni,nf})^{em}$$

$$= \int_0^\infty dE^n_f \, N_d(E^n_f,\theta_f,\phi_f)$$

$$\times \frac{2\pi}{\hbar}\frac{(n^{q_{ij}\gamma}_j+1)}{2N}|\hbar(\Xi^{\beta\gamma}_j)_{ni,nf}|^2\delta(E^n_f - E^n_i + \hbar\omega^{q_{ij}\gamma}_j)$$

$$= \frac{2\pi}{\hbar}\frac{(n^{q_{ij}\gamma}_j+1)}{2N}|\hbar(\Xi^{\beta\gamma}_j)_{ni,nf}|^2 N_d(E^n_i - \hbar\omega^{q_{ij}\gamma}_j,\theta_f,\phi_f) \quad (8.58)$$

where we require that

$$k_f(E^n_f,\theta_f,\phi_f) = k_f(E^n_i - \hbar\omega^{q_{ij}\gamma}_j,\theta_f,\phi_f) \quad (8.59)$$

8.2 CARRIER SCATTERING

For phonon absorption we insert eq. (8.35) in eq. (8.43) to find equivalently,

$$(W_i)^{abs}(\theta_f \phi_f) = \int_0^\infty dE_f^n \, N_d(E_f^n, \theta_f, \phi_f)(W_{ni,nf})^{abs}$$

$$= \frac{2\pi}{\hbar} \frac{n_j^{q_{if}\gamma}}{2N} |\hbar(\Xi_j^{\beta\gamma})_{ni,nf}|^2 N_d(E_i^n + \hbar\omega_j^{q_{if}\gamma}, \theta_f, \phi_f) \quad (8.60)$$

where we require that

$$k_f(E_f^n, \theta_f, \phi_f) = k_f(E_i^n + \hbar\omega_j^{q_{if}\gamma}, \theta_f, \phi_f) \quad (8.61)$$

In each of these equations $\beta\gamma = DO$, DA, PO or PA, so eqs (8.58) and (8.60) represent the emission and absorption rates due to just one of these four electron–phonon couplings. To obtain the differential scattering rates for each of these four cases, one should insert expressions (8.28), (8.29), (8.30) or (8.31) for $(\Xi_j^{\beta\gamma})_{ni,nf}$. Moreover, in the case of optical deformation potential and acoustical deformation potential electron–phonon coupling and of piezoelectric electron–phonon coupling all three polarization directions of the phonon modes contribute. Hence, to obtain the differential emission and absorption rates, one needs to sum eqs (8.58) and (8.60) over the index j denoting these polarization directions. Performing the summation over j and adding absorption and emission rates, the differential scattering rate for each one of the four types of electron–phonon coupling is therefore given by

$$W_i^{\beta\gamma}(\theta_f, \phi_f) = \sum_j \left[\frac{2\pi}{\hbar} \frac{(n_j^{q_{if}\gamma}+1)}{2N} |\hbar(\Xi_j^{\beta\gamma})_{ni,nf}|^2 N_d(E_i^n - \hbar\omega_j^{q_{if}\gamma}, \theta_f, \phi_f) \right.$$

$$\left. + \frac{2\pi}{\hbar} \frac{n_j^{q_{if}\gamma}}{2N} |\hbar(\Xi_j^{\beta\gamma})_{ni,nf}|^2 N_d(E_i^n + \hbar\omega_j^{q_{if}\gamma}, \theta_f, \phi_f) \right]$$

$$= \frac{2\pi}{\hbar} \sum_j \left[|\hbar(\Xi_j^{\beta\gamma})_{ni,nf}|^2 \left(\frac{(n_j^{q_{if}\gamma}+1)}{2N} N_d(E_i^n - \hbar\omega_j^{q_{if}\gamma}, \theta_f, \phi_f) \right. \right.$$

$$\left. \left. + \frac{n_j^{q_{if}\gamma}}{2N} N_d(E_i^n + \hbar\omega_j^{q_{if}\gamma}, \theta_f, \phi_f) \right) \right] \quad (8.63)$$

ISOTROPIC EFFECTIVE MASS

The final result (8.62) is very general and can be used for any shape of the band structure. As in the previous section it is instructive to consider the more special case that the effective mass formalism applies. Here we restrict ourselves to an isotropic effective mass. This is a relevant choice as it concerns the conduction band minimum of direct semiconductors as GaAs, while it also illustrates most of the features of scattering due to electon–phonon coupling. We recall eqs (8.50) and (8.56) which yield for an isotropic effective mass,

$$E_f^n = \frac{\hbar^2 k_f^2}{2m_e m^*} \quad (8.63)$$

and eq. (8.57) for the differential density of states $N_d(E_f^n, \theta_f, \phi_f)$. Then, for each of the four types of electron–phonon coupling the differential scattering rate (8.62) becomes

$$W_i^{\beta\gamma}(\theta_f, \phi_f) = \frac{2\pi}{\hbar} \sum_j \left\{ |\hbar(\Xi_j^{\beta\gamma})_{ni,nf}|^2 \left[\frac{(n_j^{q_{ij}\gamma} + 1)}{2N} N_d(E_i^n - \hbar\omega_j^{q_{ij}\gamma}, \theta_f, \phi_f) \right.\right.$$
$$\left.\left. + \frac{n_j^{q_{ij}\gamma}}{2N} N_d(E_i^n + \hbar\omega_j^{q_{ij}\gamma}, \theta_f, \phi_f) \right] \right\}$$

$$= \frac{2\pi}{\hbar} \sum_j \left\{ |\hbar(\Xi_j^{\beta\gamma})_{ni,nf}|^2 \left[\frac{(n_j^{q_{ij}\gamma} + 1)}{2N} \frac{N}{V_B} \frac{(E_i^n - \hbar\omega_j^{q_{ij}\gamma})^{1/2} (m_e m^*)^{3/2}}{\hbar^3} \right.\right.$$
$$\left.\left. + \frac{n_j^{q_{ij}\gamma}}{2N} \frac{N}{V_B} \frac{(E_i^n + \hbar\omega_j^{q_{ij}\gamma})^{1/2} (m_e m^*)^{3/2}}{\hbar^3} \right] \right\}$$

$$= \frac{\pi (m_e m^*)^{3/2}}{\hbar^2 V_B} \sum_j \left\{ |(\Xi_j^{\beta\gamma})_{ni,nf}|^2 \left[(n_j^{q_{ij}\gamma} + 1)(E_i^n - \hbar\omega_j^{q_{ij}\gamma})^{1/2} \right.\right.$$
$$\left.\left. + n_j^{q_{ij}\gamma}(E_i^n + \hbar\omega_j^{q_{ij}\gamma}) \right] \right\} \quad (8.64)$$

NON-POLAR ELECTRON–PHONON COUPLING AT THE Γ-POINT

It is now possible to continue and to insert the general expressions (8.28)–(8.31) for $(\Xi_j^{\beta\gamma})_{ni,nf}$ for the various types of electron–phonon coupling. However, as remarked above, the case of an isotropic effective mass is relevant because it corresponds to the Γ-valley of the conduction band, which is the lowest conduction band minimum in direct semiconductors like GaAs. In the remainder of this section we concentrate on this specific case and we replace the band index n by c to stress that we are treating the conduction band. In section 7.2.1 we evaluated $(\Xi_j^{\beta\gamma})_{ni,nf} = (\Xi_j^{\beta\gamma})_{ci,cf}$ for this specific case and therefore it makes more sense to insert the results of that section instead of the general expressions (8.28) to (8.31).

First we note that according to eqs (7.108) and (7.110), $(\Xi_j^{DO})_{ci,cf} = 0$, so,

$$W_i^{DO}(\theta_f, \phi_f) = 0 \quad (8.65)$$

In other words, non-polar electron–optical phonon coupling vanishes at the Γ-point and there is no scattering due to such a coupling.

Second, according to eqs (7.116) and (7.117), only longitudinal phonons contribute to non-polar electron–acoustical phonon coupling and for this coupling,

$$(\Xi^{DLA})_{ci,cf} = \frac{1}{\sqrt{\hbar\omega^{q_{if}LA}(M_1 + M_2)}} |\mathbf{k}_{q_{if}}| E \quad (8.66)$$

Here, M_1 and M_2 are the mass of the two atoms in the primitive cell and E is the only non-zero component of the effective acoustical deformation potential tensor. Moreover, as discussed in section 8.2.2, we may consider electron–acoustical phonon scattering to be quasi-elastic, i.e. we may neglect the phonon energy $\hbar\omega^{q_{if}LA}$ with

8.2 CARRIER SCATTERING

respect to the initial electron energy E_i^c. Then, the differential scattering rate due to non-polar electron–acoustical phonon scattering takes the form,

$$W_i^{DLA}(\theta_f, \phi_f) = \frac{\pi(m_e m^*)^{3/2}}{\hbar^2 V_B} \left| \frac{1}{\sqrt{\hbar \omega^{q_{ij} LA}(M_1 + M_2)}} |\mathbf{k}_{q_{ij}}| E \right|^2 (2n^{q_{ij} LA} + 1)(E_i^c)^{1/2}$$

$$= \frac{\pi(m_e m^*)^{3/2} |\mathbf{k}_{q_{ij}}|^2 E^2}{\hbar^3 \omega^{q_{ij} LA}(M_1 + M_2) V_B} (2n^{q_{ij} LA} + 1)(E_i^c)^{1/2} \tag{8.67}$$

We now note that according to eq. (3.13), the volume of a Brillouin zone $V_B = (2\pi)^3/V_C$, where V_C is the volume of a primitive cell. Now $M_1 + M_2$ is the mass of a primitive cell, so

$$(M_1 + M_2) V_B = (2\pi)^3 \frac{M_1 + M_2}{V_C} = (2\pi)^3 \rho_m \tag{8.68}$$

where ρ_m is the mass density of the crystal. Moreover, for small values of the phonon wave vector, according to eq. (6.86), $|\mathbf{k}_q| = \omega^{qLA}/c^{LA}$, where c^{LA} is the velocity of longitudinal acoustical modes. Hence,

$$W_i^{DLA}(\theta_f, \phi_f) = \frac{\pi(m_e m^*)^{3/2} \omega^{q_{ij} LA} E^2}{\hbar^3 (2\pi)^3 \rho_m (c^{LA})^2} (2n^{q_{ij} LA} + 1)(E_i^c)^{1/2}$$

$$= \frac{(m_e m^*)^{3/2} E^2}{(2\pi)^2 \hbar^4 \rho_m (c^{LA})^2} \hbar \omega^{q_{ij} LA} \left(n^{q_{ij} LA} + \frac{1}{2} \right)(E_i^c)^{1/2} \tag{8.69}$$

There is no need for an index j as there is only one LA phonon mode. Note also that this result is completely independent of the direction of the final wave vector of the electron.

A direct estimate of the differential scattering rate due to non-polar electron–acoustical phonon interaction can now be obtained immediately. For this purpose we consider the conduction band minimum at the Γ-point in GaAs. First, we note that the fraction in expression (8.69) only contains material parameters and the universal constants

$$m_e = 9.110 \cdot 10^{-31} \text{ kg}$$

$$\hbar = 1.055 \cdot 10^{-34} \text{ J s}$$

Next, we note that the remainder of this expression consists of the product of the energy

$$\hbar \omega^{q_{ij} LA} \left(n^{q_{ij} LA} + \frac{1}{2} \right)$$

of the longitudinal phonon mode with wave vector $\mathbf{k}_{q_{ij}}$ and the square root of the initial electron energy. We assume the crystal to be in thermal equilibrium at 300 K. Then both energies are equal to $k_B T$, where

$$k_B = 1.3807 \cdot 10^{-23} \text{ J K}^{-1}$$

Thus, the differential scattering rate due to non-polar electron–acoustical phonon coupling is proportional to $T^{3/2}$.

Finally, for the conduction band minimum at the Γ-point in GaAs, a temperature of 300 K and longitudinal acoustical modes in the (100)-direction [6, 24, 27],

$$m^* = 0.0665$$

$$E/e = 7.0 \text{ eV}$$

$$\rho_m = 5.3174 \text{ kg m}^{-3}$$

$$c^{LA} = 4.731 \cdot 10^3 \text{ m s}^{-1}$$

Then, we find

$$W_i^{DLA}(\theta_f, \phi_f) = 8.6 \cdot 10^9 \text{ s}^{-1} \tag{8.70}$$

Note that the differential scattering rate is defined as the probability per unit of solid angle for the electron to scatter in the (θ_f, ϕ_f)-direction. As the differential scattering rate is found to be completely isotropic, the total scattering rate is then simply the differential scattering rate multiplied by 4π. Hence, the total scattering rate due to non-polar electron–acoustical phonon coupling is estimated to be equal to

$$W_i^{DLA,tot} = 1.08 \cdot 10^{11} \text{ s}^{-1} \tag{8.71}$$

Thus, an electron is scattered about every 10 ps by this process.

POLAR ELECTRON–OPTICAL PHONON COUPLING AT THE Γ-POINT

As we will see now, polar electron–optical phonon scattering and piezoelectric electron–phonon scattering are much more efficient than the process just treated above. As we found in section 7.1.6, only longitudinal optical phonons contribute to polar electron–optical phonon coupling. According to eq. (7.118) for this coupling,

$$(\Xi^{PLO})_{ci,cf} = i \frac{ee^*}{\varepsilon_0 V_C} \sqrt{\frac{M_1 + M_2}{\hbar \omega^{q_{if} LO} M_1 M_2}} \frac{1}{|\mathbf{k}_{q_{if}}|} \tag{8.72}$$

Again M_1 and M_2 are the masses of the two atoms in the primitive cell. Now, e^* is the charge of the dipole moment induced by the relative displacement of these two atoms. Inserting this expression in eq. (8.64) we find,

$$W_i^{PLO}(\theta_f, \phi_f) = \frac{\pi(m_e m^*)^{3/2}}{\hbar^2 V_B} \left| i \frac{ee^*}{\varepsilon_0 V_C} \sqrt{\frac{M_1 + M_2}{\hbar \omega^{q_{if} LO} M_1 M_2}} \frac{1}{|\mathbf{k}_{q_{if}}|} \right|^2$$

$$\times [(n^{q_{if} LO} + 1)(E_i^c - \hbar \omega^{q_{if} LO})^{1/2} + n^{q_{if} LO}(E_i^c + \hbar \omega^{q_{if} LO})^{1/2}]$$

$$= \frac{\pi(m_e m^*)^{3/2} (ee^*)^2}{\varepsilon_0^2 \hbar^3 \omega^{q_{if} LO} V_B V_C^2} \frac{M_1 + M_2}{M_1 M_2} \frac{1}{|\mathbf{k}_{q_{if}}|^2}$$

$$\times [(n^{q_{if} LO} + 1)(E_i^c - \hbar \omega^{q_{if} LO})^{1/2} + n^{q_{if} LO}(E_i^c + \hbar \omega^{q_{if} LO})^{1/2}]$$

$$= \frac{(m_e m^*)^{1/2} e^2}{(4\pi)^2 \varepsilon_0 \hbar^2} \frac{e^{*2}}{(\omega^{LO})^2 \varepsilon_0 V_C} \frac{M_1 + M_2}{M_1 M_2} \frac{2 m_e m^*}{\hbar^2 |\mathbf{k}_{q_{if}}|^2}$$

$$\times \hbar \omega^{LO} [(n^{q_{if} LO} + 1)(E_i^c - \hbar \omega^{LO})^{1/2} + n^{q_{if} LO}(E_i^c + \hbar \omega^{LO})^{1/2}] \tag{8.73}$$

8.2 CARRIER SCATTERING

where we insert eq. (3.13) for $V_B V_C$. Also we approximate $\omega^{q_{if} LO} \approx \omega^{LO}$, because, as seen from Fig. 6.5, $\omega^{q_{if} LO}$ is nearly constant for optical phonons near the Γ-point. Finally, in the last step the various parameters are rearranged in such a way that it becomes easier to estimate the resulting scattering rate.

The first fraction in eq. (8.73) contains universal constants and material parameters only. For GaAs each of them was given above except for

$$e = 1.6022 \cdot 10^{-19} \text{ C}$$

$$\varepsilon_0 = 8.8543 \cdot 10^{-12}$$

The second fraction contains the somewhat elusive effective charge e^* of the electric dipole moment induced by an optical phonon. Without bothering to reproduce his argument, we use an estimate made by Ridley [32] yielding

$$\frac{e^{*2}}{(\omega^{LO})^2 \varepsilon_0 V_C} \frac{M_1 + M_2}{M_1 M_2} = \left[\frac{1}{\varepsilon_r(\infty)} - \frac{1}{\varepsilon_r(0)} \right] \quad (8.74)$$

where $\varepsilon_r(\infty)$ is the relative permittivity at infinite frequency and $\varepsilon_r(0)$ the relative permittivity at low frequency. For GaAs at 300 K [6],

$$\varepsilon_r(\infty) = 10.63$$

$$\varepsilon_r(0) = 12.44$$

Note that these values lead to $e^*/e = 1.02$. The final fraction in eq. (8.73) corresponds to the energy of an electron in the Γ-valley of the conduction band if it has a **k**-vector equal to $\mathbf{k}_{if} = \pm |\mathbf{k}_f - \mathbf{k}_i|$.

To estimate this final fraction and the factor in the final line of eq. (8.73), we will make a very rough assumption. Though at 300 K [6],

$$\frac{\hbar \omega^{LO}}{e} = 35.4 \text{ meV} > \frac{k_B T}{e} = 25.9 \text{ meV}$$

we ignore that $\hbar \omega^{LO}$ is finite and make the same approximations as for acoustical phonon scattering in the previous section. Then

$$\frac{2 m_e m^*}{\hbar^2 |\mathbf{k}_{q_{if}}|^2} \approx \frac{2 m_e m^*}{\hbar^2 |\mathbf{k}_i|^2} \approx (k_B T)^{-1}$$

$$\hbar \omega^{LO} [(n^{q_{if} LO} + 1)(E_i^c - \hbar \omega^{LO})^{1/2} + n^{q_{if} LO} (E_i^c + \hbar \omega^{LO})^{1/2}] \approx 2(k_B T)^{3/2}$$

So, approximately,

$$W_i^{PLO}(\theta_f, \phi_f) \approx \frac{(m_e m^*)^{1/2} e^2}{(4\pi)^2 \varepsilon_0 \hbar^2} \left[\frac{1}{\varepsilon_r(\infty)} - \frac{1}{\varepsilon_r(0)} \right] 2(k_B T)^{1/2} = 7.16 \cdot 10^{11} \text{ s}^{-1} \quad (8.75)$$

Again our result is isotropic, so the total scattering rate due to polar electron–optical phonon coupling is obtained by simply multiplying with 4π. Thus,

$$W_i^{PLO,tot} \approx 9.7 \cdot 10^{12} \text{ s}^{-1} \quad (8.76)$$

We see that an electron is scattered about every 100 fs by this process. Hence, this process is about 100 times more efficient than non-polar acoustical phonon scattering.

PIEZOELECTRIC ELECTRON–PHONON SCATTERING AT THE Γ-POINT

Without doubt piezoelectric electron–phonon coupling yields the least simple result for the scattering rate. According to eq. (7.119), even in the conduction band minimum at the Γ-point, $(\Xi_j^{PA})_{ci,cf}$ is given by a complicated expression,

$$(\Xi_j^{PA})_{ci,cf} = i \frac{2e\varepsilon^{pe}}{\varepsilon_0 \varepsilon_r} \frac{1}{\sqrt{\hbar \omega_j^{q_{if} A} (M_1 + M_2)}} \sum_\mu \left(\hat{p}_{\mu,j}^{q_{if} A} \frac{k_{q_{if}(\mu+1)} k_{q_{if}(\mu+2)}}{|\mathbf{k}_{q_{if}}|^2} \right) \quad (8.77)$$

Here ε^{pe} is the single non-zero component of the piezoelectric tensor, $\hat{p}_{\mu,j}^{q_{if} A}$ the μ component of the unit polarization vector of the jth acoustical phonon mode with wave vector $\mathbf{k}_{q_{if}}$. The complicated structure of this expression results from the sum over μ containing not only the components $\hat{p}_{\mu,j}^{q_{if} A}$ of the unit polarization vector but also the direction cosines $k_{q_{if}\mu}/|\mathbf{k}_{q_{if}}|$ of the phonon wave vector.

Here we only intend to estimate the magnitude of the scattering rate due to piezoelectric electron–phonon coupling. For this purpose we note that on the average all quantities in the sum over μ are of the order $1/\sqrt{3}$. This induces us to venture to approximate the sum over μ by $3 \times (1/\sqrt{3})^3 = 1/\sqrt{3}$. Moreover, we remember the argument of section 8.2.2 and consider electron-acoustical phonon scattering to be quasi-elastic, i.e. we neglect the phonon energy $\hbar \omega_j^{q_{if} A}$ with respect to the initial electron energy E_i^c. Then, using eq. (8.64), the differential scattering rate due to piezoelectric electron–acoustical phonon scattering takes the form,

$$W_i^{PLA}(\theta_f, \phi_f) = \frac{\pi (m_e m^*)^{3/2}}{\hbar^2 V_B} \sum_j \left| i \frac{2e\varepsilon^{pe}}{\varepsilon_0 \varepsilon_r} \frac{1}{\sqrt{\hbar \omega_j^{q_{if} A}(M_1+M_2)}} \frac{1}{\sqrt{3}} \right|^2 (2n_j^{q_{if} A} + 1)(E_i^c)^{1/2}$$

$$= \sum_j \frac{4\pi (m_e m^*)^{3/2} e^2 (\varepsilon^{pe})^2}{3 \hbar^3 \omega_j^{q_{if} A}(M_1+M_2) V_B \varepsilon_0^2 \varepsilon_r^2} (2n_j^{q_{if} A} + 1)(E_i^c)^{1/2}$$

$$= \sum_j \frac{4\pi (m_e m^*)^{3/2} e^2 (\varepsilon^{pe})^2}{3(2\pi)^3 \rho_m \hbar^4 (\omega_j^{q_{if} A})^2 \varepsilon_0^2 \varepsilon_r^2} \hbar \omega_j^{q_{if} A} (2n_j^{q_{if} A} + 1)(E_i^c)^{1/2}$$

$$= \sum_j \frac{\pi (m_e m^*)^{1/2} e^2 (\varepsilon^{pe})^2}{6\pi^3 \rho_m \hbar^2 c_j^{A 2} \varepsilon_0^2 \varepsilon_r^2} \frac{2 m_e m^*}{\hbar^2 |\mathbf{k}_{q_{if}}|^2} \hbar \omega_j^{q_{if} A} \left(n_j^{q_{if} A} + \frac{1}{2} \right) (E_i^c)^{1/2} \quad (8.78)$$

where we also insert eq. (8.68) and $\omega_j^{qA} = c_j^A |\mathbf{k}_q|$, where c_j^A is the velocity of the jth acoustical phonon mode. As in the case of non-polar electron–acoustical phonon scattering we estimate

$$\hbar \omega_j^{q_{if} A} \left(n_j^{q_{if} A} + \frac{1}{2} \right) (E_i^c)^{1/2} = (k_B T)^{3/2}$$

while equivalently to the case of polar electron–optical phonon scattering we approximate

$$\frac{2 m_e m^*}{\hbar^2 |\mathbf{k}_{q_{if}}|^2} \approx \frac{2 m_e m^*}{\hbar^2 |\mathbf{k}_i|^2} \approx (k_B T)^{-1}$$

8.2 CARRIER SCATTERING

The remaining factor in eq. (8.78) is the sum of three terms, one corresponding to longitudinal acoustical phonons with $c_j^A = c^{LA}$ and two corresponding to transverse acoustical phonons with $c_j^A = c^{TA}$. Thus, the scattering rate due to piezoelectric electron–acoustical phonon coupling can be approximated by

$$W_i^{PLA}(\theta_f, \phi_f) \approx \frac{\pi (m_e m^*)^{1/2} e^2 (\varepsilon^{pe})^2}{6\pi^3 \rho_m \hbar^2 \varepsilon_0^2 \varepsilon_r^2} \left[\frac{1}{(c^{LA})^2} + \frac{2}{(c^{TA})^2} \right] (k_B T)^{1/2} \qquad (8.79)$$

All parameters occurring in this expression have been given above while evaluating the scattering rates due to non-polar electron–acoustical phonon coupling and polar electron–optical phonon coupling, with the exception of the single component ε^{pe} of the piezoelectric tensor and the velocity of the transverse acoustical modes. Values for the former have been compiled by Ridley [32] and for the latter by Blakemore [6]. We use

$$\varepsilon^{pe} = 0.160 \text{ C m}^{-2}$$

$$\varepsilon_r = 12.8$$

$$c^{TA} = 3.345 \cdot 10^3 \text{ m s}^{-1}$$

where the latter corresponds to the $(1,0,0)$-direction. As a result we estimate the differential scattering rate due to piezoelectric electron–acoustical phonon coupling to be

$$W_i^{PLA}(\theta_f, \phi_f) \approx 0.52 \cdot 10^{11} \text{ s}^{-1} \qquad (8.80)$$

for the conduction band valley at the Γ-point of GaAs at a temperature of 300 K. Also this final result is isotropic, so the total scattering rate is given by this result multiplied by 4π. Thus,

$$W_i^{PLA,tot} \approx 0.65 \cdot 10^{12} \text{ s}^{-1} \qquad (8.81)$$

which is roughly one order of magnitude smaller than the scattering rate due to polar electron–optical phonon coupling, but also roughly one order of magnitude larger than the scattering rate due to non-polar electron–acoustical phonon coupling.

It must be stressed that the estimates made above are very rough, though the order of magnitude is surprisingly correct. This is seen from Fig. 8.3 where curves A and B are drawn using results obtained by Moglestue [27] for the total scattering rates due to polar electron–optical phonon coupling and non-polar electron–acoustical phonon coupling for the conduction band valley at the Γ-point for a temperature of 300 K and as a function of the electron energy.

8.2.6 Impurity Scattering

SHALLOW IMPURITIES

Charge carriers are not only scattered due to electron–phonon interaction, but also due to their interaction with impurities. In particular shallow donor and acceptor impurities, which are necessarily present to provide electrons and holes, are an important source of this last type of scattering. A model for such shallow impurities was

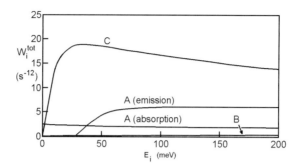

Figure 8.3 Total scattering rates for the conduction band valley at the Γ-point in GaAs: (A) due to polar electron–optical phonon coupling; (B) due to non-polar electron–acoustical phonon coupling; and (C) due to interaction with ionized impurities. $T = 300$ K and the ionized impurity concentration $N_D = 10^{17}$ cm^{-3}. The curves are drawn using data from Fig. 4.2, p. 94, Fig. 4.4, p. 100 and Fig. 4.5, p. 105 of ref. [27].

introduced in section 5.3.2. Shallow donors as well as acceptors are described using effective mass theory and taking into account the long range Coulomb potential of these impurities only. Thus, a donor is simplified to a positive elementary charge e yielding an effective potential

$$\bar{U}_d(\mathbf{r}) = \frac{e}{4\pi\varepsilon_0\varepsilon_r r} \tag{8.82}$$

where $r = |\mathbf{r}|$, is the distance to the impurity, ε_0 is the permittivity of the vacuum and ε_r is the relative permittivity of the semiconductor. Equivalently, an acceptor is reduced to a negative elementary charge $-e$ yielding an effective potential

$$\bar{U}_a(\mathbf{r}) = -\frac{e}{4\pi\varepsilon_0\varepsilon_r r} \tag{8.83}$$

In section 5.3.2 we considered the situation that an electron is bound to a donor in hydrogen like states. Now we will consider the case that not all donors are binding an electron, or equivalently, that not all acceptors are binding a hole. Such donors are generally designated as being ionized, just as atoms which have lost one or more electrons. In the first place such ionization may be caused thermally. As we calculated in section 5.3.2, the binding energy of shallow impurities is 5.5 meV in GaAs and hence much less than $kT/e = 25.9$ meV at 300K. Thus, at room temperature shallow donors in GaAs will all be ionized. In other semiconductors the binding energy may be higher, but still a considerable fraction of the impurities will be thermally excited into the ionized state. Second, even at low temperature ionization may be caused by compensation. Often, it is impossible to avoid acceptors in a material intended to contain donors only, or vice versa. Thus, if e.g. the material is intentionally doped with N_d donors, while a smaller number N_a of acceptors could not be avoided, the acceptors are said to compensate the donors in the material. Energy is gained when a hole from the latter acceptors annihilates an electron from the former donors, because the sum of the binding energies of an electron and a hole is smaller than the band gap energy. As a result, all N_a acceptors are ionized and at least N_a donors as well. Thus, such a semiconductor contains at least $2N_a$ ionized impurities.

MATRIX ELEMENTS FOR SCATTERING

Free electrons in the conduction band and free holes in the valence band experience the effective electrostatic potentials (8.82) of ionized donors and (8.83) of ionized acceptors. As these potentials are derived using effective mass theory, their effects on charge carriers can also be described using this simplifying formalism. Here we will follow the traditional method to describe the influence of ionized shallow impurities on charge carriers in terms of scattering processes. Thus, an electron or hole is supposed to approach the impurity in a state described by a plane wave (8.22),

$$e^{i\mathbf{k}_i \cdot \mathbf{r}} \tag{8.84}$$

Then it is suddenly scattered to a new state described by a plane wave (8.23),

$$e^{i\mathbf{k}_f \cdot \mathbf{r}} \tag{8.85}$$

As a result the matrix element for carrier-ionized impurity interaction is given by the simple expression,

$$(\mathcal{H}_{II})_{ni,nf} = \pm \frac{1}{NV_C} \int_{NV_C} d\mathbf{r}\, e^{-i\mathbf{k}_i \cdot \mathbf{r}} \frac{e^2}{4\pi\varepsilon_0\varepsilon_r r} e^{i\mathbf{k}_f \cdot \mathbf{r}}$$

$$= \pm \frac{e^2}{4\pi\varepsilon_0\varepsilon_r} \frac{1}{NV_C} \int_{NV_C} d\mathbf{r}\, e^{i(\mathbf{k}_f - \mathbf{k}_i)\cdot\mathbf{r}} \frac{1}{r} \tag{8.86}$$

The index n denotes the band where the electrons or holes reside. The $-$ applies for electrons scattering on ionized donors and holes scattering on ionized acceptors. The $+$ applies for electrons scattering on acceptors and holes on donors. As the wave functions are normalized over one unit of volume, the integral over the crystal with N primitive cells with volume V_C is normalized by dividing by NV_C. The integral in eqs (8.86) is treated in appendix F. We insert $V = NV_C$ and $\mathbf{k} = \mathbf{k}_f - \mathbf{k}_i$ in eq. (F.10) and find

$$(\mathcal{H}_{II})_{ni,nf} = \pm \frac{e^2}{4\pi\varepsilon_0\varepsilon_r} \frac{4\pi}{NV_C} \frac{1}{|\mathbf{k}_f - \mathbf{k}_i|^2}$$

$$= \pm \frac{1}{NV_C} \frac{e^2}{\varepsilon_0\varepsilon_r |\mathbf{k}_f - \mathbf{k}_i|^2} \tag{8.87}$$

SCATTERING RATES

The matrix element (8.87) is fully equivalent to the matrix elements (8.32) and (8.33) for phonon creation and annihilation. As in the latter case we use Fermi's golden rule (C. 67) to calculate the probability that a charge carrier is scattered on an ionized impurity. We find

$$W_{ni,nf} = \frac{2\pi}{\hbar} |(\mathcal{H}_{II})_{ni,nf}|^2 \delta(E_f^n - E_i^n)$$

$$= \frac{2\pi}{\hbar} \left| \frac{1}{NV_C} \frac{e^2}{\varepsilon_0\varepsilon_r |\mathbf{k}_f - \mathbf{k}_i|^2} \right|^2 \delta(E_f^n - E_i^n)$$

$$= \frac{2\pi}{\hbar} \frac{1}{N^2 V_C^2} \frac{e^4}{\varepsilon_0^2\varepsilon_r^2 |\mathbf{k}_f - \mathbf{k}_i|^4} \delta(E_f^n - E_i^n) \tag{8.88}$$

At this point several comments are appropriate. First, because the matrix element (8.87) has been squared, the $\pm$ signs have disappeared and hence the scattering rate is the same, whether scattering takes place on donors or on acceptors. Next, the above equation gives the probability for a charge carrier to scatter on a single ionized impurity. When there are N_{ii} ionized impurities in the crystal,

$$n_{ii} = \frac{N_{ii}}{NV_C} \qquad (8.89)$$

is the concentration of these impurities, i.e. their number per unit volume and the scattering probability can be rewritten as,

$$W_{ni,nf}^{ii} = \frac{2\pi}{\hbar} \frac{n_{ii}}{NV_C} \frac{e^4}{\varepsilon_0^2 \varepsilon_r^2 |\mathbf{k}_f - \mathbf{k}_i|^4} \delta(E_f^n - E_i^n) \qquad (8.90)$$

The third remark concerns the Dirac δ-function in eqs (8.88) and (8.90). The impurity does not take any energy from the charge carrier, nor does it yield any energy to it. Thus, this type of scattering is fully elastic. This is reflected in this Dirac δ-function which requires conservation of the energy of the carrier,

$$E_f^n = E_i^n \qquad (8.91)$$

Because of this latter property, we cannot exclude the special case that $\mathbf{k}_f = \mathbf{k}_i$. This gives rise to a final remark. Eqs (8.88) and (8.90) have the unpleasant feature that they diverge for this special case. In section 8.2.7 will consider this divergence in more detail.

DENSITY OF STATES

As in the previous section the differential density of states (8.42) is a useful device to integrate the Dirac δ-function out of eq. (8.90). The treatment is completely the same as the derivation leading from eqs (8.34) and (8.35) to eqs (8.58) and (8.60). Then, equivalently to eq. (8.58), the probability that a charge carrier in the initial state (8.84) is scattered in a unit solid angle in the (θ_f, ϕ_f)-direction, is given by

$$W_i^{ii}(\theta_f, \phi_f) = \int_0^\infty dE_f^n \, N_d(E_f^n, \theta_f, \phi_f) \, W_{ni,nf}^{ii}$$

$$= \int_0^\infty dE_f^n \, N_d(E_f^n, \theta_f, \phi_f) \times \frac{2\pi}{\hbar} \frac{n_{ii}}{NV_C} \frac{e^4}{\varepsilon_0^2 \varepsilon_r^2 |\mathbf{k}_f - \mathbf{k}_i|^4} \delta(E_f^n - E_i^n)$$

$$= \frac{2\pi}{\hbar} \frac{e^4}{\varepsilon_0^2 \varepsilon_r^2 |\mathbf{k}_f - \mathbf{k}_i|^4} \frac{n_{ii}}{NV_C} N_d(E_i^n, \theta_f, \phi_f) \qquad (8.92)$$

It is worthwhile to recall that the density of states $N_d(E_i^n, \theta_f, \phi_f)$ represents the number of Bloch states in the crystal per unit of solid angle per unit of energy. By dividing by the volume NV_C of the crystal, we obtain the density of states per unit of volume. Now, also the concentration of ionized impurities, n_{ii}, is defined per unit of volume. Hence, as expected from the physical nature of the process, the volume of the crystal does not appear in the transition probability (8.92).

8.2.7 Screening

SCATTERING RATES FOR ISOTROPIC BANDS

The final result for the differential scattering rate due to ionized impurities has an innocent look, but proves to be nasty when considered in more detail. To see this we consider the special case treated in section 8.2.4, where we assumed that effective mass theory applied, while moreover, the effective mass was assumed to be isotropic. Then, the simple expression (8.57) can be used for the differential density of states. As a result, inserting eq. (3.13),

$$V_C V_B = (2\pi)^3 \tag{8.93}$$

the differential scattering rate can be written as,

$$W_i^{ii}(\theta_f, \phi_f) = \frac{2\pi}{\hbar} \frac{e^4}{\varepsilon_0^2 \varepsilon_r^2 |\mathbf{k}_f - \mathbf{k}_i|^4} \frac{n_{ii}}{N V_C} \frac{N}{V_B} \frac{m_e m^* k_f}{\hbar^2}$$

$$= \frac{1}{(2\pi)^2} \frac{e^4}{\varepsilon_0^2 \varepsilon_r^2} \frac{m_e m^* k_f}{\hbar^3 |\mathbf{k}_f - \mathbf{k}_i|^4} n_{ii} \tag{8.94}$$

Because of isotropy, we are free to choose the axes of our frame of reference. We choose the z-axis along the $\mathbf{k}_i$-vector. Then,

$$\begin{aligned}|\mathbf{k}_f - \mathbf{k}_i|^2 &= |\mathbf{k}_f|^2 + |\mathbf{k}_i|^2 - 2|\mathbf{k}_f| \times |\mathbf{k}_i| \cos \theta_f \\ &= k_f^2 + k_i^2 - 2k_f k_i \cos \theta_f = 2k_f^2(1 - \cos \theta_f)\end{aligned} \tag{8.95}$$

This was used because of isotropy; $E_f^n = E_i^n$ implies $k_f = k_i$. Hence, the differential scattering rate is given by,

$$W_i^{ii}(\theta_f, \phi_f) = \frac{1}{(2\pi)^2} \frac{e^4}{\varepsilon_0^2 \varepsilon_r^2} \frac{m_e m^*}{4\hbar^3 k_f^3 (1 - \cos \theta_f)^2} n_{ii} \tag{8.96}$$

which is independent of ϕ_f.

TOTAL SCATTERING RATES FOR ISOTROPIC BANDS

The problem we are trying to illustrate, becomes visible when we use eq. (8.38) to calculate the total scattering rate[3]. Inserting eq. (8.96) in eq. (8.38) yields,

---

[3] Note that the total scattering rate calculated here is *not* the inverse of the relaxation rate following from analytical solutions of the Boltzmann equation. In such analytical solutions scattering events backwards count more heavily than scattering events in forward direction. Here all scattering events are counted with an equal weight.

$$W_{tot}^{ii} = \int_0^{2\pi} d\phi_f \int_0^\pi d\theta_f \sin\theta_f \, W_i^{ii}(\theta_f, \phi_f)$$

$$= \int_0^{2\pi} d\phi_f \int_0^\pi d\theta_f \sin\theta_f \frac{1}{(2\pi)^2} \frac{e^4}{\varepsilon_0^2 \varepsilon_r^2} \frac{m_e m^*}{4\hbar^3 k_f^3 (1-\cos\theta_f)^2} n_{ii}$$

$$= \frac{1}{(2\pi)^2} \frac{e^4}{\varepsilon_0^2 \varepsilon_r^2} \frac{m_e m^*}{4\hbar^3 k_f^3} n_{ii} \int_0^{2\pi} d\phi_f \int_0^\pi d\theta_f \sin\theta_f \frac{1}{(1-\cos\theta_f)^2}$$

$$= \frac{1}{(2\pi)^2} \frac{e^4}{\varepsilon_0^2 \varepsilon_r^2} \frac{m_e m^*}{4\hbar^3 k_f^3} n_{ii} \, 2\pi \int_{-1}^{+1} \frac{d\cos\theta_f}{(1-\cos\theta_f)^2}$$

$$= \frac{1}{(2\pi)} \frac{e^4}{\varepsilon_0^2 \varepsilon_r^2} \frac{m_e m^*}{4\hbar^3 k_f^3} n_{ii} \int_0^2 \frac{dx}{x^2} \tag{8.97}$$

where we inserted

$$x \equiv 1 - \cos\theta_f \tag{8.98}$$

Clearly, the integral,

$$\int_0^2 \frac{dx}{x^2} = \left[-\frac{1}{x}\right]_0^2 \to +\infty \tag{8.99}$$

diverges. Hence, according to our derivation, the total scattering rate due to interaction between charge carriers and ionized impurities *diverges*.

SCREENING EFFECTS

The problem is generally solved by taking into account that the charges of ionized donors and acceptors are never fully isolated. They are surrounded by other donors and acceptors and by electrons or holes. On a large space scale the average charge is expected to be equal to zero, so opposite charges compensate each other fully. Thus at a large distance the charge of a scattering ionized donor or acceptor is expected to be compensated by surrounding charges. This effect is called screening. It is generally accounted for by adding an exponential factor $\exp(-\alpha|\mathbf{r}|)$ to the Coulomb potential, where $1/\alpha$ is a characteristic length at which the potential of the other charges compensates the potential of the scattering ionized impurity. As a result the effective potential (8.82) of a donor is replaced by

$$\bar{U}_d(\mathbf{r}) = \frac{e}{4\pi\varepsilon_0\varepsilon_r r} e^{-\alpha r} \tag{8.100}$$

where $r = |\mathbf{r}|$ and similarly for the potential (8.83) of an acceptor. We now repeat the treatment leading to eq. (8.87) for the matrix element of this potential between two plane wave states. Eq. (8.86) is now written as

$$(\mathcal{H}_{II})_{ni,nf} = \pm \frac{e^2}{4\pi\varepsilon_0\varepsilon_r} \frac{1}{NV_C} \int_{NV_C} d\mathbf{r} \, e^{i(\mathbf{k}_f - \mathbf{k}_i)\cdot\mathbf{r}} \frac{1}{r} e^{-\alpha r} \tag{8.101}$$

8.2 CARRIER SCATTERING

The integral is treated in appendix F. Inserting $V = NV_C$ and $\mathbf{k} = \mathbf{k}_f - \mathbf{k}_i$ in eq. (F.9), we find,

$$(\mathcal{H}_{II})_{ni,nf} = \pm \frac{1}{NV_C} \frac{e^2}{\varepsilon_0 \varepsilon_r (\alpha^2 + |\mathbf{k}_f - \mathbf{k}_i|^2)} \tag{8.102}$$

Hence, we have replaced $|\mathbf{k}_f - \mathbf{k}_i|^2$ by $(\alpha^2 + |\mathbf{k}_f - \mathbf{k}_i|^2)$. This replacement can also be made in all subsequent steps leading to the differential scattering rate (9.92). Taking screening into account, this latter equation must therefore be written as,

$$W_i^{ii}(\theta_f, \phi_f) = \frac{2\pi}{\hbar} \frac{e^4}{\varepsilon_0^2 \varepsilon_r^2 (\alpha^2 + |\mathbf{k}_f - \mathbf{k}_i|^2)^2} \frac{n_{ii}}{NV_C} N_d(E_i^n, \theta_f, \phi_f) \tag{8.103}$$

Finally, making the replacement (8.95), the isotropic case yields

$$W_i^{ii}(\theta_f, \phi_f) = \frac{1}{(2\pi)^2} \frac{e^4}{\varepsilon_0^2 \varepsilon_r^2} \frac{m_e m^*}{\hbar^3 k_f^3 [(\alpha^2/k_f^2) + 2(1 - \cos\theta_f)]^2} n_{ii} \tag{8.104}$$

TOTAL SCATTERING RATES WITH SCREENING

As above we continue calculating the total scattering rate using eq. (8.38). Equivalently to eq. (8.97), we find

$$\begin{aligned}
W_{tot}^{ii} &= \int_0^{2\pi} d\phi_f \int_0^{\pi} d\theta_f \, \sin\theta_f \, W_i^{ii}(\theta_f, \phi_f) \\
&= \int_0^{2\pi} d\phi_f \int_0^{\pi} d\theta_f \, \sin\theta_f \, \frac{1}{(2\pi)^2} \frac{e^4}{\varepsilon_0^2 \varepsilon_r^2} \frac{m_e m^*}{\hbar^3 k_f^3 [(\alpha^2/k_f^2) + 2(1 - \cos\theta_f)]^2} n_{ii} \\
&= \frac{1}{(2\pi)^2} \frac{e^4}{\varepsilon_0^2 \varepsilon_r^2} \frac{m_e m^*}{4\hbar^3 k_f^3} n_{ii} \int_0^{2\pi} d\phi_f \int_0^{\pi} d\theta_f \, \sin\theta_f \, \frac{1}{[(\alpha^2/2k_f^2) + (1 - \cos\theta_f)]^2} \\
&= \frac{1}{(2\pi)^2} \frac{e^4}{\varepsilon_0^2 \varepsilon_r^2} \frac{m_e m^*}{4\hbar^3 k_f^3} n_{ii} \, 2\pi \int_{-1}^{+1} \frac{d\cos\theta_f}{[(\alpha^2/2k_f^2) + (1 - \cos\theta_f)]^2} \\
&= \frac{1}{(2\pi)} \frac{e^4}{\varepsilon_0^2 \varepsilon_r^2} \frac{m_e m^*}{4\hbar^3 k_f^3} n_{ii} \int_{\alpha^2/2k_f^2}^{2+\alpha^2/2k_f^2} \frac{dx}{x^2} \tag{8.105}
\end{aligned}$$

where we inserted

$$x \equiv (\alpha^2/2k_f^2) + 1 - \cos\theta_f \tag{8.106}$$

Now the integral,

$$\int_{\alpha^2/2k_f^2}^{2+\alpha^2/2k_f^2} \frac{dx}{x^2} = \left[-\frac{1}{x}\right]_{\alpha^2/2k_f^2}^{2+\alpha^2/2k_f^2} \approx \frac{2k_f^2}{\alpha^2} \tag{8.107}$$

does not diverge, so the total scattering rate

$$W_{tot}^{ii} = \frac{1}{(2\pi)} \frac{e^4}{\varepsilon_0^2 \varepsilon_r^2} \frac{m_e m^*}{4\hbar^3 k_f^3} n_{ii} \frac{2k_f^2}{\alpha^2} = \frac{e^4}{4\pi \varepsilon_0^2 \varepsilon_r^2} \frac{m_e m^*}{\hbar^3 k_f \alpha^2} n_{ii} \qquad (8.108)$$

remains finite.

Thus, the introduction of screening solves the problem of the divergence of the total scattering rate. Note however, that the problem is not really solved, but rather replaced by another one: the nature of screening. As is seen from eq. (8.108), the characteristic length $1/\alpha$ determining the distance at which the Coulomb potentials of ionized donors and acceptors are screened, shows up prominently in the total scattering rate. Hence, a good model of screening is imperative for a good description of ionized impurity scattering. This appears to be precisely the weakness of the present description where the influence of ionized impurities is treated as a scattering process. Various models have been suggested, each of them working well for certain conditions, but none of them has been able to give satisfactory results under all circumstances. For an overview see ref. [32]. Because of this weakness of the theory we refrain from presenting estimates of the rate of impurity scattering. On the other hand, curve C in Fig. 8.3 presents total scattering rates for ionized impurity scattering obtained by Moglestue [27].

Finally, once screening is introduced for ionized impurity scattering, one should consider introducing it into other scattering processes resulting from long range Coulomb type interactions. Otherwise, the description of transport might become inconsistent. This applies in particular to polar optical phonon scattering and piezoelectric scattering, that are both due to such long range potentials. Fortunately, in these latter two cases, the total relaxation rates do not diverge if screening is omitted. As a result, screening just constitutes a higher order correction to these processes, rather than fully determining the actual scattering rates.

9 Optical Transitions

9.1 Band Electrons in an Optical Field

In the past decades semiconductors have become the most important active media in optical devices and are used in photodetectors, incoherent light sources and lasers. Moreover, optical spectroscopy, using methods ranging from off the shelf Fourier transform spectrometers to the most sophisticated femtosecond laser systems, has proven itself as a powerful tool to investigate semiconductors. Therefore, a treatment of semiconductors is incomplete, unless the influence of rapidly varying optical fields has been covered.

Linear and non-linear optical spectroscopy is the subject of various specialized monographs (see [22], [28] and [30]). As earlier parts of this book, this chapter is an introduction to the subject, presenting the link between basic quantum mechanics and the advanced treatments given in monographs. We restrict ourselves to linear effects due to the interaction of band electrons with light in its simplest form, i.e., with a monochromatic electromagnetic plane wave. The basic structure of this chapter is very similar to the framework underlying the previous three chapters.

We start by introducing a convenient description of the optical field. For reasons of simplicity we will treat this field classically, contrary to chapter 6, where the lattice was quantized. The present approach has one clear disadvantage: spontaneous emission of light cannot be taken into account. However, the loss of this feature does not outweigh the simplicity gained by restricting to the classical picture.

Next we consider the interaction of band electrons with the optical field. In particular, following the scheme of section 7.1, we calculate transition matrix elements between Bloch states. As in chapter 7 we will see that it is sometimes possible to rewrite the result in terms of effective mass theory. Then, the single electron Hamiltonian is replaced by the *effective* Hamiltonian and transition matrix elements are obtained between two *plane* waves instead of Bloch functions. While much easier to evaluate, the effective mass approach still yields the same results numerically.

After having derived the necessary matrix elements, we continue following the scheme of chapter 8 to calculate the probabilities of various optical transitions. As mentioned above, we will refrain from non-linear optics. Still, it is impossible to account for many important linear optical effects without using second-order perturbation theory, combining matrix elements resulting from the interaction with the optical field and matrix elements resulting from electron-phonon interaction. In this respect optical transitions are more complicated than scattering processes.

Because of these second order effects, a large variety of transitions can be induced by an optical field and inevitably a choice has to be made. First we note that interband transitions play a dominant role in practical applications. Therefore, section 9.2.1 and 9.2.2 will treat interband transitions in direct and indirect semiconductors. On the other hand a particularly interesting case arises when studying optical transitions between eigenstates of an effective Hamiltonian. As an example section 9.2.2 presents optical transitions between subband levels in a quantum well. We will see that effective mass theory can be used to treat both the energy levels and the optical transitions between them. Thus, the underlying complications of the band structure have disappeared completely from the description.

However, before being able to treat these three cases, it is necessary to provide the basic formalism for the coupling between band electrons and the optical field. This will be done in sections 9.1.1–9.1.5. For simplicity we restrict our treatment of intraband transitions to non-degenerate bands.

9.1.1 The Optical Field

THE GAUGE OF AN OPTICAL FIELD

We start by reconsidering the treatment of section 5.1.2, where we derived the extra terms in the single electron Hamiltonian due to externally applied electromagnetic fields. In the special case of an electromagnetic plane wave it is possible and practical to choose a gauge where the scalar potential is equal to zero,

$$U(\mathbf{r}, t) = 0 \tag{9.1}$$

$$\mathbf{A}(\mathbf{r}, t) = \mathbf{A}_0 \cos(\mathbf{k} \cdot \mathbf{r} - \omega t) \tag{9.2}$$

Thus, the vector potential propagates as a plane wave in the **k**-direction with a velocity $c = \omega/|\mathbf{k}|$. Then $\mathbf{A}_0$ is the amplitude of this plane wave, while **k** is its wavevector and ω its frequency. To be consistent with the Coulomb gauge chosen in chapter 5, we furthermore demand

$$\begin{aligned} 0 = \nabla \cdot \mathbf{A}(\mathbf{r}, t) &= \nabla \cdot [\mathbf{A}_0 \cos(\mathbf{k} \cdot \mathbf{r} - \omega t)] \\ &= -[\nabla(\mathbf{k} \cdot \mathbf{r})] \cdot \mathbf{A}_0 \sin(\mathbf{k} \cdot \mathbf{r} - \omega t) \\ &= -(\mathbf{k} \cdot \mathbf{A}_0) \sin(\mathbf{k} \cdot \mathbf{r} - \omega t) \end{aligned} \tag{9.3}$$

Here we used

$$\nabla_\mu (\mathbf{k} \cdot \mathbf{r}) = \frac{\partial}{\partial r_\mu} \sum_\nu k_\nu r_\nu = \sum_\nu k_\nu \delta_{\nu,\mu} = k_\mu \tag{9.4}$$

Hence, the vector potential is oriented perpendicularly to the propagation direction of the plane wave,

$$\mathbf{k} \cdot \mathbf{A}_0 = 0 \tag{9.5}$$

9.1 BAND ELECTRONS IN AN OPTICAL FIELD

This choice of potentials leads to electric and magnetic fields that are given by

$$\mathbf{E}(\mathbf{r},t) = -\nabla U(\mathbf{r},t) - \frac{\partial}{\partial t}\mathbf{A}(\mathbf{r},t)$$

$$= -\mathbf{A}_0 \frac{\partial}{\partial t} \cos(\mathbf{k}\cdot\mathbf{r} - \omega t)$$

$$= \omega \mathbf{A}_0 \sin(\mathbf{k}\cdot\mathbf{r} - \omega t) \quad (9.6)$$

$$\mathbf{B}(\mathbf{r},t) = \nabla \times \mathbf{A}(\mathbf{r},t)$$

$$= \nabla \times (\mathbf{A}_0 \cos(\mathbf{k}\cdot\mathbf{r} - \omega t))$$

$$= -\{\nabla(\mathbf{k}\cdot\mathbf{r})\} \times \mathbf{A}_0 \sin(\mathbf{k}\cdot\mathbf{r} - \omega t)$$

$$= -(\mathbf{k} \times \mathbf{A}_0) \sin(\mathbf{k}\cdot\mathbf{r} - \omega t) \quad (9.7)$$

So the electric and magnetic fields propagate as plane waves in the **k**-direction with a velocity $c = \omega/|\mathbf{k}|$. Their amplitudes are given by

$$\mathbf{E}_0 = \omega \mathbf{A}_0 \quad \text{and} \quad \mathbf{B}_0 = -(\mathbf{k} \times \mathbf{A}_0) \quad (9.8)$$

Hence, remembering eq. (9.5),

$$\mathbf{E}(\mathbf{r},t) \perp \mathbf{B}(\mathbf{r},t) \quad (9.9)$$

$$\mathbf{B}(\mathbf{r},t) \perp \mathbf{k}, \quad \mathbf{E}(\mathbf{r},t) \perp \mathbf{k} \quad (9.10)$$

$$\text{and} \quad \frac{|\mathbf{E}(\mathbf{r},t)|}{|\mathbf{B}(\mathbf{r},t)|} = \frac{\omega}{|\mathbf{k}|} \quad (9.11)$$

which is the velocity of light in the crystal, where the relative permittivity is equal to ε_r. This is clearly consistent with an electromagnetic plane wave as solved from Maxwell's equations [25], provided

$$\frac{\omega}{|\mathbf{k}|} = \frac{1}{\sqrt{\mu_0 \varepsilon_0 \varepsilon_r}} = \frac{c}{\sqrt{\varepsilon_r}} \quad (9.12)$$

Here we approximate the relative permeability μ_r of the semiconductor to be equal to 1, as we did earlier in section 5.1.2.

LIGHT INTENSITY

The light intensity, i.e. the energy flux through the semiconductor, follows from the Poynting vector, which in the present gauge is equal to

$$\mathbf{P}(\mathbf{r},t) = \mathbf{E}(\mathbf{r},t) \times \mathbf{H}(\mathbf{r},t)$$

$$= \frac{1}{\mu_0}[\mathbf{E}_0(\mathbf{r},t) \times \mathbf{B}_0(\mathbf{r},t)] \sin^2(\mathbf{k}\cdot\mathbf{r} - \omega t)$$

$$= -\frac{1}{\mu_0}[\omega \mathbf{A}_0 \times (\mathbf{k} \times \mathbf{A}_0)] \sin^2(\mathbf{k}\cdot\mathbf{r} - \omega t)$$

$$= -\frac{1}{\mu_0}\omega[(\mathbf{A}_0 \cdot \mathbf{A}_0)\mathbf{k} - (\mathbf{A}_0 \cdot \mathbf{k})\mathbf{A}_0] \sin^2(\mathbf{k}\cdot\mathbf{r} - \omega t)$$

$$= -\frac{1}{\mu_0}\omega \mathbf{k} |\mathbf{A}_0|^2 \sin^2(\mathbf{k}\cdot\mathbf{r} - \omega t) \quad (9.13)$$

where we inserted eqs (9.5), (9.6) and (9.7) as well as the identity

$$\mathbf{A} \times (\mathbf{B} \times \mathbf{C}) = (\mathbf{A} \cdot \mathbf{C})\mathbf{B} - (\mathbf{A} \cdot \mathbf{B})\mathbf{C} \tag{9.14}$$

from vector calculus. We see that the energy flux is parallel to the **k**-vector, as we expect it to be for plane waves. After averaging over a single oscillation and inserting eqs (9.8) and (9.12), the light intensity is given by

$$I = \frac{\omega^2 \sqrt{\varepsilon_r}}{2\mu_0 c} |\mathbf{A}_0|^2 = \frac{\omega^2}{2} \sqrt{\frac{\varepsilon_0 \varepsilon_r}{\mu_0}} |\mathbf{A}_0|^2 = \frac{1}{2} \sqrt{\frac{\varepsilon_0 \varepsilon_r}{\mu_0}} |\mathbf{E}_0|^2 \tag{9.15}$$

9.1.2 Bloch Electrons in an Optical Field

OPTICAL INTERACTION TERMS IN THE SINGLE ELECTRON HAMILTONIAN

The next step in our treatment consists of determining the modification of the single electron Hamiltonian due to the optical field. This step is straightforward if we insert the potentials (9.1) and (9.2) in the single electron Hamiltonian (5.14)–(5.16) derived in section 5.1.2. In particular we chose the scalar potential to be equal to zero, so we may immediately write,

$$\mathcal{H}'(\mathbf{r}, t) = \mathcal{H}_0(\mathbf{r}) + \mathcal{H}_F(\mathbf{r}, t) \tag{9.16}$$

where

$$\mathcal{H}_0(\mathbf{r}) = -\frac{\hbar^2}{2m_e} \Delta + \mathcal{V}'(\mathbf{r}) \tag{9.17}$$

is the single electron Hamiltonian in the absence of the optical field and

$$\mathcal{H}_F(\mathbf{r}, t) = \frac{e}{m_e} \mathbf{A}(\mathbf{r}, t) \cdot \frac{\hbar}{i} \nabla + \frac{e^2}{m_e} |\mathbf{A}(\mathbf{r}, t)|^2 \tag{9.18}$$

is the additional term due to this field. In the present chapter we restrict ourselves to linear effects due to the optical field. Thus we will only consider terms in the Hamiltonian that are linear with $\mathbf{E}_0(\mathbf{r}, t) = \omega \mathbf{A}_0(\mathbf{r}, t)$. Among other things, this implies that we will neglect the second term in $\mathcal{H}_F(\mathbf{r}, t)$. Inserting eq. (9.2) in the first term, the interaction of a band electron with the optical field reduces to

$$\mathcal{H}_F(\mathbf{r}, t) = \frac{e}{m_e} \mathbf{A}_0 \cos(\mathbf{k} \cdot \mathbf{r} - \omega t) \cdot \frac{\hbar}{i} \nabla$$

$$= \frac{e}{2m_e} \mathbf{A}_0 e^{i\mathbf{k}_q \cdot \mathbf{r} - i\omega t} \cdot \frac{\hbar}{i} \nabla + \frac{e}{2m_e} \mathbf{A}_0 e^{-i\mathbf{k}_q \cdot \mathbf{r} + i\omega t} \cdot \frac{\hbar}{i} \nabla \tag{9.19}$$

where we split the cosine into two exponential functions, which will prove to be convenient in the following steps.

BLOCH STATES

To investigate transitions that may be induced by the optical field, we need to calculate transition matrix elements, i.e. matrix elements of $\mathcal{H}_F(\mathbf{r}, t)$ between eigen-

9.1 BAND ELECTRONS IN AN OPTICAL FIELD

states of the unperturbed Hamiltonian (9.17). As in section 7.1.1, we choose these eigenstates to be an initial Bloch state

$$\frac{1}{\sqrt{NV_C}} b_i^n(\mathbf{r}) = \frac{1}{\sqrt{NV_C}} u_i^n(\mathbf{r}) e^{i\mathbf{k}_i \cdot \mathbf{r}} \qquad (9.20)$$

and a final Bloch state

$$\frac{1}{\sqrt{NV_C}} b_f^m(\mathbf{r}) = \frac{1}{\sqrt{NV_C}} u_f^m(\mathbf{r}) e^{i\mathbf{k}_f \cdot \mathbf{r}} \qquad (9.21)$$

Here n and m are band indices while i and f indicate that the initial and final Bloch states are located at the positions $\mathbf{k}_i$ and $\mathbf{k}_f$ in the first Brillouin zone. The crystal is assumed to have a finite volume NV_C, where $N = L_1 L_2 L_3$ is the number of primitive cells and V_C the size of such a cell. As Bloch functions $b_i^n(\mathbf{r})$ are normalized over one unit of volume, an extra factor $1/\sqrt{NV_C}$ is added in order to render the functions (9.20) and (9.21) normalized over the volume of the crystal.

As a further consequence of considering a finite crystal size, we need to choose boundary conditions at the surface of the crystal. As usual we will choose periodic boundary conditions. Then, $\mathbf{k}_i$ and $\mathbf{k}_f$ may take the discrete values (3.52) only, and can be written as

$$\mathbf{k}_i = (i_1/L_1)\mathbf{b}_1 + (i_2/L_2)\mathbf{b}_2 + (i_3/L_3)\mathbf{b}_3 \qquad (9.22)$$

$$\mathbf{k}_f = (f_1/L_1)\mathbf{b}_1 + (f_2/L_2)\mathbf{b}_2 + (f_3/L_3)\mathbf{b}_3 \qquad (9.23)$$

Because we require periodic boundary conditions, the optical wave vector is also restricted to discrete values,

$$\mathbf{k}_q = (q_1/L_1)\mathbf{b}_1 + (q_2/L_2)\mathbf{b}_2 + (q_3/L_3)\mathbf{b}_3 \qquad (9.24)$$

Thus, the indices i, f and q are shorthand notations for $\mathbf{i} = (i_1, i_2, i_3)$, $\mathbf{f} = (f_1, f_2, f_3)$ and $\mathbf{q} = (q_1, q_2, q_3)$ indicating the discrete values that $\mathbf{k}_i$, $\mathbf{k}_f$ and $\mathbf{k}_q$ may take. In this notation the energy of an electron in the states $\mathbf{k}_i$ and $\mathbf{k}_f$ is given by E_i^n and E_f^m, respectively.

TRANSITION MATRIX ELEMENTS

Optical transition matrix elements are now defined as

$$(\mathcal{H}_F)_{ni,mf} = \frac{1}{NV_C} \int_{NV_C} d\mathbf{r} \, b_i^{n*}(\mathbf{r}) \mathcal{H}_F(\mathbf{r}, t) b_f^m(\mathbf{r})$$

$$= \frac{1}{NV_C} \int_{NV_C} d\mathbf{r} \, u_i^{n*}(\mathbf{r}) e^{-i\mathbf{k}_i \cdot \mathbf{r}} \frac{e}{2m_e} \left(\mathbf{A}_0 \, e^{i\mathbf{k}_q \cdot \mathbf{r} - i\omega t} \cdot \frac{\hbar}{i} \nabla \right) u_f^m(\mathbf{r}) e^{i\mathbf{k}_f \cdot \mathbf{r}}$$

$$+ \frac{1}{NV_C} \int_{NV_C} d\mathbf{r} \, u_i^{n*}(\mathbf{r}) e^{-i\mathbf{k}_i \cdot \mathbf{r}} \frac{e}{2m_e} \left(\mathbf{A}_0 \, e^{-i\mathbf{k}_q \cdot \mathbf{r} + i\omega t} \cdot \frac{\hbar}{i} \nabla \right) u_f^m(\mathbf{r}) e^{i\mathbf{k}_f \cdot \mathbf{r}} \qquad (9.25)$$

where we insert eqs (9.19)–(9.21). To evaluate this expression we recall eq. (4.4) which stood at the basis of the $\mathbf{k} \cdot \mathbf{p}$-approximation,

$$\frac{\hbar}{i} \nabla u_f^m(\mathbf{r}) e^{i\mathbf{k}_f \cdot \mathbf{r}} = e^{i\mathbf{k}_f \cdot \mathbf{r}} \left(\frac{\hbar}{i} \nabla + \hbar \mathbf{k}_f \right) u_f^m(\mathbf{r}) \qquad (9.26)$$

Then the first term in eq. (9.25) becomes

$$\frac{1}{NV_C}\int_{NV_C} d\mathbf{r}\, u_i^{n*}(\mathbf{r})e^{i(\mathbf{k}_q-\mathbf{k}_i+\mathbf{k}_f)\cdot\mathbf{r}}\frac{e}{2m_e}\mathbf{A}_0 e^{-i\omega t}\cdot\left(\frac{\hbar}{i}\nabla+\hbar\mathbf{k}_f\right)u_f^m(\mathbf{r}) \quad (9.27)$$

while a similar expression is found for the second term. We continue considering the first term. As in section 5.1.3, we split the integral over full space into a sum of integrals over the individual primitive cells. Next, in the exponential function we replace $\mathbf{r}$ by $(\mathbf{r}-\mathbf{R}_l)+\mathbf{R}_l$. As a result the first term in eq. (9.25) becomes

$$\frac{1}{N}\sum_l e^{i(\mathbf{k}_q-\mathbf{k}_i+\mathbf{k}_f)\cdot\mathbf{R}_l}\frac{1}{V_C}\int_{V_C^l} d\mathbf{r}\, u_i^{n*}(\mathbf{r})e^{i(\mathbf{k}_q-\mathbf{k}_i+\mathbf{k}_f)\cdot(\mathbf{r}-\mathbf{R}_l)}$$

$$\times\frac{e}{2m_e}\mathbf{A}_0 e^{-i\omega t}\cdot\left(\frac{\hbar}{i}\nabla+\hbar\mathbf{k}_f\right)u_f^m(\mathbf{r}) \quad (9.28)$$

where $\int_{V_C^l}$ denotes integration over the lth primitive cell. Now, as in section 5.1.3 the functions $u_i^n(\mathbf{r})$ and $u_f^m(\mathbf{r})$ obey the translational symmetry of the crystal lattice. Therefore the integral is independent of the primitive cell and we may replace the integral over the lth primitive by an integral over the primitive cell at the origin of our frame of reference, where $\mathbf{R}_l = 0$. Using also the Dirac notation defined in eqs (4.23)–(4.26), we then obtain for the first term in eq. (9.25),

$$\frac{1}{N}\sum_l e^{i(\mathbf{k}_q-\mathbf{k}_i+\mathbf{k}_f)\cdot\mathbf{R}_l}\frac{1}{V_C}\int_{V_C^0} d\mathbf{r}\, u_i^{n*}(\mathbf{r})\, e^{i(\mathbf{k}_q-\mathbf{k}_i+\mathbf{k}_f)\cdot(\mathbf{r}-\mathbf{R}_l)}$$

$$\times\frac{e}{2m_e}\mathbf{A}_0 e^{-i\omega t}\cdot\left(\frac{\hbar}{i}\nabla+\hbar\mathbf{k}_f\right)u_f^m(\mathbf{r})$$

$$=\frac{1}{N}\sum_l e^{i(\mathbf{k}_q-\mathbf{k}_i+\mathbf{k}_f)\cdot\mathbf{R}_l}\langle nk_i|e^{i(\mathbf{k}_q-\mathbf{k}_i+\mathbf{k}_f)\cdot(\mathbf{r}-\mathbf{R}_l)}$$

$$\times\frac{e}{2m_e}\mathbf{A}_0 e^{-i\omega t}\cdot\left(\frac{\hbar}{i}\nabla+\hbar\mathbf{k}_f\right)|mk_f\rangle$$

$$=\langle nk_i|\frac{e}{2m_e}\mathbf{A}_0 e^{-i\omega t}\cdot\left(\frac{\hbar}{i}\nabla+\hbar\mathbf{k}_f\right)|mk_f\rangle\, \delta_{q,q_{if}} \quad (9.29)$$

where we insert the Kronecker δ-function (3.57), which requires that

$$\mathbf{k}_q = \mathbf{k}_{q_{if}} = \mathbf{k}_i - \mathbf{k}_f \quad (9.30)$$

and subsequently use this δ-function to evaluate the exponential function within the Dirac brackets.

The second term in eq. (9.25) is evaluated in exactly the same way, except that $-\omega$ is replaced by ω and $\mathbf{k}_q$ by $-\mathbf{k}_q$. Thus, this second term is found to be given by

$$\langle nk_i|\frac{e}{2m_e}\mathbf{A}_0 e^{i\omega t}\cdot\left(\frac{\hbar}{i}\nabla+\hbar\mathbf{k}_f\right)|mk_f\rangle\, \delta_{-q,q_{if}} \quad (9.31)$$

9.1 BAND ELECTRONS IN AN OPTICAL FIELD

Then, combining both terms, we find for the optical transition matrix element,

$$(\mathcal{H}_F)_{ni,mf} = \frac{e}{2m_e} \mathbf{A}_0 \cdot \langle nk_i | \left(\frac{\hbar}{i}\nabla + \hbar\mathbf{k}_f\right) | mk_f \rangle (e^{-i\omega t} \delta_{q,q_{if}} + e^{i\omega t} \delta_{-q,q_{if}}) \quad (9.32)$$

At this point we note that the wavelengths corresponding to optical fields are extremely long compared to the size of the primitive cells. This is even valid for ultraviolet light with a wavelength of 100 nm corresponding to a photon energy of 12.4 eV, because the primitive cells of semiconductors have dimensions of typically 0.5 nm only. Hence, the wave vector $\mathbf{k}_q$ of an optical field may generally be neglected with respect to the wave vectors $\mathbf{k}_i$ and $\mathbf{k}_f$ describing the initial and final state of a band electron. As a result, the optical transition matrix element $(\mathcal{H}_F)_{ni,mf}$ simplifies to

$$\begin{aligned}(\mathcal{H}_F)_{ni,mf} &= \frac{e}{2m_e} \mathbf{A}_0 \cdot \langle nk_i | \left(\frac{\hbar}{i}\nabla + \hbar\mathbf{k}_f\right) | mk_f \rangle (e^{-i\omega t} \delta_{i,f} + e^{i\omega t} \delta_{i,f}) \\ &= \frac{e}{2m_e} \mathbf{A}_0 \cdot \langle nk_i | \left(\frac{\hbar}{i}\nabla + \hbar\mathbf{k}_f\right) | mk_f \rangle (e^{-i\omega t} + e^{i\omega t}) \delta_{i,f} \\ &= \frac{e}{m_e} \mathbf{A}_0 \cdot \langle nk_i | \left(\frac{\hbar}{i}\nabla + \hbar\mathbf{k}_f\right) | mk_f \rangle \cos(\omega t) \, \delta_{i,f}\end{aligned} \quad (9.33)$$

where the Kronecker δ-function $\delta_{i,f}$ requires that $\mathbf{k}_f = \mathbf{k}_i$. Thus, the optical transition matrix element $(\mathcal{H}_F)_{ni,mf}$ is zero unless the wave vectors $\mathbf{k}_i$ and $\mathbf{k}_f$ of the two Bloch functions are equal. This particular property of optical transition matrix elements has a physical interpretation, which is found by taking into account that $\hbar\mathbf{k}_i$ and $\hbar\mathbf{k}_f$ represent crystal momentum for each of the two Bloch states. Then the optical transition matrix element is non-zero only, if the two Bloch states have equal crystal momentum, i.e. the optical field does not change crystal momentum. In a quantum description of the optical field we say that the photon absorbed or emitted by the electron has negligible momentum.

9.1.3 k·p Treatment of Transition Matrix Elements

Often the initial Bloch state is located near the Γ-point. Then we can exploit the finding above, that optical transition matrix elements vanish unless the wave vectors of the initial and final Bloch functions are equal. Thus the final Bloch state should lie near the Γ-point as well and the **k·p**-approximation offers a convenient way to evaluate eq. (9.33).

LOWEST ORDER TRANSITION MATRIX ELEMENTS

In lowest order approximation we simply approximate

$$|nk_i\rangle \approx |n0\rangle \quad (9.34)$$

$$|mk_f\rangle \approx |m0\rangle \quad (9.35)$$

So the optical transition matrix element (9.33) can be written as

$$(\mathcal{H}_F)_{ni,mf} \approx \frac{e}{m_e} \mathbf{A}_0 \cdot \langle n0| \left(\frac{\hbar}{i}\nabla + \hbar \mathbf{k}_f\right) |m0\rangle \cos(\omega t)\, \delta_{i,f}$$

$$= \frac{e}{m_e} \mathbf{A}_0 \cdot \left(\langle n0|\frac{\hbar}{i}\nabla|m0\rangle + \hbar \mathbf{k}_f \langle n0|m0\rangle\right) \cos(\omega t)\, \delta_{i,f}$$

$$= \frac{e}{m_e} \sum_\mu A_{0\mu} \left(\frac{\hbar}{i} P_\mu^{nm} + \hbar k_{f\mu} \delta_{n,m}\right) \cos(\omega t)\, \delta_{i,f}$$

$$= \frac{\hbar e}{i m_e} \sum_\mu A_{0\mu}\, P_\mu^{nm} \cos(\omega t)\, \delta_{i,f} + \frac{\hbar e}{m_e} \mathbf{A}_0 \cdot \mathbf{k}_f \cos(\omega t)\, \delta_{n,m}\, \delta_{i,f} \qquad (9.36)$$

where $\mu = x, y$ or z and where we insert the δ-function (4.25) and the matrix element P_μ^{nm} defined by eq. (4.36).

It is instructive to note the close resemblance of the resulting two terms with those resulting from the matrix representation of the perturbation in the $\mathbf{k}\cdot\mathbf{p}$-approximation given in section 4.1.2. To see this resemblance we ignore the factor $\cos(\omega t)\,\delta_{i,f}$. Then the first term in eq. (9.36) is completely equivalent to the first term in eq. (4.28), provided we replace the factor $e\mathbf{A}_0$ by $\hbar\mathbf{k}_f$. Equivalently, the second term in eq. (9.36) is completely equivalent to the second term in eq. (4.28), as long as we replace $e\mathbf{A}_0$ by $(1/2)\hbar\mathbf{k}_f$. Thus, also eqs (4.29) and (4.32) apply. Then the first term is non-zero only for $n \neq m$, while the second term is non-zero only in the opposite case that $n = m$.

In sections 9.2.1 and 9.2.2 we direct our attention to interband transitions, so $n \neq m$. Then the optical transition matrix element reduces to

$$(\mathcal{H}_F)_{ni,mf} = \frac{\hbar e}{i m_e} \sum_\mu A_{0\mu}\, P_\mu^{nm} \cos(\omega t)\, \delta_{i,f} \qquad (9.37)$$

FIRST-ORDER TRANSITION MATRIX ELEMENTS

At first sight the second term in eq. (9.36) is exactly what is needed to treat intraband transitions, i.e. transitions such that $n = m$. However, an estimate of its order of magnitude shows that this second term is very small, so we need to go one order higher in the $\mathbf{k}\cdot\mathbf{p}$-approximation to obtain a reliable result. To obtain such an estimate we refer to section 4.2.2. There we estimated P_μ^{nm} for the special case that $|n0\rangle$ corresponds to the bottom of the conduction band at the Γ-point and $|m0\rangle$ corresponds to the top of the valence band. We found P_μ^{nm} to be roughly equal to the length $|\mathbf{b}_\beta| = 2\sqrt{3}\pi/c$ of a vector $\mathbf{b}_\beta$ spanning a Brillouin zone, where c is the length of the cubic unit cell. On the other hand, the $\mathbf{k}\cdot\mathbf{p}$-approximation only applies for $|\mathbf{k}_i|$ and $|\mathbf{k}_f| \ll |\mathbf{b}_\beta|$. Hence, the second term in eq. (9.36) is much smaller than the first term.

To treat intraband coupling to first order in the $\mathbf{k}\cdot\mathbf{p}$ approximation we apply non-degenerate perturbation theory as developed in section C.1.1 of appendix C. We expand the final state $|mk_f\rangle$ and the initial state $\langle nk_i|$ using eq. (C.12) and insert

9.1 BAND ELECTRONS IN AN OPTICAL FIELD

them in the transition matrix element (9.33). According to section 4.1.1 the unperturbed Hamiltonian $\mathcal{H}^{(0)}$ and the perturbation $\mathcal{H}^{(1)}$ are given by eqs (4.7) and (4.8). Inserting eqs (4.27) to (4.29) for their matrix elements and eigenstates, the final state $|mk_f\rangle$ is found to be given by,

$$|mk_f\rangle = |m0\rangle + \frac{\hbar^2}{im_e} \sum_{p \neq m} |p0\rangle \frac{\langle p0|\mathcal{H}^{(1)}|m0\rangle}{E_0^m - E_0^p}$$

$$= |m0\rangle + \frac{\hbar^2}{im_e} \sum_{p \neq m} |p0\rangle \frac{\langle p0|(\mathbf{k}_f \cdot \nabla)|m0\rangle}{E_0^m - E_0^p}$$

$$= |m0\rangle + \frac{\hbar^2}{im_e} \sum_\mu k_{f\mu} \sum_{p \neq m} |p0\rangle \frac{\langle p0|\nabla_\mu|m0\rangle}{E_0^m - E_0^p} \quad (9.38)$$

where $\mu = x, y$ or z. Equivalently for the initial state $\langle nk_i|$ we find,

$$\langle nk_i| = \langle n0| + \frac{\hbar^2}{im_e} \sum_\mu k_{i\mu} \sum_{p \neq n} \frac{\langle n0|\nabla_\mu|p0\rangle}{E_0^n - E_0^p} \langle p0| \quad (9.39)$$

Note that for intraband coupling the initial and final states are in the same band. Hence, when inserting these expansions in eq. (9.33) we may set $n = m$. Then, up to first order the matrix element in eq. (9.33) becomes,

$$\langle nk_i|\left(\frac{\hbar}{i}\nabla_\nu + \hbar k_{f\nu}\right)|nk_f\rangle = \left(\langle n0| + \frac{\hbar^2}{im_e} \sum_\mu k_{i\mu} \sum_{p \neq n} \frac{\langle n0|\nabla_\mu|p0\rangle}{E_0^n - E_0^p} \langle p0|\right)\left(\frac{\hbar}{i}\nabla_\nu + \hbar k_{f\nu}\right)$$

$$\times \left(|n0\rangle + \frac{\hbar^2}{im_e} \sum_\mu k_{f\mu} \sum_{p \neq n} |p0\rangle \frac{\langle p0|\nabla_\mu|n0\rangle}{E_0^n - E_0^p}\right)$$

$$= \langle n0|\left(\frac{\hbar}{i}\nabla_\nu + \hbar k_{f\nu}\right)|n0\rangle$$

$$+ \frac{\hbar^2}{im_e} \sum_\mu \left(k_{f\mu} \sum_{p \neq n} \frac{\langle n0|\left(\frac{\hbar}{i}\nabla_\nu + \hbar k_{f\nu}\right)|p0\rangle\langle p0|\nabla_\mu|n0\rangle}{E_0^n - E_0^p}\right.$$

$$\left. + k_{i\mu} \sum_{p \neq n} \frac{\langle n0|\nabla_\mu|p0\rangle\langle p0|\left(\frac{\hbar}{i}\nabla_\nu + \hbar k_{f\nu}\right)|n0\rangle}{E_0^n - E_0^p}\right) \quad (9.40)$$

Equivalently to eq. (9.36), in this expression,

$$\langle n0|\left(\frac{\hbar}{i}\nabla_\nu + \hbar k_{f\nu}\right)|p0\rangle = \frac{\hbar}{i}P_\nu^{np} + \hbar k_{f\nu}\delta_{n,p} \quad (9.41)$$

where $P_\nu^{np} = 0$ if $n = p$. Hence, remembering eq. (4.36),

$$\langle nk_i | \left(\frac{\hbar}{i} \nabla_\nu + \hbar k_{f\nu} \right) | nk_f \rangle = \hbar k_{f\nu} - \frac{\hbar^3}{m_e} \sum_\mu \left(k_{f\mu} \sum_{p \neq n} \frac{P_\nu^{np} P_\mu^{pn}}{E_0^n - E_0^p} + k_{i\mu} \sum_{p \neq n} \frac{P_\mu^{np} P_\nu^{pn}}{E_0^n - E_0^p} \right)$$

$$= \frac{1}{2} \sum_\mu \hbar k_{f\mu} \left(\delta_{\mu,\nu} - \frac{2\hbar^2}{m_e} \sum_{p \neq n} \frac{P_\nu^{np} P_\mu^{pn}}{E_0^n - E_0^p} \right)$$

$$+ \frac{1}{2} \sum_\mu \hbar k_{f\mu} \left(\delta_{\mu,\nu} - \frac{2\hbar^2}{m_e} \sum_{p \neq n} \frac{P_\mu^{np} P_\nu^{pn}}{E_0^n - E_0^p} \right) \quad (9.42)$$

where we insert $k_{f\mu} = k_{i\mu}$. This insertion makes sense because we will insert the present result in eq. (9.33), and the latter equation contains a δ-function requiring $k_{f\mu} = k_{i\mu}$.

The attraction of the result (9.42) lies in the fact that the expressions between square brackets are precisely the effective mass tensors defined in eq. (4.35). Thus, we obtain the extremely simple expression,

$$\langle nk_i | \left(\frac{\hbar}{i} \nabla_\nu + \hbar k_{f\nu} \right) | nk_f \rangle = \sum_\mu \left(\frac{1}{2m_{\mu\nu}^*} + \frac{1}{2m_{\nu\mu}^*} \right) \hbar k_{f\mu} \quad (9.43)$$

We may now evaluate eq. (9.33) for the optical transition matrix element for intraband coupling. It is found to be given by

$$(\mathcal{H}_F)_{ni,nf} = \frac{\hbar e}{m_e} \sum_{\nu,\mu} \left(\frac{1}{2m_{\mu\nu}^*} + \frac{1}{2m_{\nu\mu}^*} \right) A_{0\nu} k_{f\mu} \cos(\omega t) \delta_{i,f} \quad (9.44)$$

9.1.4 Effective Mass Treatment of Optical Coupling

EFFECTIVE MASS HAMILTONIAN IN AN OPTICAL FIELD

Thus we obtained a surprisingly simple result for the intraband optical transition matrix element. In particular it contains the effective mass tensor only and no other material parameters. This suggests that the same result could also have been obtained invoking effective mass theory. In the present section we will investigate whether this suggestion is correct.

Eq. (9.44) gives the optical transition matrix element for the case of a non-degenerate band described by an effective mass tensor with elements $m_{\mu\nu}^*$. According to eq. (5.64) the effective Hamiltonian is then given by

$$\mathcal{H}_{eff} = \frac{1}{2m_e} \sum_{\mu,\nu} \left\{ -\frac{\hbar^2}{m_{\mu\nu}^*} \nabla_\mu \nabla_\nu + \frac{\hbar e}{i m_{\mu\nu}^*} [\bar{A}_\mu(\mathbf{r}, t) \nabla_\nu + \bar{A}_\nu(\mathbf{r}, t) \nabla_\mu] + \frac{e^2}{m_{\mu\nu}^*} \bar{A}_\mu(\mathbf{r}, t) \bar{A}_\nu(\mathbf{r}, t) \right\}$$

(9.45)

where we ignored the constant term E_0^n and took account of our choice of (9.1)–(9.3) for the gauge, so $\bar{U}(\mathbf{r}, t) = 0$. We recall that we have averaged the vector potential over each primitive cell, which is denoted by the bar over $\bar{A}_\mu(\mathbf{r}, t)$. As discussed above,

9.1 BAND ELECTRONS IN AN OPTICAL FIELD

optical wavelengths are always much longer than the size of primitive cells, so any effects of such averaging are negligible.

Note that in our present gauge the term (5.65) is not equal to zero, so it makes no sense to rewrite this effective Hamiltonian in the more elegant shape of the second line of eq. (5.64). This poses no problems however. Because we restrict ourselves to linear optics, we ignore all terms quadratic in $\bar{A}_\mu$. Then, eq. (9.45) is the most convenient form anyhow.

We split the effective Hamiltonian (9.45) in the same way as we split the single electron Hamiltonian in eqs (9.16)–(9.18). We obtain

$$\mathcal{H}_{eff} = \mathcal{H}_{0,eff} + \mathcal{H}_{F,eff} \tag{9.46}$$

where the unperturbed Hamiltonian is given by

$$\mathcal{H}_{0,eff} = -\frac{\hbar^2}{2m_e} \sum_{\mu,\nu} \frac{1}{m^*_{\mu\nu}} \nabla_\mu \nabla_\nu \tag{9.47}$$

and the perturbation due to the optical fields is equal to

$$\begin{aligned}
\mathcal{H}_{F,eff} &= \frac{\hbar e}{im_e} \sum_{\mu,\nu} \frac{1}{2m^*_{\mu\nu}} \left(\bar{A}_\mu \nabla_\nu + \bar{A}_\nu \nabla_\mu \right) \\
&= \frac{e}{m_e} \sum_{\mu,\nu} \left(\frac{1}{2m^*_{\mu\nu}} + \frac{1}{2m^*_{\nu\mu}} \right) \bar{A}_\nu \left(\frac{\hbar}{i} \nabla_\mu \right) \\
&= \frac{e}{m_e} \sum_{\mu,\nu} \left(\frac{1}{2m^*_{\mu\nu}} + \frac{1}{2m^*_{\nu\mu}} \right) \bar{A}_{0\nu} \cos(\mathbf{k}_q \cdot \mathbf{r} - \omega t) \left(\frac{\hbar}{i} \nabla_\mu \right) \\
&= \frac{e}{2m_e} \sum_{\mu,\nu} \left(\frac{1}{2m^*_{\mu\nu}} + \frac{1}{2m^*_{\nu\mu}} \right) \times \bar{A}_{0\nu} \left(e^{i\mathbf{k}_q \cdot \mathbf{r} - i\omega t} + e^{-i\mathbf{k}_q \cdot \mathbf{r} + i\omega t} \right) \left(\frac{\hbar}{i} \nabla_\mu \right) \quad (9.48)
\end{aligned}$$

Here we first inserted eq. (9.2), subsequently we inverted the indices μ and ν in the second term and finally we split the cosine into two exponential functions. Note also that the optical wave vector $\mathbf{k}_q$ is required to obey eq. (9.24).

PLANE WAVE STATES AND TRANSITION MATRIX ELEMENTS

In effective mass theory the initial and final states are simple plane waves

$$\frac{1}{\sqrt{NV_C}} e^{i\mathbf{k}_i \cdot \mathbf{r}} \tag{9.49}$$

and

$$\frac{1}{\sqrt{NV_C}} e^{i\mathbf{k}_f \cdot \mathbf{r}} \tag{9.50}$$

rather than the Bloch states (9.20) and (9.21). Note that we have added the same normalization factors as in eqs (9.20) and (9.21). Thus, in effective mass theory the optical transition matrix element is obtained by calculating the matrix element of the perturbation (9.48) between these two plane waves,

$$(\mathcal{H}_{F,eff})_{i,f} = \frac{1}{NV_C} \int_{NV_C} d\mathbf{r}\, e^{-i\mathbf{k}_i \cdot \mathbf{r}}\, \mathcal{H}_{F,eff}\, e^{i\mathbf{k}_f \cdot \mathbf{r}} \tag{9.51}$$

For this purpose we evaluate

$$\frac{1}{NV_C} \int_{NV_C} d\mathbf{r}\, e^{-i\mathbf{k}_i \cdot \mathbf{r}} e^{i\mathbf{k}_q \cdot \mathbf{r}} \left(\frac{\hbar}{i}\nabla_\mu\right) e^{i\mathbf{k}_f \cdot \mathbf{r}} = \frac{1}{NV_C} \int_{NV_C} d\mathbf{r}\, e^{i(\mathbf{k}_q - \mathbf{k}_i + \mathbf{k}_f)\cdot \mathbf{r}}\, \hbar k_{f\mu}$$

$$= \hbar k_{f\mu}\, \delta_{q,q_{if}} \tag{9.52}$$

where we insert a Kronecker δ-function equivalent to eq. (3.35). This δ-function requires that

$$\mathbf{k}_q = \mathbf{k}_i - \mathbf{k}_f \tag{9.53}$$

just as in eq. (9.30). Similarly,

$$\frac{1}{NV_C} \int_{NV_C} d\mathbf{r}\, e^{-i\mathbf{k}_i \cdot \mathbf{r}} e^{-i\mathbf{k}_q \cdot \mathbf{r}} \left(\frac{\hbar}{i}\nabla_\mu\right) e^{i\mathbf{k}_f \cdot \mathbf{r}} = \hbar k_{f\mu}\, \delta_{-q,q_{if}} \tag{9.54}$$

Then the matrix element of $\mathcal{H}_{F,eff}$ between the two plane waves (9.49) and (9.50) is found to be given by

$$(\mathcal{H}_{F,eff})_{i,f} = \frac{\hbar e}{2m_e} \sum_{\mu,\nu} \left(\frac{1}{2m^*_{\mu\nu}} + \frac{1}{2m^*_{\nu\mu}}\right) \bar{A}_{0\nu} k_{f\mu} \times (e^{-i\omega t}\, \delta_{q,q_{if}} + e^{i\omega t}\, \delta_{-q,q_{if}}) \tag{9.55}$$

Again we note that the optical wave vector is generally much smaller than the wave vector of the electron. Hence we may again neglect $\mathbf{k}_q$ with respect to $\mathbf{k}_i$ and $\mathbf{k}_f$, so $\delta_{q,q_{if}} \approx \delta_{i,f}$. Then,

$$(\mathcal{H}_{F,eff})_{i,f} = \frac{\hbar e}{2m_e} \sum_{\mu,\nu} \left(\frac{1}{2m^*_{\mu\nu}} + \frac{1}{2m^*_{\nu\mu}}\right) \bar{A}_{0\nu} k_{f\mu} (e^{-i\omega t} + e^{i\omega t}) \delta_{i,f}$$

$$= \frac{\hbar e}{m_e} \sum_{\mu,\nu} \left(\frac{1}{2m^*_{\mu\nu}} + \frac{1}{2m^*_{\nu\mu}}\right) \bar{A}_{0\nu} k_{f\mu} \cos(\omega t)\, \delta_{i,f} \tag{9.56}$$

which is precisely the same expression as eq. (9.44), showing that such a result can also be obtained with effective mass theory.

Below we will treat optical transitions between levels that result from quantization of the effective Hamiltonian. Eq. (9.56) will then allow us to calculate the corresponding transition probabilities using effective mass theory only, i.e. both for the states between which the transitions take place, and for the transition matrix elements themselves.

9.1.5 Optical Coupling Near the Γ-point

BASIS STATES NEAR THE Γ-POINT

A model for the basis states for a $\mathbf{k} \cdot \mathbf{p}$ treatment of a semiconductor near the Γ-point was presented in section 4.2.1. There we found the conduction band to be non-degenerate, yielding a single basis state $|c0\rangle$, while the valence band is threefold

9.1 BAND ELECTRONS IN AN OPTICAL FIELD

degenerate with basis states $|vx0\rangle$, $|vy0\rangle$ and $|vz0\rangle$. In the latter case, the eigenstates $|nk\rangle$ are linear combinations

$$|nk\rangle = \sum_\mu |v\mu 0\rangle \langle v\mu 0|nk\rangle \tag{9.57}$$

of the basis states $|v\mu 0\rangle$ at $\mathbf{k}=0$, where $\mu = x$, y or z. The proportionality constants $\langle v\mu o|nk\rangle$ in these linear combinations are the components of the eigenvectors of the matrix consisting of the sum of the matrices (4.72) and (4.105). Here $n = hh$, lh or λh, denote the heavy holes, the light holes and the spin-orbit band, respectively.

We remember that even in lowest order these eigenvectors depend on $\mathbf{k}$ contrary to the basis state $|c0\rangle$ of the conduction band. Note also that the proportionality constants $\langle v\mu o|nk\rangle$, etc., are actually binary vectors because of the spin of the valence electrons. Equivalently, the conduction band state $|c0\rangle$ should actually be multiplied by a binary vector denoting the spin state. In the following treatment we ignore these spin effects for two reasons. First, the optical field does not act directly on the electron spin. Second, we are only interested in the total coupling between all initial spin states to all final spin states. Under those two conditions spin effects do not influence the final result.

OPTICAL COUPLING BETWEEN THE CONDUCTION AND VALENCE BANDS

An optical field coupling the valence band and the conduction band yields a basic example of interband coupling due to the zero-order transition matrix element (9.36). To estimate it we use the extremely simple model introduced in section 4.2.1. We recall that according to our findings in section 4.2.2, the only non-zero matrix elements of ∇ between the valence and the conduction band are

$$\langle vx0|\nabla_x|c0\rangle = \langle vy0|\nabla_y|c0\rangle = \langle vz0|\nabla_z|c0\rangle = P \tag{9.58}$$

Furthermore, in this special case the second term in eq. (9.36) is equal to zero, because the Kronecker δ-function yields zero. So the non-zero optical transition matrix elements between each of the valence bands and the conduction band have the simple shape,

$$(\mathcal{H}_F)_{ni,cf} = \frac{\hbar e}{im_e} \mathbf{A}_0 \cdot \langle nk_i|\nabla|c0\rangle \cos(\omega t)\, \delta_{i,f}$$

$$= \frac{\hbar e}{im_e} \left(\sum_\nu A_{0\nu} \sum_\mu \langle nk_i|v\mu 0\rangle\langle v\mu 0|\nabla_\nu|c0\rangle \right) \cos(\omega t)\, \delta_{i,f}$$

$$= \frac{\hbar e}{im_e} \left(\sum_\nu A_{0\nu} \sum_\mu \langle nk_i|v\mu 0\rangle P\, \delta_{\nu,\mu} \right) \cos(\omega t)\, \delta_{i,f}$$

$$= \frac{\hbar e}{im_e} \left(\sum_\mu \langle nk_i|v\mu 0\rangle A_{0\mu} \right) P \cos(\omega t)\, \delta_{i,f} \tag{9.59}$$

Note that eq. (9.59) can be evaluated with reasonable precision using values for P that are extracted from Table 4.1.

OPTICAL INTRABAND COUPLING IN THE CONDUCTION BAND

Expression (9.44) can be directly used to obtain optical transition matrix elements for transitions within the conduction band. Then, according to the simple model of section 4.2.2 as well as in the more refined case treated in section 4.2.6, the effective mass in the conduction band minimum at the Γ-point is isotropic,

$$\frac{1}{m^*_{\mu\nu}} = \frac{1}{m^*}\delta_{\mu,\nu} \qquad (9.60)$$

Then, eq. (9.44), reduces to the even more simple shape,

$$(\mathcal{H}_F)_{ni,nf} = \frac{\hbar e}{m_e}\sum_{\nu,\mu}\frac{A_{0\nu}k_{f\mu}}{m^*}\delta_{\mu,\nu}\cos(\omega t)\delta_{i,f}$$

$$= \frac{\hbar e}{m_e}\sum_{\mu}\frac{A_{0\mu}k_{f\mu}}{m^*}\cos(\omega t)\delta_{i,f}$$

$$= \frac{\hbar e}{m_e m^*}\mathbf{A}_0 \cdot \mathbf{k}_f \cos(\omega t)\delta_{i,f} \qquad (9.61)$$

This simple shape would also have been obtained with effective mass theory. Thus, the coupling between a free electron with mass $m_e m^*$ and an optical field with frequency ω yields precisely the same result (9.61).

9.2 Optical Absorption

Optical transition matrix elements as treated in the previous section are used to calculate the probability of a variety of transitions between energy levels in semiconductors. Here we consider a few of those, just to show the method of calculation and to demonstrate a few basic types of transitions. For more extensive treatments the reader is referred to specialized texts on optics in semiconductors (see [22], [28] and [30]).

In semiconductors the fundamental optical absorption is across the bandgap. When discussing this type of transition it is essential to take into account that the optical field barely changes the crystal momentum of a band electron. This property is reflected in eq. (9.33) as a δ-function requiring crystal momentum to be the same for the initial and final states that are coupled by the optical field. Thus, in a band structure diagram the initial and final states need to be vertically above each other. As a result direct and indirect semiconductors form two clearly distinct cases. Fig. 9.1a and 9.1b show the band structures of GaAs and Ge as examples of the direct and indirect case respectively. In the former case the conduction band minimum occurs at the Γ-point and the optical field may couple states near the valence band maximum and the conduction band minimum directly. An example of such direct optical coupling is shown as arrow A in Fig. 9.1a. In the case of the indirect semiconductor germanium, the lowest conduction band minima occur at the four L-points and conservation of crystal momentum inhibits direct optical coupling of the type B in Fig. 9.1b.

Of course, in the case of germanium, direct optical coupling still exists between the valence band maximum and the higher conduction band valley at the Γ-point as

9.2 OPTICAL ABSORPTION

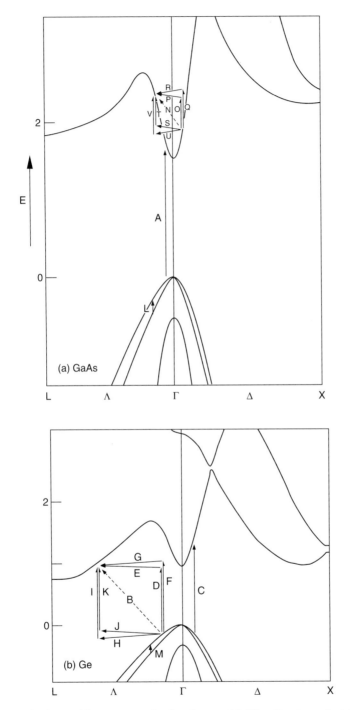

Figure 9.1 Optical transitions across the band gap. (a) The direct semiconductor GaAs (band structure adapted from Fig. 8.21, p. 103, ref. [10], © Springer Verlag). (b) The indirect semiconductor Ge (band structure adapted from Fig. 8.11, p. 92, ref. [10], © Springer Verlag). For a further explanation see the text.

indicated by arrow C in Fig. 9.1b. However, second order processes are needed to transfer electrons from the valence band maximum to a conduction band valley at an L-point. In these processes a photon is absorbed, while simultaneously a phonon is absorbed or emitted as indicated by the arrows $D+E$, $F+G$, $H+I$ and $J+K$ in Fig. 9.1b. Here the phonon carries the crystal momentum which the photon lacks. Clearly, because these so-called indirect transitions are second order, the corresponding transition probabilities are much smaller than those for direct transitions. It is precisely the much higher transition probability of direct transitions across the bandgap which makes direct semiconductors the logical choice for semiconductor lasers. Below in sections 9.2.1 and 9.2.2 we will consider direct and indirect transitions across the band gap in more detail.

Optical coupling within the valence band or within the conduction band suffers the same restrictions as coupling across the band gap. Because crystal momentum has to be conserved, direct optical coupling between different valence bands—arrows L and M in Fig. 9.1a and 9.1b—is allowed. However, direct optical coupling within the conduction band, as indicated by arrow N, is forbidden. Here, one again has to resort to indirect optical coupling where a photon is absorbed, while simultaneously a phonon is absorbed or emitted as indicated by the arrows $O+P$, $Q+R$, $S+T$ and $U+V$. For a further treatment of such indirect intraband transitions the reader is referred to ref. [35].

A strongly different situation occurs however, when such bands are further quantized into subbands due to externally applied fields, impurity potentials or when the semiconductor is part of an artificial structure like a quantum well. Then, individual Bloch functions do not represent the eigenstates of the band electrons. The externally applied fields or the potentials imposed by the impurity or artificial structure mix many Bloch functions into new eigenstates. As *different* new eigenstates may contain the *same* Bloch function, direct optical coupling between these two eigenstates is fully allowed, even when they originate in the same band. As an example we will treat optical absorption between subbands in a quantum well in section 9.2.3.

9.2.1 Direct Semiconductors

TRANSITION RATES

The purpose of the present section is to calculate the optical absorption coefficient due to direct optical transitions across the bandgap, i.e. transitions of type A in Fig 9.1a. For simplicity we single out one of the valence bands, the heavy holes band. However, without much problem the treatment is repeated for direct optical transitions originating in the light holes or spin-orbit bands. Adding the contributions due to all three valence bands yields the total absorption coefficient. As a further simplification we consider the semiconductor to be in its electronic ground state, so the conduction band is empty and the valence band is full. Thus, the optical field may only induce transitions from the valence band up to the conduction band and not vice versa. The extension to the opposite case is straightforward, except that transitions will only take place when some conduction band levels are occupied and some valence band levels are empty. Hence, one needs to take the actual occupation of conduction band levels and valence band levels into account. Thus, one needs to consider semi-

9.2 OPTICAL ABSORPTION

conductor statistics which is beyond the scope of this book. Also, spontaneous emission due to transitions from the conduction band to the valence band are beyond the present treatment, as the understanding of such transitions requires the optical field to be quantized.

In section 9.1.5 we obtained the matrix element (9.59) which is responsible for direct optical transitions from the heavy holes band to the conduction band. Setting $n = hh$,

$$(\mathcal{H}_F)_{hhi,cf} = \frac{\hbar e}{im_e}\left(\sum_\mu \langle hhk_i | v\mu 0\rangle A_{0\mu}\right) P \cos(\omega t)\, \delta_{i,f} \quad (9.62)$$

where $\mu = x$, y or z and P is defined in eq. (9.58).

As we consider the crystal to be in its ground state, the valence band is full, so all states in the heavy holes band may serve as an initial state. Moreover, the conduction band is empty, so all conduction band states may in principle serve as a final state. However, selection rules may provide restrictions. In eq. (9.62) the first of these selection rules emerges as the Kronecker δ-function $\delta_{i,f}$. This δ-function requires that the initial wave vector $\mathbf{k}_i$ is equal to the final wave vector $\mathbf{k}_f$. Hence, given an initial state in the heavy holes band, in practice only a single conduction band state may serve as the final state. The matrix element which is responsible for this allowed transition is given by

$$(\mathcal{H}_F)_{hhi,ci} = \frac{\hbar e}{im_e}\left(\sum_\mu \langle hhk_i | v\mu 0\rangle A_{0\mu}\right) P \cos(\omega t)$$

$$= \frac{\hbar e}{2im_e} P\left(\sum_\mu \langle hhk_i | v\mu 0\rangle A_{0\mu}\right) e^{i\omega t} + \frac{\hbar e}{2im_e} P\left(\sum_\mu \langle hhk_i | v\mu 0\rangle A_{0\mu}\right) e^{-i\omega t} \quad (9.63)$$

We now use time dependent perturbation theory as developed in appendix C, section C.2.2. The second term in eq. (9.63) has the shape of the perturbation term of eq. (C.73), so one may immediately apply Fermi's golden rule (C.79) to find the transition probability due to this second term. We find,

$$W^+_{hhi,c} = \frac{\pi\hbar e^2}{2m_e^2}\left|P\sum_\mu \langle hhk_i | v\mu 0\rangle A_{0\mu}\right|^2 \delta(E_i^c - E_i^{hh} - \hbar\omega) \quad (9.64)$$

The δ-function represents conservation of energy and provides a second selection rule. Note that $\omega > 0$, while the conduction band energy E_i^c is always larger than the valence band energy E_i^{hh}, so it is possible to satisfy the second δ-function.

The first term in eq. (9.63) also has the shape of the oscillating term in eq. (C.73), except that $-\omega$ is replaced by $+\omega$. Therefore, Fermi's golden rule (C.79) may be applied again, provided we make the same replacement. Thus, this first term yields a transition probability

$$W^-_{hhi,ci} = \frac{\pi\hbar e^2}{2m_e^2}\left|P\sum_\mu \langle hhk_i | v\mu 0\rangle A_{0\mu}\right|^2 \delta(E_i^c - E_i^{hh} + \hbar\omega) = 0 \quad (9.65)$$

because the second δ-function can never be satisfied. Thus, only the second term in the matrix element (9.63) contributes to a finite transition probability.

SELECTION OF POSSIBLE INITIAL AND FINAL STATES

We continue considering the total probability that the optical field causes a transition across the bandgap. At this point one must realise that we consider the crystal to be in its ground state, so the valence band is full. As a result all states in the heavy holes band may serve as an initial state, i.e. all positions in the first Brillouin zone may serve as initial $\mathbf{k}_i$-vectors. Therefore, the problem of calculating the total transition probability consists of summing eq. (9.64) over all these initial $\mathbf{k}_i$-vectors,

$$W_{hh,c}^{tot} = \sum_i^N W_{hhi,ci}^+ = \frac{\pi \hbar e^2}{2m_e^2} \sum_i^N \left| P \sum_\mu \langle hhk_i | v\mu 0 \rangle A_{0\mu} \right|^2 \delta(E_i^c - E_i^{hh} - \hbar\omega) \quad (9.66)$$

To perform this summation over the initial wave vector $\mathbf{k}_i$ we need to follow a similar procedure as used in section 8.2.3, where we determined the total scattering rate. First, in order to suppress the δ-function in eq. (9.66), we require summation over one variable only, while in our three-dimensional world the sum over $\mathbf{k}_i$ is a threefold sum. Choosing to sum over one of the components of $\mathbf{k}_i$, the resulting total scattering rate still contains sums over the other two components. As in section 8.2.3 it is practical to choose spherical coordinates (k_i, θ_i, ϕ_i) and to sum over the length $k_i = |\mathbf{k}_i|$ of the initial wave vector to suppress the δ-function $\delta(E_i^c - E_i^{hh} - \hbar\omega)$.

A second problem arises because the remaining δ-function in eq. (9.66) is actually a Dirac δ-function which requires *integration* instead of summation. As in section 8.2.3 we solve this problem by approximating the sum over $\mathbf{k}_i$ by an integral, so

$$W_{hh,c}^{tot} \approx \frac{N}{V_B} \int_{V_B} d\phi_i d\theta_i \, \sin\theta_i \, dk_i \, k_i^2 \, W_{hhi,ci}^+$$

$$= \frac{N}{V_B} \int_\infty d\phi_i d\theta_i \, \sin\theta_i \, dk_i \, k_i^2 \, W_{hhi,ci}^+$$

$$= \int_0^{2\pi} d\phi_i \int_0^\pi d\theta_i \, \sin\theta_i \left(\frac{N}{V_B} \int_0^\infty dk_i \, k_i^2 \, W_{hhi,ci}^+ \right)$$

$$= \int_0^{2\pi} d\phi_i \int_0^\pi d\theta_i \, \sin\theta_i \, W_{hh,c}^{diff}(\theta_i, \phi_i) \quad (9.67)$$

Here, the factor

$$\frac{N}{V_B} = \frac{NV_C}{(2\pi)^3} \quad (9.68)$$

is included to account for the fact that the original sum contains N initial wave vectors $\mathbf{k}_i$ and the resulting integral is performed over a volume V_B. Note that V_B is expressed in V_C using eq. (3.13), while NV_C is the volume of the crystal. In the first step the integral is extended to infinite space, because the Dirac δ-function is non-zero for $\mathbf{k}_i$

9.2 OPTICAL ABSORPTION

near the Γ-point only. So, without loss of generality, $W^+_{hhi,c}$ may be assumed to vanish for $\mathbf{k}_i$ outside the first Brillouin zone. In eq. (9.67), $d\phi_i d\theta_i \sin \theta_i$ is an element of solid angle and hence

$$W^{diff}_{hh,c}(\theta_i, \phi_i) = \frac{N}{V_B} \int_0^\infty dk_i \, k_i^2 \, W^+_{hhi,ci} \tag{9.69}$$

represents the probability per unit of solid angle that an electron in the heavy holes band with a wave vector $\mathbf{k}_i$ in the (θ_i, ϕ_i)-direction is excited to the conduction band. Thus this quantity is very similar to the differential scattering rate (8.41).

JOINT DIFFERENTIAL DENSITIES OF STATE FUNCTIONS

A remaining problem is that eq. (9.69) contains an integral over k_i, while the integrand $W^+_{hhi,ci}$ contains a Dirac δ-function requiring integration over the energy difference $(E_i^c - E_i^{hh})$ between the conduction band and the heavy holes band at position $\mathbf{k}_i$ in reciprocal space. Therefore we need to transform the integral in eq. (9.69) to an integral over this energy difference. The Jacobian of this transformation is the partial derivative of $(E_i^c - E_i^{hh})$ with respect to k_i, while keeping θ_i and ϕ_i constant,

$$\left[\frac{\partial}{\partial k_i} (E_i^c - E_i^{hh}) \right]_{\theta_i, \phi_i} \tag{9.70}$$

So, eq. (9.69) may be written as

$$W^{diff}_{hh,c}(\theta_i, \phi_i) = \frac{N}{V_B} \int_0^\infty dk_i \, k_i^2 \, W^+_{hhi,ci}$$

$$= \frac{N}{V_B} \int_0^\infty d(E_i^c - E_i^{hh}) \, \frac{k_i^2}{\left[\dfrac{\partial (E_i^c - E_i^{hh})}{\partial k_i} \right]_{\theta_i, \phi_i}} W^+_{hhi,ci} \tag{9.71}$$

where k_i is a function of $(E_i^c - E_i^{hh})$, θ_i and ϕ_i. Note, the value of k_i is such that at $\mathbf{k}_i = (k_i, \theta_i, \phi_i)$ the energy difference between the conduction band and the heavy holes band is precisely equal to $(E_i^c - E_i^{hh})$.

Introducing the joint differential density of states,

$$N_d^{hh,c}[(E_i^c - E_i^{hh}), \theta_i, \phi_i] = \frac{N}{V_B} \frac{k_i^2[(E_i^c - E_i^{hh}), \theta_i, \phi_i]}{\left[\dfrac{\partial (E_i^c - E_i^{hh})}{\partial k_i} \right]_{\theta_i, \phi_i}} \tag{9.72}$$

we can shorten eq. (9.71) to

$$W_{hh,c}^{diff}(\theta_i, \phi_i) = \int_0^\infty d(E_i^c - E_i^{hh}) \, N_d^{hh,c}[(E_i^c - E_i^{hh}), \theta_i, \phi_i] \, W_{hhi,ci}^+$$

$$= \int_0^\infty d(E_i^c - E_i^{hh}) \, N_d^{hh,c}[(E_i^c - E_i^{hh}), \theta_i, \phi_i]$$

$$\times \frac{\pi \hbar e^2}{2m_e^2} \left| P \sum_\mu \langle hhk_i | v\mu 0 \rangle A_{0\mu} \right|^2 \delta(E_i^c - E_i^{hh} - \hbar\omega)$$

$$= \frac{\pi \hbar e^2}{2m_e^2} N_d^{hh,c}(\hbar\omega, \theta_i, \phi_i) \left| P \sum_\mu \langle hhk_i | v\mu 0 \rangle A_{0\mu} \right|^2 \quad (9.73)$$

where we also inserted eq. (9.66) for $W_{hhi,ci}^+$ and furthermore performed the integration cancelling the Dirac δ-function. Furthermore,

$$N_d^{hh,c}(\hbar\omega, \theta_i, \phi_i) = \{N_d^{hh,c}[(E_i^c - E_i^{hh}), \theta_i, \phi_i]\}_{(E_i^c - E_i^{hh} = \hbar\omega)} \quad (9.74)$$

is the joint differential density of states at a position $\mathbf{k}_i$ in reciprocal space defined by the angles θ_i and ϕ_i and the requirement that the energy difference $(E_i^c - E_i^{hh})$ between the conduction band and the heavy holes band is precisely equal to $\hbar\omega$. Note that the matrix element $\langle hhk_i | v\mu 0 \rangle$ is also evaluated at this position $\mathbf{k}_i$.

The total probability (9.67) that the optical field induces a transition from the valence band to the conduction band can now finally be written as

$$W_{hh,c}^{tot} = \int_0^{2\pi} d\phi_i \int_0^\pi d\theta_i \sin\theta_i \, W_{hh,c}^{diff}(\theta_i, \phi_i)$$

$$= \frac{\pi \hbar e^2}{2m_e^2} \int_0^{2\pi} d\phi_i \int_0^\pi d\theta_i \sin\theta_i \, N_d^{hh,c}(\hbar\omega, \theta_i, \phi_i) \left| P \sum_\mu \langle hhk_i | v\mu 0 \rangle A_{0\mu} \right|^2 \quad (9.75)$$

Before continuing we recall our discussion of section 8.2.3 where we derived a direct physical interpretation for differential densities of state. Translating that discussion to the present context, we find that the joint differential density of states corresponds to the number of states per unit of solid angle and per unit of energy for $\mathbf{k}$ oriented in the (θ, ϕ)-direction, while its length k is such that the energy difference between the heavy holes band and the conduction band is equal to $E^c - E^{hh}$.

OPTICAL ABSORPTION

$W_{hh,c}^{tot}$ as given by eq. (9.75) gives the number of transitions per second in the crystal, provided the valence band is full and the conduction band empty. However, to interpret optical absorption experiments one prefers to know the optical absorption coefficient $\alpha_{hh,c}$ defined by

$$\alpha_{hh,c} = -\frac{1}{I}\frac{\partial I}{\partial z} = -\frac{\partial}{\partial z}(\ln I) \quad (9.76)$$

where I is the light intensity defined in eq. (9.15) and z the position along the propagation direction of the light. Note that this definition implies that a constant value of $\alpha_{hh,c}$ incurs an exponential decay of the light intensity,

$$I(z) = I(0) \, e^{-\alpha_{hh,c} z} \quad (9.77)$$

9.2 OPTICAL ABSORPTION

To obtain the absorption coefficient we first multiply the total transition probability with the transition energy $\hbar\omega$. Thus we obtain the energy absorbed per second from the optical field. Dividing subsequently by the volume NV_C of the crystal provides us with the absorbed energy per second per unit of volume. A final division by the light intensity I gives the absorption coefficient $\alpha_{hh,c}$ of the light due to the transitions described by eq. (9.75). We find

$$\alpha_{hh,c} = \frac{W^{tot}_{hh,c} \hbar\omega}{NV_C I} = \frac{\pi\hbar e^2}{2m_e^2} \int_0^{2\pi} d\phi_i \int_0^{\pi} d\theta_i \sin\theta_i$$

$$\times \frac{N_d^{hh,c}(\hbar\omega, \theta_i, \phi_i)\, \hbar\omega}{NV_C I} \left| P \sum_\mu \langle hhk_i | v\mu 0 \rangle A_{0\mu} \right|^2 \quad (9.78)$$

This expression can be simplified by taking into account that the vector potential $\mathbf{A}_0$ and the light intensity are related by eq. (9.15). For this purpose we define the optical polarization vector $\mathbf{e}_0$ as the unit vector parallel to the electric field component $\mathbf{E}_0$ of the optical field. As, according to eq. (9.8), the vector potential $\mathbf{A}_0$ is also parallel to this electric field component, we may then write

$$\left| P \sum_\mu \langle hhk_i | v\mu 0 \rangle A_{0\mu} \right|^2 = |P|^2 \left| \sum_\mu \langle hhk_i | v\mu 0 \rangle e_{0\mu} \right|^2 |\mathbf{A}_0|^2$$

$$= |P|^2 \left| \sum_\mu \langle hhk_i | v\mu 0 \rangle e_{0\mu} \right|^2 \frac{2}{\omega^2} \sqrt{\frac{\mu_0}{\varepsilon_0 \varepsilon_r}} I \quad (9.79)$$

As a result the optical absorption coefficient is given by

$$\alpha_{hh,c} = \frac{\pi\hbar e^2}{2m_e^2} \int_0^{2\pi} d\phi_i \int_0^{\pi} d\theta_i \sin\theta_i$$

$$\times \frac{N_d^{hh,c}(\hbar\omega, \theta_i, \phi_i)\, \hbar\omega}{NV_C I} |P|^2 \left| \sum_\mu \langle hhk_i | v\mu 0 \rangle e_{0\mu} \right|^2 \frac{2}{\omega^2} \sqrt{\frac{\mu_0}{\varepsilon_0 \varepsilon_r}} I$$

$$= \frac{\pi\hbar^3 e^2}{m_e^2} \sqrt{\frac{\mu_0}{\varepsilon_0 \varepsilon_r}} \frac{|P|^2}{NV_C \hbar\omega} \int_0^{2\pi} d\phi_i \int_0^{\pi} d\theta_i \sin\theta_i$$

$$\times N_d^{hh,c}(\hbar\omega, \theta_i, \phi_i) \left| \sum_\mu \langle hhk_i | v\mu 0 \rangle e_{0\mu} \right|^2 \quad (9.80)$$

We see that $\alpha_{hh,c}$ is independent of the light intensity, so it obeys eqs (9.76) and (9.77). As eq. (9.76) is a linear differential equation, the optical absorption process described in the present section represents a linear optical process.

ISOTROPIC PARABOLIC BANDS

To illustrate the result (9.80) for the optical absorption coefficient, we consider the simple case that the effective mass of the heavy holes is so large that we may approximate it to be infinite. Such an approximation is not unreasonable. In many direct semiconductors the effective mass of the heavy holes is an order of magnitude larger than the the effective mass of the electrons. For example, in GaAs, for electrons at the Γ-point, $m_c^* = 0.0665$, while for heavy holes $m_{hh}^* = 0.50$. Note that m_{hh}^* is actually anisotropic, so this represents an average over all directions. Then the heavy holes band energy is independent of the position $\mathbf{k}_i$ in reciprocal space and the joint density of states simplifies to

$$N_d^{hh,c}(\hbar\omega, \theta_i, \phi_i) = \frac{N}{V_B} \left\{ \frac{k_i^2[(E_i^c - E_0^{hh}), \theta_i, \phi_i]}{\left(\frac{\partial E_i^c}{\partial k_i}\right)_{\theta_i, \phi_i}} \right\}_{(E_i^c - E_0^{hh} = \hbar\omega)} \quad (9.81)$$

Such an approximation entails a further simplification, because near the Γ-point, the conduction band is isotropic. Hence, this expression can be evaluated using eq. (8.57). We find,

$$N_d^{hh,c}(\hbar\omega, \theta_f, \phi_f) = \frac{N}{V_B} \frac{\sqrt{2(\hbar\omega - E_g)}(m_e m_c^*)^{3/2}}{\hbar^3} \quad (9.82)$$

which is independent of the angle θ_i, ϕ_i. Here we take into account that eq. (8.57) chooses the zero point of energy at the band minimum at $\mathbf{k}_i = 0$, while now the zero point is shifted downwards by an amount $E_g = E_0^c - E_0^{hh}$. As we also require $E_i^c - E_0^{hh} = \hbar\omega$, we need to replace E_f^n in eq. (8.57) by $\hbar\omega - E_g$ in eq. (9.82).

Then the optical absorption coefficient is found to be given by

$$\alpha_{hh,c} = \frac{\pi\hbar^3 e^2}{m_e^2} \sqrt{\frac{\mu_0}{\varepsilon_0\varepsilon_r}} \frac{N_d^{hh,c}(\hbar\omega, \theta_f, \phi_f)}{NV_C \hbar\omega} |P|^2 \times \int_0^{2\pi} d\phi_i \int_0^\pi d\theta_i \sin\theta_i \left|\sum_\mu \langle hhk_i | v\mu 0 \rangle e_{0\mu}\right|^2$$

$$= \frac{\pi\hbar^3 e^2}{m_e^2} \sqrt{\frac{\mu_0}{\varepsilon_0\varepsilon_r}} \frac{1}{NV_C \hbar\omega} \frac{N}{V_B} \frac{\sqrt{2(\hbar\omega - E_g)}(m_e m_c^*)^{3/2}}{\hbar^3} |P|^2$$

$$\times \int_0^{2\pi} d\phi_i \int_0^\pi d\theta_i \sin\theta_i \left|\sum_\mu \langle hhk_i | v\mu 0 \rangle e_{0\mu}\right|^2$$

$$= \frac{e^2}{16\pi^2\hbar^2 m_e} \sqrt{\frac{\mu_0}{\varepsilon_0\varepsilon_r}} \sqrt{2(\hbar\omega - E_g)}(m_e m_c^*)^{3/2} \frac{2\hbar^2|P|^2}{m_e\hbar\omega}$$

$$\times \int_0^{2\pi} d\phi_i \int_0^\pi d\theta_i \sin\theta_i \left|\sum_\mu \langle hhk_i | v\mu 0 \rangle e_{0\mu}\right|^2 \quad (9.85)$$

where we insert $V_B V_C = (2\pi)^3$. A further simplification of this expression is obtained by inserting expression (4.55) for the electron effective mass m_c^*. Note that we approximate $m_{hh}^* \gg m_c^*$, which generally entails $m_c^* \ll 1$. So approximately,

$$\frac{1}{m_c^*} \approx \frac{2\hbar^2 P^2}{m_e E_g} \approx \frac{2\hbar^2 P^2}{m_e \hbar\omega} \quad (9.84)$$

9.2 OPTICAL ABSORPTION

as for absorption near the bandgap, also $E_g \approx \hbar\omega$. Then,

$$\alpha_{hh,c} = \frac{e^2}{16\pi^2 \hbar^2 m_e} \sqrt{\frac{\mu_0}{\varepsilon_0 \varepsilon_r}} \sqrt{2(\hbar\omega - E_g)} (m_e m_c^*)^{3/2} \frac{1}{m_c^*}$$

$$\times \int_0^{2\pi} d\phi_i \int_0^{\pi} d\theta_i \sin\theta_i \left| \sum_\mu \langle hhk_i | v\mu 0 \rangle e_{0\mu} \right|^2$$

$$= \left(\frac{e}{4\pi\hbar}\right)^2 \sqrt{\frac{\mu_0}{\varepsilon_0 \varepsilon_r}} \sqrt{2(\hbar\omega - E_g) m_e m_c^*}$$

$$\times \int_0^{2\pi} d\phi_i \int_0^{\pi} d\theta_i \sin\theta_i \left| \sum_\mu \langle hhk_i | v\mu 0 \rangle e_{0\mu} \right|^2 \quad (9.85)$$

Thus, the optical absorption varies as $\sqrt{(\hbar\omega - E_g)}$, which is clearly illustrated in experimental results for GaAs shown in Fig. 9.2. An estimate of the absorption coefficient is now also easily obtained for GaAs. We set

$$e = 1.602 \cdot 10^{-19} \text{ C}$$
$$\hbar = 1.055 \cdot 10^{-34} \text{ J s}$$
$$\mu_0 = 4\pi \cdot 10^{-7}$$
$$\varepsilon_0 = c^{-2} \mu_0^{-1}$$
$$c = 2.998 \cdot 10^8 \text{ m s}^{-1}$$
$$m_e = 9.11 \cdot 10^{-31} \text{ kg}$$
$$\varepsilon_r = 12.8$$
$$m_c^* = 0.0665 \quad (9.86)$$

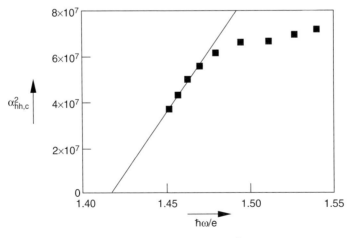

Figure 9.2 The squared optical absorption coefficient $\alpha_{hh,c}^2$ for GaAs as a function of the frequency of the optical field. The data are from ref. [37].

Furthermore we assume the integrand in eq. (9.85) to be approximately equal to one, corresponding to a random orientation of the polarization direction $\mathbf{e}_0$ with respect to the initial wave vector $\mathbf{k}_i$ of the electron. Hence, the integral over θ_i and ϕ_i yields 4π. Then we find

$$\alpha_{hh,c} = 2.69 \cdot 10^6 \times \sqrt{\frac{\hbar\omega - E_g}{e}} \text{ m}^{-1}$$

Hence, for $(\hbar\omega - E_g)/e = 0.1$ eV, we find $\alpha_{hh,c}^2 = 7.25 \cdot 10^7$ cm^{-2}. We may compare this result with a fit to the experimental results shown in Fig. 9.2. Then, $E_g = 1.417$ eV, falling within the range of values found in the literature [24], while the experiment shows $\alpha_{hh,c}^2 = 10.5 \cdot 10^7$ cm^{-2} at $(\hbar\omega - E_g)/e = 0.1$ eV. However, note the many approximations that were made and the small number of data used in the fit.

9.2.2 Indirect Semiconductors

THE NATURE OF INDIRECT OPTICAL TRANSITIONS

Above we found that direct optical transitions from the valence band maximum to the conduction band minimum are forbidden in indirect semiconductors. Only indirect optical transitions where simultaneously a phonon is emitted or absorbed are allowed. Fig. 9.1b shows a number of those indirect transitions for germanium, while Fig. 9.3a singles one of these transitions out as the arrows A + B. Both figures depict these processes in the customary way. Arrow A denotes the absorption of a photon with energy $\hbar\omega$, while arrow B corresponds the simultaneous emission of a longitudinal optical phonon with energy $\hbar\omega^{qLO}$ and wave vector $\mathbf{k}_q$. Often one says that an electron is transferred from the initial state 1 via an intermediate state 2 to a final state 3. As the intermediate state does not seem to exist as a real state in the band structure, this intermediate state is called 'virtual'.

Here we wish to describe this process using second order time dependent perturbation theory as developed in appendix C. At this point it is very important to realize

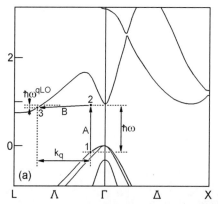
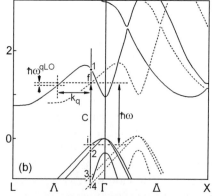

Figure 9.3 (a) The standard way to depict an indirect transition. (b) The way an indirect transition needs to be depicted in order to able to use eq. (C.83). The band structure of Ge is adapted from Fig. 8.11, p. 92, ref. [10], © Springer Verlag.

9.2 OPTICAL ABSORPTION

that the optical field is treated differently from the lattice vibrations. The former is treated classically and the energy of the optical field is *not* part of the Hamiltonian. Only its interaction with the band electrons figures in this Hamiltonian and the interaction term oscillates with the optical frequency ω. Because of this time dependence, the energy of the crystal may increase or decrease upon absorption or emission of light. On the other hand the lattice vibrations are treated quantum mechanically and give rise to the lattice term $\mathcal{H}_L$ in the Hamiltonian. The electron–phonon coupling is time-independent and does *not* change the total energy of the crystal. Energy emitted by band electrons is absorbed by phonons and vice versa.

To implement these details it is better to depict an indirect transition as done in Fig. 9.3b. We consider excitation of an electron in the heavy holes band with a wave vector $\mathbf{k}_i$. We choose the zero-point of energy and the zero-point of the total wave vector of the crystal to coincide with the ground state of this particular electron. This point is indicated in Fig. 9.3b by the index i. The drawn lines denote the energy and the wave vector of the whole crystal when our selected electron is transferred to another electron state, while the lattice is unperturbed. This new state may be in the valence band, but more importantly for our present discussion, it may also be in one of the conduction bands. As the optical field carries negligible momentum, the final electron state should be on a vertical line through the point i, i.e. on the vertical axis. Hence, only the points indicated by 1 and 2 may serve as final states. These are precisely the allowed direct transitions.

Thus far we kept the number of phonons constant. However, in an indirect optical transition also a phonon is created or annihilated. Let us assume that an optical phonon is created and that it has an energy $\hbar\omega^{qLO}$ and a wave vector $\mathbf{k}_q$. Then the total energy of the crystal increases by an extra amount $\hbar\omega^{qLO}$, while the total wave vector of the crystal is shifted by $\mathbf{k}_q$. In Fig. 9.3b the resulting energy spectrum of the crystal is indicated by dashed lines. So, when simultaneously with an excitation of the electron, such a phonon is created, the final energy and the final wave vector of the crystal are somewhere on these dashed lines.

Now we argue again that the optical field carries negligible momentum, so an optical excitation of our selected electron should keep this electron on a vertical line through the point i. Hence, in such an indirect optical transition only the states indicated by the points f, 3 and 4 are allowed as final states. We see that now an indirect optical transition can be depicted by a single arrow 'C'. Clearly, in this way of depicting the transition, the energy difference between the final and initial states is precisely the photon energy $\hbar\omega$, while the total wave vector of the crystal is conserved. Thus, it shows how this transition obeys conservation of energy and momentum as required for optical transitions.

Before continuing a warning should be given. Fig. 9.3b gives the impression that conservation of energy and momentum allows a single final state only. This impression is incorrect because Fig. 9.3b has singled out one phonon wave vector $\mathbf{k}_q$. In principle *all* phonon wave vectors are available for indirect optical transitions. Then, even taking conservation of energy into account, a large number of final electron states is possible. In this respect, indirect optical transitions differ from direct optical transitions, as the latter allow a single final state only.

THE INTERMEDIATE STATE

Still, though the transition shown in Fig. 9.3b now has a 'direct' appearance, it is forbidden in first-order perturbation theory. The reason is that the optical field as well as the electron–phonon coupling have vanishing matrix elements between the initial and final state. Hence, one needs to go to second order to calculate the transition probability. Then one invokes a third state, such that the optical field has a non-zero matrix element between the initial state and this third state, while the electron–phonon coupling has a non-zero matrix element between this third state and the final state. It is important to realize that this intermediate state is a real existing state of the crystal as a whole, so it is *not* 'virtual' at all!

The complexity of indirect optical transitions originates in the fact that a variety of states may serve as intermediate states. Examples are numbered as 1, 2, 3,..., in Fig. 9.3b. Moreover, not only longitudinal optical phonons but also the other phonon modes may participate in the process. On top of that, the final conduction band valleys are degenerate, in germanium there are four L-valleys, while in silicon we have six Δ-valleys. Also, we have to consider three initial bands, the heavy holes band, the light holes band and the spin-orbit band. Clearly, calculating the total transition probability at a given optical frequency, may easily evolve to an tedious administrative task masking the basic principles. Therefore, here we will single out one of all these processes, where the initial state is a heavy holes state, the intermediate state is in the Γ-valley of the conduction band (intermediate state 1 in Fig. 9.3b) and the final state is in one of the L-valleys. The optical field is supposed to couple the initial heavy holes state to intermediate state, while the electron–phonon coupling couples the intermediate state to the final state in the L-valley. Finally, we consider longitudinal optical phonons only in a non-polar semiconductor, e.g., germanium, thus excluding polar optical phonon coupling.

Unfortunately, the treatment is still somewhat tedious due to the length of formulae originating in second-order time dependent perturbation theory.

MATRIX ELEMENTS

With the above restrictions the initial electron state is a heavy holes state

$$\frac{1}{\sqrt{NV_C}}|hhi\rangle e^{i\mathbf{k}_i \cdot \mathbf{r}} \tag{9.86}$$

with band energy E_i^{hh} and wave vector $\mathbf{k}_i$ near the Γ-point. The intermediate electron state is a conduction band state

$$\frac{1}{\sqrt{NV_C}}|cp\rangle e^{i\mathbf{k}_p \cdot \mathbf{r}} \tag{9.87}$$

with band energy E_p^c and wave vector $\mathbf{k}_p$, also near the Γ-point. The final electron state is a conduction band state

$$\frac{1}{\sqrt{NV_C}}|cf\rangle e^{i\mathbf{k}_f \cdot \mathbf{r}} \tag{9.88}$$

9.2 OPTICAL ABSORPTION

at a band energy E_f^c and with a wave vector $\mathbf{k}_f$ in one of the L-valleys. Note that all these states are normalized on the volume NV_C of the crystal.

The initial and intermediate state are coupled by the optical field, while the intermediate and final state are coupled via interaction with LO phonons. The intermediate state is in the Γ-valley of the conduction band, so we may use eq. (9.62) for the transition matrix element between the initial and intermediate state,

$$(\mathcal{H}_F)_{hhi,cp} = \frac{\hbar e}{im_e} \left(\sum_\mu \langle hhk_i | v\mu 0 \rangle A_{0\mu} \right) P \cos(\omega t) \, \delta_{i,p}$$

$$= (\mathcal{H}_F)^0_{hhi,cp} e^{i\omega t} + (\mathcal{H}_F)^0_{hhi,cp} e^{-i\omega t} \quad (9.89)$$

where

$$(\mathcal{H}_F)^0_{hhi,cp} = \frac{\hbar e}{2im_e} P \left(\sum_\mu \langle hhk_i | v\mu 0 \rangle A_{0\mu} \right) \delta_{i,p}$$

and $\mu = x$, y or z and P defined in eq. (9.58). Note that we split the cosine into two exponential functions as in eq. (9.63).

The transition matrix element coupling the intermediate and final state is more difficult to obtain. In principle we would like to use the longitudinal optical phonon term in eq. (7.73),

$$\mathcal{H}^{LO}_{cf,cp} = \frac{\hbar}{\sqrt{2N}} \left(\Xi^{DLO} \right)_{cf,cp} (a^{q_{pf} LO} + a^{q_{pf} LO\dagger}) \quad (9.90)$$

Here the index cf denotes the conduction band state (9.88) at wave vector $\mathbf{k}_f$, which is in the chosen L-valley. The index cp denotes the conduction band state (9.87) at wave vector $\mathbf{k}_p$ in the Γ-valley. Furthermore $a^{q_{pf} LO}$ and $a^{q_{pf} LO\dagger}$ are phonon annihilation and creation operators for LO phonons with a wave vector

$$\mathbf{k}_{q_{pf}} = \mathbf{k}_p - \mathbf{k}_f \quad (9.91)$$

It is indeed possible to use eq. (9.90), but we *cannot* simply use the original definition (7.54) for $(\Xi^{DLO})_{cf,cp}$. The reason is that eq. (7.54) is calculated for the case that $|\mathbf{k}_p - \mathbf{k}_f|$ is small. Thus it contains the optical deformation potential (7.38),

$$\left(D^O_\mu \right)_{c,c} = \lim_{\mathbf{k}_{q_{pf}} \to 0} \langle ck_p | d^{q_{pf} O}_\mu(\mathbf{r}) | ck_f \rangle \quad (9.92)$$

The solution to this problem is to redefine $(\Xi^{DLO})_{cf,cp}$ in such a way that this approximation is avoided. So we set

$$(\Xi^{DLO})_{cf,cp} = \sqrt{\frac{M_1 + M_2}{\hbar \omega^{q_{pf} LO} M_1 M_2}} \sum_\mu p^{q_{pf} LO}_\mu \langle ck_p | d^{q_{pf} O}_\mu(\mathbf{r}) | ck_f \rangle \quad (9.93)$$

Of course, this quantity may now differ significantly from its value used to calculate optical deformation potential scattering rates.

The matrix element we are looking for, couples the intermediate electron state with the final electron state while simultaneously a phonon is emitted. Assuming $n^{q_{pf}LO}$ LO phonons at wave vector $\mathbf{k}_{pf}$ in the initial state and using eq. (7.68) we find

$$\langle n^{q_{pf}LO} | \mathcal{H}^{LO}_{cf,cp} | n^{q_{pf}LO} + 1 \rangle$$

$$= \frac{\hbar}{\sqrt{2N}} \left(\Xi^{DLO} \right)_{cf,cp}$$

$$\times \left(\langle n^{q_{pf}LO} | a^{q_{pf}LO} | n^{q_{pf}LO} + 1 \rangle + \langle n^{q_{pf}LO} | a^{q_{pf}LO\dagger} | n^{q_{pf}LO} + 1 \rangle \right)$$

$$= \frac{\hbar}{\sqrt{2N}} \left(\Xi^{DLO} \right)_{cf,cp} \sqrt{n^{q_{pf}LO} + 1} \qquad (9.94)$$

TRANSITION PROBABILITIES

Now we have found the matrix elements responsible for coupling the initial state with the intermediate state and the intermediate state with the final state, we may directly use eq. (C.85) to calculate the transition probability. We note that all Bloch states in the Γ-valley of the conduction band may serve as intermediate state, though selection rules will restrict their number. So the transition probability contains a sum over the intermediate wave vector $\mathbf{k}_p$ and δ-functions representing these selection rules. Thus, inserting eqs (9.89) and (9.94),

$$W_{hhi,cf} = \frac{2\pi}{\hbar} \left| \sum_p \frac{(\mathcal{H}_F)^0_{hhi,cp} \langle n^{q_{pf}LO} | \mathcal{H}^{LO}_{cf,cp} | n^{q_{pf}LO} + 1 \rangle}{(E^c_p - E^c_f - \hbar\omega^{q_{pf}LO})} \right|^2 \times \delta(E^c_f + \hbar\omega^{q_{pf}LO} - E^{hh}_i - \hbar\omega)$$

$$= \frac{2\pi}{\hbar} \left| \sum_p \frac{1}{E^c_p - E^c_f - \hbar\omega^{q_{pf}LO}} \times \left\{ \frac{\hbar e}{2im_e} P \left(\sum_\mu \langle hhk_i | v\mu 0 \rangle A_{0\mu} \right) \delta_{i,p} \right\} \right.$$

$$\left. \times \left\{ \frac{\hbar}{\sqrt{2N}} \left(\Xi^{DLO} \right)_{cf,cp} \sqrt{n^{q_{pf}LO} + 1} \right\} \right|^2 \times \delta(E^c_f + \hbar\omega^{q_{pf}LO} - E^{hh}_i - \hbar\omega)$$

$$= \frac{\pi \hbar^3 e^2}{4 m_e^2} \frac{n^{q_{if}LO} + 1}{N} \left| \frac{P(\sum_\mu \langle hhk_i | v\mu 0 \rangle A_{0\mu}) \left(\Xi^{DLO} \right)_{cf,ci}}{\hbar\omega} \right|^2$$

$$\times \delta(E^c_f + \hbar\omega^{q_{if}LO} - E^{hh}_i - \hbar\omega) \qquad (9.95)$$

We see that the sum over intermediate wave vectors has been cancelled against the δ-function $\delta_{i,p}$ requiring conservation of crystal momentum for optical transition matrix elements. To establish the remaining Dirac δ-function, we took into account that, because of the emission of a phonon, the energy of the crystal in the final state is equal to $E^c_f + \hbar\omega^{q_{if}LO} - E^{hh}_i$, while it is zero in the initial state and $E^c_i - E^{hh}_i$ in the intermediate state. Moreover, then only the second term in eq. (9.89) contributes to our result, as the first term yields a Dirac δ-function which can never be satisfied. Finally, in the last step, we used the Dirac δ-function to simplify the denominator.

9.2 OPTICAL ABSORPTION

POSSIBLE INITIAL AND FINAL STATES

A major difference between direct and indirect optical transitions lies in the possible choices of initial and final states. While for direct optical transition we have the stringent requirement that the initial and final wave vector should be equal, the requirements for indirect optical transitions are much more relaxed. The reason is that phonons may now take up the missing crystal momentum. Thus, to obtain the total transition probability, one has to sum eq. (9.95) over all initial $\mathbf{k}_i$-vectors *and* over all final $\mathbf{k}_f$-vectors. Only the δ-function represents some restriction in these summations. As a result, the total transition probability is given by,

$$W_{hh,c}^{tot} = \sum_i^N \sum_f^N W_{hhi,cf}$$

$$= \frac{\pi \hbar^3 e^2}{4m_e^2} \sum_i^N \sum_f^N \frac{n^{q_{if}LO} + 1}{N} \left| \frac{P(\sum_\mu \langle hhk_i | v\mu 0 \rangle A_{0\mu})(\Xi^{DLO})_{cf,ci}}{\hbar \omega} \right|^2$$

$$\times \delta(E_f^c + \hbar \omega^{q_{if}LO} - E_i^{hh} - \hbar \omega) \tag{9.96}$$

To continue we follow the procedure for direct optical transitions. We write the initial wave vector directly in spherical coordinates (k_i, θ_i, ϕ_i). Before doing the same with the final wave vector $\mathbf{k}_f$, we note that $\mathbf{k}_f$ lies near the minimum of an L-valley of the conduction band, rather than near the Γ-point. Therefore it is more practical to choose the origin at this L-point rather than in the Γ-point. Thus, we write

$$\mathbf{k}_f' = (k_f', \theta_f', \phi_f') \equiv \mathbf{k}_f - \mathbf{k}_L \tag{9.97}$$

where $\mathbf{k}_L$ is the position of this L-point in reciprocal space.

Next we approximate the sums over i and f by integrals. The latter step is required because the suppression of the Dirac δ-function needs integration instead of summation, while the former step allows a suitable choice of the parameter over which we integrate in order to achieve this suppression. Thus,

$$W_{hh,c}^{tot} \approx \frac{N}{V_B} \int_{V_B} d\phi_i\, d\theta_i\, \sin\theta_i\, dk_i\, k_i^2 \; \frac{N}{V_B} \int_{V_B} d\phi_f'\, d\theta_f'\, \sin\theta_f'\, dk_f'\, k_f'^2\, W_{hhi,cf}$$

$$= \frac{N}{V_B} \int_\infty d\phi_i\, d\theta_i\, \sin\theta_i\, dk_i\, k_i^2 \; \frac{N}{V_B} \int_\infty d\phi_f'\, d\theta_f'\, \sin\theta_f'\, dk_f'\, k_f'^2\, W_{hhi,cf}$$

$$= \int_0^{2\pi} d\phi_i \int_0^\pi d\theta_i\, \sin\theta_i \int_0^{2\pi} d\phi_f' \int_0^\pi d\theta_f'\, \sin\theta_f'$$

$$\times \frac{N}{V_B} \int_0^\infty dk_i\, k_i^2 \; \frac{N}{V_B} \int_0^\infty dk_f'\, k_f'^2\, W_{hhi,cf} \tag{9.98}$$

DENSITIES OF STATES

Subsequently we transform the integrals over k_i and k_f into integrations over the initial and final energies E_i^{hh} and E_f^c. The Jacobians of these transformations are

$$\left(\frac{\partial E_i^{hh}}{\partial k_i} \right)_{\theta_i, \phi_i} \quad \text{and} \quad \left(\frac{\partial E_f^c}{\partial k_f'} \right)_{\theta_f', \phi_f'} \tag{9.99}$$

Furthermore, we recall the differential density of states (8.42), which can now be written as

$$N_d^{hh}(E_i^{hh}, \theta_i, \phi_i) = -\frac{N}{V_B} \frac{k_i^2}{\left(\frac{\partial E_i^{hh}}{\partial k_i}\right)_{\theta_i, \phi_i}} \qquad (9.100)$$

$$N_d^c(E_f^c, \theta_f', \phi_f') = \frac{N}{V_B} \frac{k_f'^2}{\left(\frac{\partial E_f^c}{\partial k_f'}\right)_{\theta_f', \phi_f'}} \qquad (9.101)$$

for the initial and final states, respectively. As usual we choose the zero point of the energy at the top of the valence band so $E_0^{hh} = 0$. Note the minus sign in the differential density of states for the heavy holes band. For the valence band the derivative (9.99) is negative, so this minus sign is needed to render $N_d^{hh}(E_i^{hh}, \theta_i, \phi_i)$ positive. Then, the total transition probability is immediately found to be given by

$$W_{hh,c}^{tot} \approx \int_0^{2\pi} d\phi_i \int_0^{\pi} d\theta_i \sin\theta_i \int_0^{2\pi} d\phi_f' \int_0^{\pi} d\theta_f' \sin\theta_f'$$

$$\times \frac{N}{V_B} \int_0^{\infty} dk_i \, k_i^2 \, \frac{N}{V_B} \int_0^{\infty} dk_f' \, k_f'^2 \, W_{hhi,cf}$$

$$= \int_0^{2\pi} d\phi_i \int_0^{\pi} d\theta_i \sin\theta_i \int_0^{2\pi} d\phi_f' \int_0^{\pi} d\theta_f' \sin\theta_f'$$

$$\times \int_{-\infty}^0 dE_i^{hh} \, N_d^{hh}(E_i^{hh}, \theta_i, \phi_i) \int_{E_g}^{\infty} dE_f^c \, N_d^c(E_f^c, \theta_f', \phi_f') \, W_{hhi,cf}$$

$$= \int_0^{2\pi} d\phi_i \int_0^{\pi} d\theta_i \sin\theta_i \int_0^{2\pi} d\phi_f' \int_0^{\pi} d\theta_f' \sin\theta_f'$$

$$\times \int_{E_g}^{\infty} dE_f^c \int_{-\infty}^0 dE_i^{hh} \, N_d^{hh}(E_i^{hh}, \theta_i, \phi_i) \, N_d^c(E_f^c, \theta_f', \phi_f')$$

$$\times \frac{\pi \hbar^3 e^2}{4 m_e^2} \frac{n^{q_{if}LO} + 1}{N} \left| \frac{P(\sum_\mu \langle hhk_i | v\mu 0 \rangle A_{0\mu})(\Xi^{DLO})_{cf,ci}}{\hbar\omega} \right|^2$$

$$\times \delta(E_f^c + \hbar\omega^{q_{if}LO} - E_i^{hh} - \hbar\omega)$$

$$= \int_0^{2\pi} d\phi_i \int_0^{\pi} d\theta_i \sin\theta_i \int_0^{2\pi} d\phi_f' \int_0^{\pi} d\theta_f' \sin\theta_f'$$

$$\times \frac{1}{N} \int_{E_g}^{\hbar\omega - \hbar\omega^{q_{if}LO}} dE_f^c \, N_d^c(E_f^c, \theta_f', \phi_f') N_d^{hh}(\hbar\omega - \hbar\omega^{q_{if}LO} - E_f^c, \theta_i, \phi_i)$$

$$\times \frac{\pi \hbar^3 e^2}{4 m_e^2} (n^{q_{if}LO} + 1) \left| \frac{P(\sum_\mu \langle hhk_i | v\mu 0 \rangle A_{0\mu})(\Xi^{DLO})_{cf,ci}}{\hbar\omega} \right|^2 \qquad (9.102)$$

9.2 OPTICAL ABSORPTION

where we integrated over E_i^{hh} in order to eliminate the Dirac δ-function. Note also that the conduction band density of states (9.101) is equal to zero if E_f^c is less than the band gap E_g, so the integral over E_f^c is naturally restricted to $E_f^c > E_g$. Similarly, the valence band density of states (9.100) is equal to zero if $E_i^{hh} > 0$, so in the last step the integral over E_f^c is restricted to $E_f^c < \hbar\omega - \hbar\omega^{q_{if}LO}$ as well.

COMPARISON BETWEEN DIRECT AND INDIRECT TRANSITIONS

No doubt the final result (9.102) for the transition probability for indirect optical transitions has a rather intimidating look, due to its many integrals and factors in the integrand. However, if we allow ourselves some well-chosen approximations we are able to obtain some insight in its analogies and differences with the result (9.75) for direct optical transitions.

First, we observe that the differential density of states is now replaced by the convolution

$$I = \frac{1}{N}\int_{E_g}^{\hbar\omega - \hbar\omega^{q_{if}LO}} dE_f^c\, N_d^c(E_f^c, \theta_f', \phi_f') N_d^{hh}(\hbar\omega - \hbar\omega^{q_{if}LO} - E_f^c, \theta_i, \phi_i) \quad (9.103)$$

of the individual densities of states for each band. Though grossly incorrect, we will simplify this integral by assuming both the conduction band and the valence band to be isotropic. Then, according to eq. (8.57),

$$N_d^c = \frac{N}{V_B} \frac{\sqrt{2(E_f^c - E_g)}(m_e m_c^*)^{3/2}}{\hbar^3} \quad (9.104)$$

$$N_d^{hh} = \frac{N}{V_B} \frac{\sqrt{-2E_i^{hh}}(m_e m_{hh}^*)^{3/2}}{\hbar^3} \quad (9.105)$$

Note the change of sign for the heavy holes band where the density of states increases with decreasing energy. As a result the convolution (9.103) becomes simply

$$I = \frac{1}{N}\int_{E_g}^{\hbar\omega - \hbar\omega^{q_{if}LO}} dE_f^c$$

$$\times \frac{N}{V_B}\frac{\sqrt{2(E_f^c - E_g)}(m_e m_c^*)^{3/2}}{\hbar^3} \frac{N}{V_B}\frac{\sqrt{2(\hbar\omega - \hbar\omega^{q_{if}LO} - E_f^c)}(m_e m_{hh}^*)^{3/2}}{\hbar^3}$$

$$= \frac{2N}{V_B^2}\frac{(m_e m_c^*)^{3/2}}{\hbar^3}\frac{(m_e m_{hh}^*)^{3/2}}{\hbar^3} \times \int_{E_g}^{\hbar\omega - \hbar\omega^{q_{if}LO}} dE_f^c \sqrt{(E_f^c - E_g)(\hbar\omega - \hbar\omega^{q_{if}LO} - E_f^c)}$$

$$= \frac{2N}{V_B^2}\frac{(m_e m_c^*)^{3/2}}{\hbar^3}\frac{(m_e m_{hh}^*)^{3/2}}{\hbar^3}(\hbar\omega - \hbar\omega^{q_{if}LO} - E_g)^2 \int_0^1 dx\,\sqrt{x(1-x)}$$

$$= \frac{\pi N}{4 V_B^2}\frac{m_e^3(m_c^* m_{hh}^*)^{3/2}}{\hbar^6}(\hbar\omega - \hbar\omega^{q_{if}LO} - E_g)^2 \quad (9.106)$$

To obtain a further simplification we observe that three quantities depending on the angles θ_f' and ϕ_f' remain in the integrand in (9.102). The occupation number $n^{q_{if}LO}$ of the LO phonon mode at wave number $\mathbf{k}_{q_{if}}$, the coupling between this phonon mode and the excited electron $(\Xi^{DLO})_{cf,ci}$ and the phonon energy $\hbar\omega^{q_{if}LO}$.

We start with the last of these three quantities. For low values of $\hbar\omega$, the energy conservation expressed by the Dirac δ-function in eqs (9.96) and (9.102), entails that the excited electron ends up near the bottom of the chosen L-valley. Hence all phonons involved in the process have wave vectors near $\mathbf{k}_L$, which is the wave vector connecting the Γ-point with the bottom of the chosen L-valley. We remember that we encountered a similar situation in chapter 7, in our treatment of deformation potential electron-optical phonon coupling. Also there only phonons in a small region of the Brillouin zone were involved and we approximated the deformation potential to be constant. Therefore, here we will do the same and approximate

$$(\Xi^{DLO})_{cf,ci} \approx (\Xi^{DLO})_{\Gamma,L} \qquad (9.107)$$

Moreover, we will again approximate the phonon energy to have an approximate value $\hbar\omega^{q_{if}LO} = \hbar\omega^{\Gamma L,LO}$.

It is not so straightforward to handle the phonon occupation number $n^{q_{if}LO}$. However, if we restrict our comparison between direct and indirect optical transitions to low temperatures, then we may simply set $n^{q_{if}LO} = 0$, thus avoiding any further complicated discussions. One actually does not need to go to very low temperatures, as $\hbar\omega_{LO}^{q_{if}}/k_B \approx 350$ K for germanium.

As a result of these arguments, the integrand in eq. (9.102) has been made independent of θ'_f and ϕ'_f, so the integrals over these angles are trivial and yield 4π. Thus,

$$W_{hh,c}^{tot} \approx \frac{\pi\hbar^3 e^2}{4m_e^2} \frac{\pi N}{4V_B^2} \frac{m_e^3 (m_c^* m_{hh}^*)^{3/2}}{\hbar^6} (\hbar\omega - \hbar\omega^{\Gamma L,LO} - E_g)^2 \, 4\pi \left|\frac{(\Xi^{DLO})_{\Gamma,L}}{\hbar\omega}\right|^2$$

$$\times \int_0^{2\pi} d\phi_i \int_0^{\pi} d\theta_i \, \sin\theta_i \left| P\sum_\mu \langle hhk_i|v\mu 0\rangle A_{0\mu}\right|^2 \qquad (9.108)$$

We will not try a quantitative comparison of this result with the direct case (9.75). Such a comparison makes little sense as we have not taken into account all possible processes leading to indirect optical transitions. However, eq. (9.108) shows three features meriting attention. First, the optical absorption vanishes when

$$\hbar\omega < \hbar\omega^{\Gamma L,LO} - E_g \qquad (9.109)$$

instead of $\hbar\omega < E_g$. This is clearly due to fact that now also a phonon is emitted, while in direct transitions this is not the case. Next, we see that the indirect case differs from the direct case by various factors, among which

$$\left|\frac{(\Xi^{DLO})_{\Gamma,L}}{\omega}\right|^2 \qquad (9.110)$$

This factor represents the fact that indirect transitions are second order instead of first order. Now, $(\hbar/e)(\Xi^{DLO})_{\Gamma,L}$ is typically of the order 50 meV, while $(\hbar/e)\omega$ is rather 1 eV, so this factor yields a reduction in optical absorption of more than two orders of magnitude. Finally, we see that the optical absorption now depends on the *square* of $(\hbar\omega - \hbar\omega^{\Gamma L,LO} - E_g)$, while it depends on the *square root* of $(\hbar\omega - E_g)$ in

9.2 OPTICAL ABSORPTION

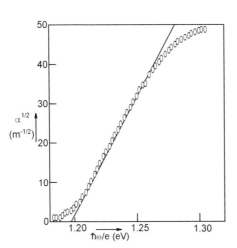

Figure 9.4 The square root of the optical absorption coefficient of Si *versus* photon energy at 6 K.

the direct case. Thus, the frequency dependence is profoundly different for indirect optical absorption compared to direct optical absorption. Fig. 9.4 shows unpublished experiments by the author on Si at 6 K illustrating this quadratic dependence nicely.

9.2.3 Quantum Wells

INTRABAND TRANSITIONS

Our final treatment concerns intraband transitions. For simplicity we restrict ourselves to the simple situation of a non-degenerate band with an isotropic effective mass m^* as e.g. the Γ-valley of the conduction band in GaAs. In section 9.1.4 we saw that this situation can be fully handled using effective mass theory only. Then the unperturbed Hamiltonian is given by eq. (9.47) which for an isotropic effective mass reduces to

$$\mathcal{H}_{0,\text{eff}} = -\frac{\hbar^2}{2m_e m^*} \Delta \quad (9.111)$$

The eigenstates of this unperturbed Hamiltonian are plane waves. Thus, the optical field couples initial and final electron states (9.49) and (9.50),

$$\phi_i(\mathbf{r}) = \frac{1}{\sqrt{NV_c}} e^{i\mathbf{k}_i \cdot \mathbf{r}} \quad (9.112)$$

$$\phi_f(\mathbf{r}) = \frac{1}{\sqrt{NV_c}} e^{i\mathbf{k}_f \cdot \mathbf{r}} \quad (9.113)$$

with energy,

$$E_i^n = \frac{\hbar^2 |\mathbf{k}_i|^2}{2m_e m^*} \quad (9.114)$$

$$E_f^n = \frac{\hbar^2 |\mathbf{k}_f|^2}{2m_e m^*} \quad (9.115)$$

This coupling is described by the transition matrix element (9.61),

$$(\mathcal{H}_F)_{ni,nf} = \frac{\hbar e}{m_e m^*} \mathbf{A}_0 \cdot \mathbf{k}_f \cos(\omega t)\delta_{i,f} \qquad (9.116)$$

Here $\mathbf{A}_0$ is the amplitude of the vector potential describing the optical field and ω its frequency.

We investigate the transitions which the optical field may induce between the plane waves (9.112) and (9.113). Up to first order, the transition rate is obtained using eq. (C.79). Then,

$$W_{ni,nf} = \frac{2\pi}{\hbar}|(\mathcal{H}_F)_{ni,nf}|^2 \delta(E_i^n - E_f^n - \hbar\omega)$$

$$= \frac{2\pi}{\hbar}\left|\frac{\hbar e}{m_e m^*}(\mathbf{A}_0 \cdot \mathbf{k}_f)\delta_{i,f}\right|^2 \delta\left(\frac{\hbar^2|\mathbf{k}_i|^2}{2m_e m^*} - \frac{\hbar^2|\mathbf{k}_f|^2}{2m_e m^*} - \hbar\omega\right) \qquad (9.117)$$

Two δ-functions restrict the number of possible transitions severely. The function $\delta_{i,f}$ requires

$$\mathbf{k}_i = \mathbf{k}_f \qquad (9.118)$$

However, then $E_i^n = E_f^n$ and the Dirac δ-function reduces to the requirement

$$\omega = 0 \qquad (9.119)$$

Thus, in first order only *time-independent* electromagnetic fields are able to act on electrons in a non-degenerate band.

In second order the situation is different. Indirect optical transitions involving phonons *and* the optical field are capable of inducing transitions between states with different wave vectors. In general they are a dominant source of optical absorption due to interaction with band electrons. For a more detailed treatment see ref. [35].

QUANTUM WELLS

As mentioned earlier, a strongly different situation occurs when a band is further quantized due to externally applied fields, impurity potentials or potentials occurring in an artificial structure like a quantum well. The essential point is that the extra potential mixes plane waves, thus breaking the severe selection rules described above. As an example we will consider optical absorption between subbands in a square quantum well. This example has the further beauty that both the energy levels and the coupling with the optical field can be described using effective mass theory only. Furthermore the treatment is given in such a way that it shows clearly how plane waves mix and thus allow for breaking the severe selection rules.

We refer to section 5.3.3 for the treatment of an infinitely deep square quantum well. The effective mass is assumed to be isotropic, so $m^*_{\mu\nu} = m^*\delta_{\mu,\nu}$. Then, according to eqs (5.129)–(5.132) the effective Hamiltonian is written as

$$\mathcal{H}_{eff} = \mathcal{H}^x_{eff} + \mathcal{H}^y_{eff} + \mathcal{H}^z_{eff} \qquad (9.120)$$

9.2 OPTICAL ABSORPTION

where

$$\mathcal{H}_{eff}^x = -\frac{\hbar^2}{2m_e m^*}\frac{\partial^2}{\partial x^2} \qquad (9.121)$$

$$\mathcal{H}_{eff}^y = -\frac{\hbar^2}{2m_e m^*}\frac{\partial^2}{\partial y^2} \qquad (9.122)$$

$$\mathcal{H}_{eff}^z = -\frac{\hbar^2}{2m_e m^*}\frac{\partial^2}{\partial z^2} - eU(z) \qquad (9.123)$$

Thus the first two terms (9.121) and (9.122) describe a free motion in the x- and y-direction. The last term (9.123) corresponds to the problem of a one-dimensional potential well, yielding bound states. We consider an infinitely deep square well with a width $2z_0$, so $eU(z) = 0$ in the interval $-z_0 < z < +z_0$ and $eU(z) = \infty$ outside this interval. According to eq. (5.134) it has eigenfunctions

$$F_n(\mathbf{r}) = \frac{1}{\sqrt{z_0}}\frac{1}{(NV_C)^{1/3}} \sin k_n(z+z_0)\, e^{i(k_x x + k_y y)} \quad \text{if} \quad -z_0 \le z \le z_0 \qquad (9.124)$$

while $F_n(\mathbf{r}) = 0$ for $z < -z_0$ and $z > z_0$. Furthermore the factors $1/\sqrt{z_0}$ and $1/(NV_C)^{1/3}$ assure that the wave functions are normalized over the volume of the well. The quantum numbers n are given by

$$n = 1, 2, 3, \ldots \qquad (9.125)$$

and k_n is quantized to values

$$k_n = \frac{n\pi}{2z_0} \qquad (9.126)$$

while the energy takes discrete values

$$E_n = \frac{\hbar^2 k_n^2 + \hbar^2 k_x^2 + \hbar^2 k_y^2}{2m_e m^*} \qquad (9.127)$$

Fig. 9.5 shows the wave functions as a function of z and the bottoms of the subbands following from eqs (9.126) and (9.127).

EXPANSION IN PLANE WAVES

We expand the eigenfunctions (9.124) in plane waves $\exp(i\mathbf{k}\cdot\mathbf{r})$. Though this is not the shortest way to a result for the optical absorption, it shows us why optical absorption is now allowed in first order while it is forbidden in an unperturbed non-degenerate band. For this purpose we introduce the function $B(z)$ such that

$$B(z) = 1 \quad \text{if} \quad -z_0 \le z \le z_0$$

$$B(z) = 0 \quad \text{if} \quad z > z_0 \quad \text{or} \quad z < -z_0 \qquad (9.128)$$

which allows us to write the eigenfunctions (9.124) without the restriction $-z_0 \le z \le z_0$. Also splitting the sine in these equations one finds for all values of z,

$$F_n(\mathbf{r}) = \frac{1}{\sqrt{z_0}}\frac{1}{(NV_C)^{1/3}} B(z) \frac{1}{i\sqrt{2}}\left(e^{ik_n(z+z_0)} - e^{-ik_n(z+z_0)}\right)e^{i(k_x x + k_y y)} \qquad (9.129)$$

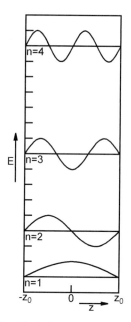

Figure 9.5 The infinitely deep square well.

Next we note that $B(z)$ can be expanded as

$$B(z) = \int_{-\infty}^{+\infty} dk' \, \frac{\sin k' z_0}{\pi k'} e^{ik'z} \tag{9.130}$$

Then, eq. (9.129) becomes,

$$\begin{aligned}
F_n(\mathbf{r}) &= \frac{1}{\sqrt{z_0}} \frac{1}{(NV_C)^{1/3}} \int_{-\infty}^{+\infty} dk' \, \frac{\sin k' z_0}{\pi k'} e^{ik'z} \times \frac{1}{i\sqrt{2}} (e^{ik_n z} - e^{-ik_n z}) e^{i(k_x x + k_y y)} \\
&= \frac{(NV_C)^{1/6}}{i\sqrt{2 z_0}} \int_{-\infty}^{+\infty} dk' \, \frac{\sin k' z_0}{\pi k'} \times \left(e^{ik_n z_0} \frac{1}{\sqrt{NV_C}} e^{i(k_x x + k_y y + (k' + k_n)z)} \right. \\
&\quad \left. - e^{-ik_n z_0} \frac{1}{\sqrt{NV_C}} e^{i(k_x x + k_y y + (k' - k_n)z)} \right)
\end{aligned} \tag{9.131}$$

which shows that the eigenfunctions $F_n(\mathbf{r})$ are actually linear combinations of plane waves

$$\frac{1}{\sqrt{NV_C}} e^{i(k_x x + k_y y + (k' + k_n)z)} \tag{9.132}$$

$$\frac{1}{\sqrt{NV_C}} e^{i(k_x x + k_y y + (k' - k_n)z)} \tag{9.133}$$

with wave vectors $\mathbf{k}_\pm = (k_x, k_y, (k' \pm k_n))$, amplitude

$$\pm \frac{(NV_C)^{1/3}}{i\sqrt{2 z_0}} \frac{\sin k' z_0}{\pi k'} \tag{9.134}$$

and phase $\pm k_n z_0$.

SELECTION RULES

We now recall the most important feature of eq. (9.116), which is the δ-function requiring the initial and final wave vectors of the electron to be the same. As a result, when no external potentials are applied, the optical field cannot induce any transitions. However, if an external potential is switched on, as is the case in our quantum well, then the situation is different. Eq. (9.131) clearly shows that each eigenstate is constructed from many plane waves. Hence, the optical field may have non-zero matrix elements between *different* eigenstates $F_n(\mathbf{r})$ and $F_m(\mathbf{r})$ of the electron, provided $F_n(\mathbf{r})$ contains a component with wave vector

$$\mathbf{k} = [k_x, k_y, (k' \pm k_n)] \tag{9.135}$$

and $F_m(\mathbf{r})$ a component with wave vector

$$\mathbf{k} = [k_x, k_y, (k'' \pm k_m)] \tag{9.136}$$

such that

$$k' \pm k_n = k'' \pm k_m \tag{9.137}$$

Note that we have encountered a phenomenom extending far beyond the special case of the square quantum well. An optical field may induce transitions between quantum states that originate in the *same* band, provided an *extra* potential is switched on. This extra potential combines Bloch states into wave packets and if the same Bloch states participate in two different wave packets, then optical transitions between these two wave packets are allowed.

Continuing the present treatment towards the optical absorption coefficient is direct though lengthy. For this purpose one needs to insert eq. (9.131) to expand the matrix elements

$$\int_\infty d\mathbf{r}\, F_n^*(\mathbf{r})\, \mathcal{H}_F\, F_m(\mathbf{r}) \tag{9.138}$$

in matrix elements (9.116) of $\mathcal{H}_F$ between plane waves. Next, one uses eq. (C.79) to obtain the transition probability. Subsequently, to obtain the total absorption rate, one needs to derive the differential density of states for the case that an electron is free in two directions but bound in the third direction and finally one has to rewrite the transition rate into an absorption coefficient as done previously in section 9.2.1.

However, the present treatment is not the shortest route towards the desired result. For this purpose it is better to start again in the earlier stage treated in section 9.1.4. Thus, we immediately add the external potential to the effective Hamiltonian (9.45). As a result, the unperturbed Hamiltonian (9.47) contains a potential term due to this externally applied potential and the eigenstates (9.49) and (9.50) need to be replaced by eigenstates of a Hamiltonian including such a potential term. Then the matrix elements of the optical coupling (9.48) are calculated between these new eigenstates and immediately transition rates can be obtained by invoking eq. (C.79). Thus, we calculate the matrix elements of the optical coupling (9.48) between two eigenstates (9.124) and immediately find the desired transition matrix elements. This shorter route will be left to the reader as an exercise.

Appendix A
The Hydrogen Atom

The description of atomic structure is based on the wave functions of the hydrogen atom. This simplest of all atoms is treated in most textbooks on quantum mechanics. Therefore we limit ourselves to repeating the results. The hydrogen atom consists of an electron and a proton having mutual electrostatic interaction. Our first approximation corresponds to replacing the nucleus by a central force field. Then, neglecting the electron spin, the Hamiltonian of the hydrogen atom is given by

$$\mathcal{H} = -\frac{\hbar^2}{2m_e}\Delta + \mathcal{V} \tag{A.1}$$

Here, Δ is the Laplace operator,

$$\Delta = \nabla \cdot \nabla \quad \text{where} \quad \nabla = \left(\frac{\partial}{\partial x}, \frac{\partial}{\partial y}, \frac{\partial}{\partial z}\right)$$

and $\mathcal{V}$ is the electrostatic attraction between the nucleus and the electron

$$\mathcal{V} = -\frac{e^2}{4\pi\varepsilon_0 r} \tag{A.2}$$

In these expressions, $\mathbf{r} = (x, y, z)$ is the position of the electron with respect to the proton, so $r = \sqrt{x^2 + y^2 + z^2}$ is the distance between these two particles. Furthermore m_e is the mass of the electron and $-e$ its charge while $h = 2\pi\hbar$ is Planck's constant. We use rationalized Giorgi (SI) units. Then

$$4\pi\varepsilon_0 = c^{-2} \cdot 10^7$$

$$c = 2.998 \cdot 10^8 \text{ m s}^{-1} = \text{the velocity of light}$$

$$e = 1.602 \cdot 10^{-19} \text{ C}$$

$$m_e = 9.110 \cdot 10^{-31} \text{ kg}$$

$$\hbar = 1.055 \cdot 10^{-34} \text{ J s}$$

The eigenstates Ψ_{nlm} of the hydrogen atom are characterized by three quantum numbers, the principal quantum number n and two quantum numbers, l and m, that determine total angular momentum and its z-component. They take values

$$n = 1, 2, 3, \ldots, \infty$$

$$l = 0, 1, 2, \ldots, n-1$$

$$m = -l, -l+1, \ldots, l-1, l$$

The energy levels of the hydrogen atom are determined by the principal quantum number n only, so they are n^2-fold degenerate with respect to the quantum numbers l and m:

$$E_n = -\frac{Ry}{n^2} \tag{A.3}$$

where Ry is the Rydberg, the splitting between the lowest and highest energy levels E_1 and E_∞. $-Ry$ corresponds to the electrostatic potential (A.2) at a distance $r = 2a$, where a is the Bohr radius

$$a = \frac{4\pi\varepsilon_0 \hbar^2}{m_e e^2} = 0.5292 \cdot 10^{-10} \text{ m} \tag{A.4}$$

So

$$\frac{Ry}{e} = \frac{e}{8\pi\varepsilon_0 a} = 13.606 \text{ eV} \tag{A.5}$$

where 1 eV corresponds to $1.602 \cdot 10^{-19}$ J. The energy levels of the hydrogen atom are shown in Fig. A.1.

The eigenstates of the hydrogen atom can be written as the product of a radial part $\rho_{nl}(r)r^l$ depending on the quantum numbers n and l and a spherical harmonic function $Y_l^m(\theta, \phi)$ determined by the quantum numbers l and m describing the state of angular momentum:

$$\Psi_{nlm} = \rho_{nl}(r)r^l Y_l^m(\theta, \phi) \tag{A.6}$$

where (r, θ, ϕ) are spherical coordinates defined by

$$x = r \sin\theta \cos\phi$$

$$y = r \sin\theta \sin\phi$$

$$z = r \cos\theta$$

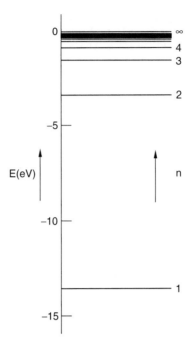

Figure A.1 The energy levels of the hydrogen atom.

Table A.1 The factor $\rho_{nl}(r)$ for the nine hydrogen wave functions with the lowest energy.

n	l	m	$\rho_{nl}(r)$
1	0	0	$\dfrac{2}{\sqrt{a^3}} \cdot \exp\left(-\dfrac{r}{a}\right)$
2	0	0	$\dfrac{1}{\sqrt{2a^3}} \cdot \left[1 - \dfrac{r}{2a}\right] \exp\left(-\dfrac{r}{2a}\right)$
2	1	0	$\dfrac{1}{2\sqrt{6a^5}} \cdot \exp\left(-\dfrac{r}{2a}\right)$
2	1	1	$\dfrac{1}{2\sqrt{6a^5}} \cdot \exp\left(-\dfrac{r}{2a}\right)$
2	1	-1	$\dfrac{1}{2\sqrt{6a^5}} \cdot \exp\left(-\dfrac{r}{2a}\right)$
3	0	0	$\dfrac{2}{3\sqrt{3a^3}} \cdot \left[1 - \dfrac{2r}{3a} + \dfrac{2r^2}{27a^2}\right] \cdot \exp\left(-\dfrac{r}{3a}\right)$
3	1	0	$\dfrac{8}{27\sqrt{6a^5}} \cdot \left[1 - \dfrac{r}{6a}\right] \cdot \exp\left(-\dfrac{r}{3a}\right)$
3	1	1	$\dfrac{8}{27\sqrt{6a^5}} \cdot \left[1 - \dfrac{r}{6a}\right] \cdot \exp\left(-\dfrac{r}{3a}\right)$
3	1	-1	$\dfrac{8}{27\sqrt{6a^5}} \cdot \left[1 - \dfrac{r}{6a}\right] \cdot \exp\left(-\dfrac{r}{3a}\right)$

THE HYDROGEN ATOM

Table A.2 The factor $r^l Y_l^m(\theta,\phi)$ for the nine hydrogen wave functions with the lowest energy.

n	l	m	$r^l Y_l^m(\theta,\phi)$
1	0	0	$\dfrac{1}{\sqrt{4\pi}} = \dfrac{1}{\sqrt{4\pi}}$
2	0	0	$\dfrac{1}{\sqrt{4\pi}} = \dfrac{1}{\sqrt{4\pi}}$
2	1	0	$r\sqrt{\dfrac{3}{4\pi}}\cos\theta = \sqrt{\dfrac{3}{4\pi}}z$
2	1	1	$-r\sqrt{\dfrac{3}{8\pi}}\sin\theta e^{i\phi} = -\sqrt{\dfrac{3}{8\pi}}(x+iy)$
2	1	-1	$r\sqrt{\dfrac{3}{8\pi}}\sin\theta e^{-i\phi} = \sqrt{\dfrac{3}{8\pi}}(x-iy)$
3	0	0	$\dfrac{1}{\sqrt{4\pi}} = \dfrac{1}{\sqrt{4\pi}}$
3	1	0	$r\sqrt{\dfrac{3}{4\pi}}\cos\theta = \sqrt{\dfrac{3}{4\pi}}z$
3	1	1	$-r\sqrt{\dfrac{3}{8\pi}}\sin\theta e^{i\phi} = -\sqrt{\dfrac{3}{8\pi}}(x+iy)$
3	1	-1	$r\sqrt{\dfrac{3}{8\pi}}\sin\theta e^{-i\phi} = \sqrt{\dfrac{3}{8\pi}}(x-iy)$

Table A.3 The factor $F_{l\mu}(\theta,\phi)$ in the real linear combinations of the nine hydrogen wave functions with the lowest energy.

$l\mu$		$\Psi_{nl\mu}$	$F_{l\mu}(\theta,\phi)$
00	s	Ψ_{ns}	$Y_0^0(\theta,\phi) = \dfrac{1}{\sqrt{4\pi}}$
1z	p_z	Ψ_{np_z}	$rY_1^0(\theta,\phi) = \sqrt{\dfrac{3}{4\pi}}z$
1x	p_x	Ψ_{np_x}	$\dfrac{1}{\sqrt{2}}\cdot[-rY_1^1(\theta,\phi)+rY_1^{-1}(\theta,\phi)] = \sqrt{\dfrac{3}{4\pi}}x$
1y	p_y	Ψ_{np_y}	$\dfrac{i}{\sqrt{2}}\cdot[rY_1^1(\theta,\phi)+rY_1^{-1}(\theta,\phi)] = \sqrt{\dfrac{3}{4\pi}}y$

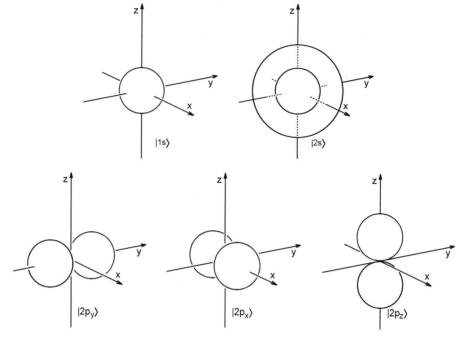

Figure A.2 Artist's impression of some hydrogen wave functions.

Tables A.1 and A.2 show the functions $\rho_{nl}(r)$ and $r^l Y_l^m(\theta,\phi)$ for the nine lowest states. Further tables for $\rho_{nl}(r)$ and $r^l Y_l^m(\theta,\phi)$ up to $n=4$ and $l,m=3$ can be found in ref. [3]. Apparently, the wave functions are complex for $m=1,-1$. However, because all eigenstates with equal principal quantum number n are degenerate, one may replace the eigenstates Ψ_{nlm} by real linear combinations,

$$\Psi_{nl\mu} = \rho_{nl}(r) F_{l\mu}(\theta,\phi) \qquad \mu = x,y,z, \tag{A.7}$$

where the functions $F_{l\mu}(\theta,\phi)$ are given in Table A.3 for $l=0,1$. In Table A.3 both current notations $l=0,1,\ldots$ and $l=s,p,\ldots$ are given in columns 1 and 2, respectively. An artist's impression of some of the real hydrogen wave functions thus obtained are depicted in Fig. A.2.

Appendix B
The Harmonic Oscillator

The Hamiltonian

The one-dimensional harmonic oscillator is described by the Hamiltonian

$$\mathcal{H} = -\frac{\hbar^2}{2m}\frac{\partial^2}{\partial x^2} + \frac{1}{2}C_0 x^2 \tag{B.1}$$

where m is the mass of the particle, x the coordinate and C_0 the force constant. For a stable oscillator C_0 is real and positive, so it may be written as

$$C_0 = m\omega_0^2 \tag{B.2}$$

We note that ω_0 has the dimension $[t]^{-1}$ or frequency. Actually, it is the classical oscillation frequency of the harmonic oscillator. Then

$$\mathcal{H} = -\frac{\hbar^2}{2m}\frac{\partial^2}{\partial x^2} + \frac{1}{2}m\omega_0^2 x^2 \tag{B.3}$$

To simplify the calculation of the eigenvalues and eigenstates one introduces a reduced coordinate

$$q = \sqrt{\frac{m}{\hbar}}\, x \quad \text{so} \quad x = \sqrt{\frac{\hbar}{m}}\, q \tag{B.4}$$

Note that q has the dimension $[t]^{1/2}$. The corresponding reduced momentum is defined as

$$p = \frac{1}{i}\frac{\partial}{\partial q} = \sqrt{\frac{1}{\hbar m}}\frac{\hbar}{i}\frac{\partial}{\partial x} \quad \text{so} \quad \frac{\hbar}{i}\frac{\partial}{\partial x} = \sqrt{\hbar m}\,\frac{1}{i}\frac{\partial}{\partial q} = \sqrt{\hbar m}\, p \tag{B.5}$$

So p has dimension $[t]^{-1/2}$. Then $\mathcal{H}$ reduces to

$$\mathcal{H} = \frac{\hbar}{2}(p^2 + \omega_0^2 q^2) = \frac{\hbar}{2}\left(-\frac{\partial^2}{\partial q^2} + \omega_0^2 q^2\right) \tag{B.6}$$

There are various ways to calculate the eigenvalues and eigenfunction of $\mathcal{H}$. The most elegant method ([11], [15]) introduces two operators, the so-called annihilation operator

$$a = \frac{1}{\sqrt{2}}\left(\sqrt{\omega_0}\, q + i\frac{1}{\sqrt{\omega_0}}\, p\right) = \frac{1}{\sqrt{2}}\left(\sqrt{\omega_0}\, q + \frac{1}{\sqrt{\omega_0}}\frac{\partial}{\partial q}\right) \qquad (B.7)$$

and the so-called creation operator

$$a^\dagger = \frac{1}{\sqrt{2}}\left(\sqrt{\omega_0}\, q - i\frac{1}{\sqrt{\omega_0}}\, p\right) = \frac{1}{\sqrt{2}}\left(\sqrt{\omega_0}\, q - \frac{1}{\sqrt{\omega_0}}\frac{\partial}{\partial q}\right) \qquad (B.8)$$

So inversely

$$q = \sqrt{\frac{1}{2\omega_0}}(a + a^\dagger) \qquad (B.9)$$

$$\frac{\partial}{\partial q} = \sqrt{\frac{\omega_0}{2}}(a - a^\dagger) \qquad (B.10)$$

Then, after some calculation the Hamiltonian can be written as

$$\mathcal{H} = \hbar\omega_0\left(a^\dagger a + \frac{1}{2}\right) = \hbar\omega_0\left(aa^\dagger - \frac{1}{2}\right) \qquad (B.11)$$

Thus, as a by-product we got the commutation relation

$$a^\dagger a - aa^\dagger = 1 \qquad (B.12)$$

The calculation of the eigenvalues and eigenstates of the Hamiltonian (B.11) is now reduced to the determination of the eigenvalues and eigenfunctions of $a^\dagger a$.

EIGENVALUES AND EIGENSTATES

We will not pursue this calculation as it can be found in many books on quantum mechanics (see [11], [15]). The eigenvalues of $a^\dagger a$ are found to be equal to

$$n = 0, 1, 2, 3, \ldots \qquad (B.13)$$

So $\mathcal{H}$ has eigenvalues

$$E_n = \hbar\omega_c\left(n + \frac{1}{2}\right) \qquad (B.14)$$

The eigenfunctions of $a^\dagger a$, and hence of $\mathcal{H}$, corresponding to the quantum numbers n are denoted as

$$|0\rangle, |1\rangle, \ldots, |n\rangle, \ldots \qquad (B.15)$$

They are defined by the recursion relation

$$|n\rangle = \frac{1}{\sqrt{n}}\, a^\dagger |n-1\rangle \qquad (B.16)$$

We now consider some of the consequences of the solutions (B.13) to (B.16). First, the recursion relation (B.16) explains why $a^\dagger$ is called a creation operator. As seen from eq. (B.16), $a^\dagger$ 'creates' an excitation from state $|n-1\rangle$ to state $|n\rangle$. Using $\langle n|a^\dagger a|n\rangle = n$, and eqs (B.12) and (B.16) one may derive a second recursion relation,

$$|n-1\rangle = \frac{1}{\sqrt{n}} a |n\rangle \tag{B.17}$$

This relation shows that the annihilation operator is appropriately named because it 'annihilates' the excitation created by $a^\dagger$.

Next, inserting eq. (B.16) repetitively yields,

$$|n\rangle = \frac{1}{\sqrt{n!}} (a^\dagger)^n |0\rangle \tag{B.18}$$

so, once an explicit expression for $|0\rangle$ is obtained, all other eigenfunctions can be calculated using eqs (B.4), (B.5), (B.8) and (B.18). Again we refer to textbooks on quantum mechanics like [11], where $|0\rangle$ is found to be given by

$$|0\rangle = \left(\frac{m\omega_0}{\pi\hbar}\right)^{1/4} \exp\left(-\frac{m\omega_0}{2\hbar} x^2\right) \tag{B.19}$$

Furthermore, we observe that the eigenstates $|n\rangle$ are orthonormal, so $\langle n|m\rangle = \delta_{m,n}$. Therefore the matrix elements of the creation and annihilation operators on the basis of these eigenstates can be directly calculated from the recursion relations (B.16) and (B.17). One finds

$$\langle n|a|m\rangle = \sqrt{n+1}\, \delta_{m,n+1} \tag{B.20}$$

$$\langle n|a^\dagger|m\rangle = \sqrt{n}\, \delta_{m,n-1} \tag{B.21}$$

SECOND QUANTIZATION

Finally, the solutions (B.13)–(B.16) allow an interesting way to look at a harmonic oscillator. We see that its energy is quantized and can only be increased or decreased with quanta equal to $\hbar\omega_0$, where ω_0 is its classical oscillation frequency. We may compare this outcome with the description of monochromatic light which is quantized into quanta equal to $\hbar\omega$ where ω is the frequency of the light. In the case of light we are accustomed to interpret these quanta as particles that are called photons. This induces us to interpret the quanta $\hbar\omega_0$ of the harmonic oscillator as particles as well. In this language, a harmonic oscillator in the state $|n\rangle$ is said to contain n quanta of energy $\hbar\omega_0$. Furthermore, $a^\dagger$ is now really a *creation operator* because it creates a quantum with energy $\hbar\omega_0$, while a is now really an *annihilation operator* because it annihilates such a quantum.

Moreover, when the oscillator is excited into state $|n\rangle$, one could say that n quanta *occupy* this oscillator. Hence, in a language where quanta are interpreted as particles, one may interpret the oscillator as the state of these particles where they exist. Thus, the role of particles and states has been reversed. This reversal of the role of particles and eigenstates is often denoted as second quantization. One may wonder whether the quanta $\hbar\omega_0$ are bosons or fermions. For this purpose we observe that many phonons may occupy the same harmonic oscillator. This excludes the quanta $\hbar\omega_0$ to be fermions for which multiple occupation is not allowed because of the Pauli exclusion principle. For a more elaborate discussion of second quantization the reader is referred to the still hard to surpass treatment of photons by Dirac [15].

Appendix C
Perturbation Theory

This appendix is devoted to a derivation of perturbation theory up to second order. The results of time-independent perturbation theory are needed for the development of the **k·p**-approximation in chapter 5. Its derivation is given in two steps. We start by considering the case that the unperturbed states are non-degenerate. This case is treated in most standard textbooks on quantum mechanics and repeated for convenience in section C.1.1. In section C.1.2 we continue with the more complicated situation that the unperturbed states are degenerate. This treatment is needed in the **k·p**-approximation because the basis states for the valence band are degenerate.

Time-dependent perturbation theory is needed up to first order to describe the transition probabilities due to electron–phonon interaction treated in chapter 7 and of direct optical transitions described in chapter 8. To understand indirect optical transitions—also described in chapter 8—we need to extend time-dependent perturbation theory to second order however. Such an extension will be given in sections C.2.1 and C.2.2, though restricted to non-degenerate unperturbed levels.

C.1 Time-independent Perturbation Theory

C.1.1 Non-degenerate Unperturbed States

SERIES EXPANSION

Time-independent perturbation theory is used to solve the eigenvalue problem

$$[\mathcal{H}^{(0)} + \mathcal{H}^{(1)}] |n\rangle = E_n |n\rangle \tag{C.1}$$

where $\mathcal{H}^{(0)}$ is the unperturbed Hamiltonian and $\mathcal{H}^{(1)}$ the perturbation. To determine the eigenvalues E_n and eigenfunctions $|n\rangle$ we develop

$$E_n = E_n^{(0)} + E_n^{(1)} + E_n^{(2)} + \cdots \tag{C.2}$$

$$|n\rangle = |n\rangle^{(0)} + |n\rangle^{(1)} + |n\rangle^{(2)} + \cdots \tag{C.3}$$

PERTURBATION THEORY

and insert these expansions in the eigen value problem (C.1), yielding

$$[\mathcal{H}^{(0)} + \mathcal{H}^{(1)}][|n\rangle^{(0)} + |n\rangle^{(1)} + |n\rangle^{(2)} + \cdots]$$
$$= [E_n^{(0)} + E_n^{(1)} + E_n^{(2)} + \cdots][|n\rangle^{(0)} + |n\rangle^{(1)} + |n\rangle^{(2)} + \cdots]$$

Next we sort the various terms according to their order of magnitude:

$$[\mathcal{H}^{(0)}|n\rangle^{(0)}] + [\mathcal{H}^{(0)}|n\rangle^{(1)} + \mathcal{H}^{(1)}|n\rangle^{(0)}] + [\mathcal{H}^{(0)}|n\rangle^{(2)} + \mathcal{H}^{(1)}|n\rangle^{(1)}] + \cdots$$
$$= [E_n^{(0)}|n\rangle^{(0)}] + [E_n^{(0)}|n\rangle^{(1)} + E_n^{(1)}|n\rangle^{(0)}]$$
$$+ [E_n^{(0)}|n\rangle^{(2)} + E_n^{(1)}|n\rangle^{(1)} + E_n^{(2)}|n\rangle^{(0)}] + \cdots$$

EQUATIONS OF INCREASING ORDER

Subsequently we equalize terms of the same order. Then,

$$\mathcal{H}^{(0)}|n\rangle^{(0)} = E_n^{(0)}|n\rangle^{(0)} \tag{C.4}$$

$$\mathcal{H}^{(0)}|n\rangle^{(1)} + \mathcal{H}^{(1)}|n\rangle^{(0)} = E_n^{(0)}|n\rangle^{(1)} + E_n^{(1)}|n\rangle^{(0)} \tag{C.5}$$

$$\mathcal{H}^{(0)}|n\rangle^{(2)} + \mathcal{H}^{(1)}|n\rangle^{(1)} = E_n^{(0)}|n\rangle^{(2)} + E_n^{(1)}|n\rangle^{(1)} + E_n^{(2)}|n\rangle^{(0)} \tag{C.6}$$

etc.

The zero-order equation (C.4) corresponds to the unperturbed eigenvalue problem. In perturbation theory the solutions of the higher order equations are expressed in the eigenvalues and eigenfunctions of this zero-order equation. Here we assume that such zero order solutions have been found and that the unperturbed eigenvalues are non-degenerate. We denote latter formally by E_n^0 and the unperturbed eigenstates formally by $|n0\rangle$.

To solve the higher order equations we insert the formal zero-order solutions in eqs (C.5) and (C.6). Next, we multiply these equations by $\langle m0|$ on the left-hand side. We obtain

$$\langle m0|\mathcal{H}^{(0)}|n\rangle^{(1)} + \langle m0|\mathcal{H}^{(1)}|n0\rangle = E_n^0\langle m0|n\rangle^{(1)} + E_n^{(1)}\langle m0|n0\rangle$$

$$\langle m0|\mathcal{H}^{(0)}|n\rangle^{(2)} + \langle m0|\mathcal{H}^{(1)}|n\rangle^{(1)} = E_n^0\langle m0|n\rangle^{(2)} + E_n^{(1)}\langle m0|n\rangle^{(1)} + E_n^{(2)}\langle m0|n0\rangle$$

etc.

Inserting the complex conjugate of the zero order equation,

$$\langle m0|\mathcal{H}^{(0)} = \langle m0|E_m^0$$

we find

$$[E_m^0 - E_n^0]\langle m0|n\rangle^{(1)} + \langle m0|\mathcal{H}^{(1)}|n0\rangle = E_n^{(1)}\langle m0|n0\rangle = E_n^{(1)}\delta_{m,n} \tag{C.7}$$

$$[E_m^0 - E_n^0]\langle m0|n\rangle^{(2)} + \langle m0|\mathcal{H}^{(1)}|n\rangle^{(1)} = E_n^{(1)}\langle m0|n\rangle^{(1)} + E_n^{(2)}\langle m0|n0\rangle$$
$$= E_n^{(1)}\langle m0|n\rangle^{(1)} + E_n^{(2)}\delta_{m,n} \tag{C.8}$$

etc.

where we use the orthonormality of the unperturbed eigenstates,

$$\langle m0|n0\rangle = \delta_{m,n}$$

FIRST-ORDER SOLUTIONS

First we solve the first order equation (C.7). For $n = m$ it yields the first-order corrections on the energy levels,

$$E_n^{(1)} = \langle n0|\mathcal{H}^{(1)}|n0\rangle \tag{C.9}$$

For $n \neq m$ we get the first order corrections on the eigenstates,

$$\langle m0|n\rangle^{(1)} = \frac{\langle m0|\mathcal{H}^{(1)}|n0\rangle}{E_n^0 - E_m^0} \tag{C.10}$$

Thus, these first-order corrections are due to mixing between unperturbed states $|m0\rangle$ and $|n0\rangle$. We note that the first-order correction $\langle n0|n\rangle^{(1)}$ is undefined because we excluded $n = m$ in the derivation of eq. (C.10). However, the number of solutions (C.9) and (C.10) is already equal to the number of independent first-order equations (C.7), so we are free to choose the value of this remaining unknown quantity. We choose,

$$\langle n0|n\rangle^{(1)} = 0 \tag{C.11}$$

As a result, up to first order the eigenstates are given by

$$|n0\rangle + |n\rangle^{(1)} = |n0\rangle + \sum_m |m0\rangle\langle m0|n\rangle^{(1)}$$

$$= |n0\rangle + \sum_{m \neq n} |m0\rangle\langle m0|n\rangle^{(1)}$$

$$= |n0\rangle + \sum_{m \neq n} |m0\rangle \frac{\langle m0|\mathcal{H}^{(1)}|n0\rangle}{E_n^0 - E_m^0} \tag{C.12}$$

where we inserted the closure relation,

$$\sum_m |m0\rangle\langle m0| = 1 \tag{C.13}$$

for the unperturbed eigenstates.

SECOND-ORDER SOLUTIONS

Next we calculate the second-order corrections on the energy levels. Using the result (C.9) for the first-order energy shift and inserting $n = m$ in eq. (C.8) we find

$$E_n^{(2)} = \langle n0|\mathcal{H}^{(1)}|n\rangle^{(1)} - \langle n0|\mathcal{H}^{(1)}|n0\rangle\langle n0|n\rangle^{(1)}$$

PERTURBATION THEORY

We now insert the closure relation (C.13) in the first term and subsequently eq. (C.10) for $\langle n0|n\rangle^{(1)}$. Then,

$$\begin{aligned}
E_n^{(2)} &= \sum_m \langle n0|\mathcal{H}^{(1)}|m0\rangle\langle m0|n\rangle^{(1)} - \langle n0|\mathcal{H}^{(1)}|n0\rangle\langle n0|n\rangle^{(1)} \\
&= \sum_{m\neq n} \langle n0|\mathcal{H}^{(1)}|m0\rangle\langle m0|n\rangle^{(1)} \\
&= \sum_{m\neq n} \frac{\langle n0|\mathcal{H}^{(1)}|m0\rangle\langle m0|\mathcal{H}^{(1)}|n0\rangle}{E_n^0 - E_m^0}
\end{aligned} \quad (C.14)$$

Note that our arbitrary choice of $\langle n0|n\rangle^{(1)}$ does not influence the result. Thus, up to second order, the energy is given by

$$\begin{aligned}
E_n &= E_n^0 + E_n^{(1)} + E_n^{(2)} \\
&= E_n^0 + \langle n0|\mathcal{H}^{(1)}|n0\rangle + \sum_{m\neq n} \frac{\langle n0|\mathcal{H}^{(1)}|m0\rangle\langle m0|\mathcal{H}^{(1)}|n0\rangle}{E_n^0 - E_m^0}
\end{aligned} \quad (C.15)$$

C.1.2 Degenerate Unperturbed States

Now we extend time-independent perturbation theory to include degenerate unperturbed energy levels. That this extension is not obvious, is directly seen from eqs (C.10) to (C.12). When two states $|n0\rangle$ and $|m0\rangle$ are degenerate, clearly the first-order correction to the wave function and the second-order correction to the energy diverge.

DEGENERATE MULTIPLETS

To avoid such divergence we need to modify the treatment given in the previous section. In this modification we consider the situation that the unperturbed, zero-order equation yields degenerate energy levels E_n^0. Each of these levels corresponds to a multiplet of degenerate states $|na0\rangle$ where $a = 1, 2, \ldots, g_n$, where the degeneracy g_n may depend on the quantum number n. The perturbation $\mathcal{H}^1$ has two effects. First, its matrix elements between *different* multiplets modify the states $|na0\rangle$ in a similar way as described in the previous section, so they change them by a small amount into $|na\rangle$. Second, it lifts part or all of the degeneracy by completely mixing these modified states $|na\rangle$ *within* each multiplet. We include this latter effect by writing the exact eigenfunctions $|\psi\rangle$ as linear combinations of the functions $|na\rangle$ within a multiplet

$$|\psi\rangle = \sum_{a=1}^{g_n} |na\rangle\langle na|\psi\rangle \quad (C.16)$$

Furthermore we suppose that the coefficients $\langle na|\psi\rangle$ and energy levels E of the full Hamiltonian are solutions of matrix equations

$$\sum_{b=1}^{g_n} H_{ab}^n \langle nb|\psi\rangle = E\langle na|\psi\rangle \quad (C.17)$$

To obtain the matrix elements H_{ab}^n we consider the full Schrödinger equation

$$[\mathcal{H}^{(0)} + \mathcal{H}^{(1)}]|\psi\rangle = E|\psi\rangle$$

Next, we insert eqs (C.16) and (C.17):

$$\sum_{b=1}^{g_n}[\mathcal{H}^{(0)} + \mathcal{H}^{(1)}]|nb\rangle\langle nb|\psi\rangle = \sum_{c=1}^{g_n}|nc\rangle E\langle nc|\psi\rangle$$

$$= \sum_{c=1}^{g_n}|nc\rangle \sum_{b=1}^{g_n} H_{cb}^n \langle nb|\psi\rangle$$

$$= \sum_{b=1}^{g_n}\left[\sum_{c=1}^{g_n}|nc\rangle H_{cb}^n\right]\langle nb|\psi\rangle$$

and we find the matrix elements H_{ab}^n by requiring each term on the left-hand side to be equal to each term on the right-hand side,

$$[\mathcal{H}^{(0)} + \mathcal{H}^{(1)}]|nb\rangle = \sum_{c=1}^{g_n}|nc\rangle H_{cb}^n \qquad (C.18)$$

and multiplying on the left by $\langle na|$, so

$$\langle na|[\mathcal{H}^{(0)} + \mathcal{H}^{(1)}]|nb\rangle = \sum_{c=1}^{g_n}\langle na|nc\rangle H_{cb}^n = \sum_{c=1}^{g_n}\delta_{a,c}H_{cb}^n = H_{ab}^n \qquad (C.19)$$

SERIES EXPANSION

Now we are ready to continue in the same way as in non-degenerate perturbation theory. We develop

$$H_{ab}^n = (H_{ab}^n)^{(0)} + (H_{ab}^n)^{(1)} + (H_{ab}^n)^{(2)} + \cdots$$

$$|na\rangle = |na\rangle^{(0)} + |na\rangle^{(1)} + |na\rangle^{(2)} + \cdots \qquad (C.20)$$

and insert these expansions in eq. (C.18) to obtain

$$[\mathcal{H}^{(0)} + \mathcal{H}^{(1)}][|nb\rangle^{(0)} + |nb\rangle^{(1)} + |nb\rangle^{(2)} + \cdots]$$

$$= \sum_{a=1}^{g_n}[|na\rangle^{(0)} + |na\rangle^{(1)} + |na\rangle^{(2)} + \cdots][(H_{ab}^n)^{(0)} + (H_{ab}^n)^{(1)} + (H_{ab}^n)^{(2)} + \cdots]$$

Next we sort the various terms according to order of magnitude:

$$[\mathcal{H}^{(0)}|nb\rangle^{(0)}] + [\mathcal{H}^{(0)}|nb\rangle^{(1)} + \mathcal{H}^{(1)}|nb\rangle^{(0)}] + [\mathcal{H}^{(0)}|nb\rangle^{(2)} + \mathcal{H}^{(1)}|nb\rangle^{(1)}] + \cdots$$

$$= \sum_{a=1}^{g_n}[|na\rangle^{(0)}(H_{ab}^n)^{(0)}] + \sum_{a=1}^{g_n}[|na\rangle^{(1)}(H_{ab}^n)^{(0)} + |na\rangle^{(0)}(H_{ab}^n)^{(1)}]$$

$$+ \sum_{a=1}^{g_n}[|na\rangle^{(2)}(H_{ab}^n)^{(0)} + |na\rangle^{(1)}(H_{ab}^n)^{(1)} + |na\rangle^{(0)}(H_{ab}^n)^{(2)}] + \cdots$$

EQUATIONS OF INCREASING ORDER

Subsequently we equalize terms of the same order. Then

$$\mathcal{H}^{(0)}|nb\rangle^{(0)} = \sum_{a=1}^{g_n} |na\rangle^{(0)} (H^n_{ab})^{(0)} \tag{C.21}$$

$$\mathcal{H}^{(0)}|nb\rangle^{(1)} + \mathcal{H}^{(1)}|nb,\rangle^{(0)} = \sum_{a=1}^{g_n} |na\rangle^{(1)} (H^n_{ab})^{(0)} + \sum_{a=1}^{g_n} |na\rangle^{(0)} (H^n_{ab})^{(1)} \tag{C.22}$$

$$\mathcal{H}^{(0)}|nb\rangle^{(2)} + \mathcal{H}^{(1)}|nb\rangle^{(1)} = \sum_{a=1}^{g_n} |na\rangle^{(2)} (H^n_{ab})^{(0)} + \sum_{a=1}^{g_n} |na\rangle^{(1)} (H^n_{ab})^{(1)}$$

$$+ \sum_{a=1}^{g_n} |na\rangle^{(0)} (H^n_{ab})^{(2)} \tag{C.23}$$

etc.

As in the previous section we consider the solutions of the zero-order equation (C.21) to be known. We note that the multiplet with basis states $|na\rangle^{(0)}$, where $a = 1, 2, \ldots, g_n$ is degenerate in zero order. This allows us to write these solutions as

$$(H^n_{ab})^{(0)} = E^n_0 \delta_{ab} \tag{C.24}$$

$$|na\rangle^{(0)} = |na0\rangle \tag{C.25}$$

To solve the higher order equations we multiply them on the left hand side by $\langle mc0|$ and insert the zero-order results (C.24) and (C.25). Then

$$\langle mc0|\mathcal{H}^{(0)}|nb\rangle^{(1)} + \langle mc0|\mathcal{H}^{(1)}|nb0\rangle = E^n_0 \langle mc0|nb\rangle^{(1)} + \sum_{a=1}^{g_n} \langle mc0|na0\rangle (H^n_{ab})^{(1)}$$

$$\langle mc0|\mathcal{H}^{(0)}|nb\rangle^{(2)} + \langle mc0|\mathcal{H}^{(1)}|nb\rangle^{(1)} = E^n_0 \langle mc0|nb\rangle^{(2)} + \sum_{a=1}^{g_n} \langle mc0|na\rangle^{(1)} (H^n_{ab})^{(1)}$$

$$+ \sum_{a=1}^{g_n} \langle mc0|na0\rangle (H^n_{ab})^{(2)}$$

etc.

Inserting the complex conjugate of the zero-order equation,

$$\langle mc0|\mathcal{H}^{(0)} = \langle mc0|E^m_0$$

we find

$$[E^m_0 - E^n_0]\langle mc0|nb\rangle^{(1)} + \langle mc0|\mathcal{H}^{(1)}|nb0\rangle$$

$$= \sum_{a=1}^{g_n} \langle mc0|na0\rangle (H^n_{ab})^{(1)} = (H^n_{cb})^{(1)} \delta_{m,n} \tag{C.26}$$

$$[E_0^m - E_0^n]\langle mc0|nb\rangle^{(2)} + \langle mc0|\mathcal{H}^{(1)}|nb\rangle^{(1)}$$

$$= \sum_{a=1}^{g_n} \langle mc0|na\rangle^{(1)} (H_{ab}^n)^{(1)} + \sum_{a=1}^{g_n} \langle mc0|na0\rangle (H_{ab}^n)^{(2)}$$

$$= \sum_{a=1}^{g_n} \langle mc0|na\rangle^{(1)} (H_{ab}^n)^{(1)} + (H_{cb}^n)^{(2)} \delta_{m,n} \tag{C.27}$$

etc.
where we use the orthonormality relation

$$\langle mc0|na0\rangle = \delta_{m,n}\delta_{c,a}$$

FIRST-ORDER SOLUTIONS

We first solve the first-order equation (C.26). For $n = m$ we get the first order corrections on the energy levels,

$$(H_{cb}^n)^{(1)} = \langle nc0|\mathcal{H}^{(1)}|nb0\rangle \tag{C.28}$$

For $n \neq m$ we obtain the first order corrections on the eigenstates,

$$\langle mc0|nb\rangle^{(1)} = \frac{\langle mc0|\mathcal{H}^{(1)}|nb0\rangle}{E_0^n - E_0^m} \tag{C.29}$$

Thus, these first-order corrections are due to mixing between states $|mc0\rangle$ and $|nb0\rangle$ by $\mathcal{H}^{(1)}$. It is important to note that the first-order corrections $\langle nc0|nb\rangle^{(1)}$ are undefined because eq. (C.29) is obtained by solving eq. (C.26) for $n \neq m$. Moreover, for $n = m$, eq. (C.29) would diverge. However, the case $n = m$ is taken care of exactly by solving the matrix equation (C.17). So we need not worry about the fact eq. (C.29) cannot be used if $n = m$ and we may choose

$$\langle nc0|nb\rangle^{(1)} = 0 \tag{C.30}$$

for any b and c within the nth multiplet. As a result, up to first order the basis states are given by

$$|nb0\rangle + |nb\rangle^{(1)} = |nb0\rangle + \sum_m \sum_{c=1}^{g_m} |mc0\rangle\langle mc0|nb\rangle^{(1)}$$

$$= |nb0\rangle + \sum_{m \neq n} \sum_{c=1}^{g_m} |mc0\rangle\langle mc0|nb\rangle^{(1)}$$

$$= |nb0\rangle + \sum_{m \neq n} \sum_{c=1}^{g_m} |mc0\rangle \frac{\langle mc0|\mathcal{H}^{(1)}|nb0\rangle}{E_n^0 - E_m^0} \tag{C.31}$$

where we inserted the closure relation

$$\sum_m \sum_{c=1}^{g_m} |mc0\rangle\langle mc0| = 1 \tag{C.32}$$

for the unperturbed eigenstates.

SECOND-ORDER SOLUTIONS

Next we calculate the second-order corrections on the energy levels. We insert $n = m$ in eq. (C.27) and use eq. (C.30) to find

$$(H_{cb}^n)^{(2)} = \langle nc0 | \mathcal{H}^{(1)} | nb \rangle^{(1)}$$

In this expression we insert the closure relation (C.32) for the unperturbed eigenstates. Then inserting eqs (C.29) and (C.30), we find

$$(H_{cb}^n)^{(2)} = \sum_m \sum_{a=1}^{g_m} \langle nc0 | \mathcal{H}^{(1)} | ma0 \rangle \langle ma0 | nb \rangle^{(1)}$$

$$= \sum_{m \neq n} \sum_{a=1}^{g_m} \frac{\langle nc0 | \mathcal{H}^{(1)} | ma0 \rangle \langle ma0 | \mathcal{H}^{(1)} | nb0 \rangle}{E_0^n - E_0^m} \quad \text{(C.34)}$$

Thus, up to second order the matrix elements are given by

$$H_{cb}^n = (H_{cb}^n)^{(0)} + (H_{cb}^n)^{(1)} + (H_{cb}^n)^{(2)}$$

$$= E_0^n \delta_{c,b} + \langle nc0 | \mathcal{H}^{(1)} | nb0 \rangle + \sum_{m \neq n} \sum_{a=1}^{g_m} \frac{\langle nc0 | \mathcal{H}^{(1)} | ma0 \rangle \langle ma0 | \mathcal{H}^{(1)} | nb0 \rangle}{E_0^n - E_0^m} \quad \text{(C.35)}$$

Now the matrix elements H_{cb}^n are known up to second order, the energy levels E are obtained up to the same order by solving eq. (C.17). Thus the energy levels E are the eigenvalues of the $g_n \times g_n$ matrix $\mathbf{H}^n$ with elements H_{cb}^n and are obtained by solving

$$\text{Det} | \mathbf{H}^n - E\mathbf{1} | = 0 \quad \text{(C.35)}$$

where $\mathbf{1}$ is the $g_n \times g_n$ unit matrix with elements $\delta_{c,b}$.

C.2 Time-dependent Perturbation Theory

C.2.1 Time-independent Perturbations

TRANSITIONS BETWEEN UNPERTURBED EIGENSTATES

Time-dependent perturbation theory is used to solve the time-dependent Schrödinger equation

$$i\hbar \frac{\partial}{\partial t} | \psi(t) \rangle = [\mathcal{H}^{(0)} + \mathcal{H}^{(1)}] | \psi(t) \rangle \quad \text{(C.36)}$$

where $\mathcal{H}^{(0)}$ is the unperturbed Hamiltonian and $\mathcal{H}^{(1)}$ the perturbation. In particular we are interested to know the probability that the perturbation $\mathcal{H}^{(1)}$ induces a transition from eigenstate $|n\rangle$ to eigenstate $|m\rangle$ of the unperturbed Hamiltonian $\mathcal{H}^{(0)}$. In the present section we will assume the perturbation to be time independent. The extension to time-dependent perturbations will be given in the following section.

To formulate the problem more precisely, we start by considering that the latter eigenstates are solutions of the zero-order time-independent Schrödinger equation (C.4),

$$\mathcal{H}^{(0)} | p \rangle = E_p | p \rangle \quad \text{(C.37)}$$

Note that we have omitted indices 0 and (0) in the eigenstates and eigenvalues, as they are redundant in the present derivation. Next, we expand the solution $|\psi(t)\rangle$ of eq. (C.36) in eigenstates $|p\rangle$,

$$|\psi(t)\rangle = \sum_p |p\rangle\langle p|\psi(t)\rangle \tag{C.38}$$

Finally, we demand the system to be in state $|m\rangle$ at $t = 0$, so we put

$$\langle p|\psi(0)\rangle = \delta_{m,p} \tag{C.39}$$

The probability to find the system in the eigenstate $|n\rangle$ at a later time t is given by $|\langle n|\psi(t)\rangle|^2$. As a result, the probability per unit time that the perturbation induces a transition from eigenstate $|m\rangle$ to eigenstate $|n\rangle$ is equal to $(\partial/\partial t)\,|\langle n|\psi(t)\rangle|^2$. Hence, the problem basically consists of obtaining and solving a differential equation for $\langle n|\psi(t)\rangle$.

THE BASIC DIFFERENTIAL EQUATION

The required equation is found by inserting the expansion (C.38) in the time-dependent Schrödinger equation (C.36),

$$i\hbar \sum_p \frac{\partial}{\partial t} |p\rangle\langle p|\psi(t)\rangle = \sum_p [\mathcal{H}^{(0)} + \mathcal{H}^{(1)}]|p\rangle\langle p|\psi(t)\rangle \tag{C.40}$$

and by multiplying on the left-hand side by $\langle n|$. Then, we find

$$i\hbar \sum_p \frac{\partial}{\partial t} \langle n|p\rangle\langle p|\psi(t)\rangle = \sum_p \langle n|[\mathcal{H}^{(0)} + \mathcal{H}^{(1)}]|p\rangle\langle p|\psi(t)\rangle$$

$$= \sum_p [\langle n|\mathcal{H}^{(0)}|p\rangle + \langle n|\mathcal{H}^{(1)}|p\rangle]\langle p|\psi(t)\rangle$$

$$= \sum_p [E_p\langle n|p\rangle + \langle n|\mathcal{H}^{(1)}|p\rangle]\langle p|\psi(t)\rangle \tag{C.41}$$

where we insert the time-independent Schrödinger equation (C.37) in the last step. Now, we take into account that the eigenstates of the unperturbed Hamiltonian $\mathcal{H}^{(0)}$ are orthonormal and obey

$$\langle n|p\rangle = \delta_{n,p} \tag{C.42}$$

As a result, eq. (C.41) reduces to the required differential equation, which reads as,

$$i\hbar \frac{\partial}{\partial t} \langle n|\psi(t)\rangle = \sum_p [E_p \delta_{n,p} + \langle n|\mathcal{H}^{(1)}|p\rangle]\langle p|\psi(t)\rangle$$

$$= E_n \langle n|\psi(t)\rangle + \sum_p \langle n|\mathcal{H}^{(1)}|p\rangle\langle p|\psi(t)\rangle \tag{C.43}$$

As a first step to solve eq. (C.43) we write

$$\langle n|\psi(t)\rangle = a_n(t)\, e^{-(i/\hbar)E_n t} \tag{C.44}$$

PERTURBATION THEORY

Then, differentiating both sides of this equation with respect to time, we may write,

$$i\hbar \frac{\partial}{\partial t} \langle n|\psi(t)\rangle = i\hbar \frac{\partial}{\partial t} [a_n(t)\, e^{-(i/\hbar)E_n t}]$$

$$= e^{-(i/\hbar)E_n t} \left[i\hbar \frac{\partial}{\partial t} a_n(t) - i\hbar a_n(t)(i/\hbar)E_n \right]$$

$$= e^{-(i/\hbar)E_n t} i\hbar \frac{\partial}{\partial t} a_n(t) + E_n \langle n|\psi(t)\rangle \tag{C.45}$$

Inserting this expression in the left-hand side of eq. (C.43) and dividing both sides by $\exp(-i/\hbar E_n t)$, we obtain a differential equation for $a_n(t)$,

$$i\hbar \frac{\partial}{\partial t} a_n(t) = \sum_p a_p(t) \langle n|\mathcal{H}^{(1)}|p\rangle e^{(i/\hbar)(E_n - E_p)t} \tag{C.46}$$

In fact, solving the latter equation is just as useful as solving eq. (C.43) itself. The reason is that phase factors $\exp[-(i/\hbar)E_n t]$ are not relevant for the calculation of the probability

$$|\langle n|\psi(t)\rangle|^2 = |a_n(t)|^2 \tag{C.47}$$

to find the system in the eigenstate $|n\rangle$ at a time t.

SERIES EXPANSION

Eq. (C.46) is solved by developing

$$a_n(t) = a_n^{(0)}(t) + a_n^{(1)}(t) + a_n^{(2)}(t) + \cdots \tag{C.48}$$

so it is written as

$$i\hbar \frac{\partial}{\partial t} [a_n^{(0)}(t) + a_n^{(1)}(t) + a_n^{(2)}(t) + \cdots]$$

$$= \sum_p [a_p^{(0)}(t) + a_p^{(1)}(t) + a_p^{(2)}(t) + \cdots] \langle n|\mathcal{H}^{(1)}|p\rangle\, e^{(i/\hbar)(E_n - E_p)t} \tag{C.49}$$

and subsequently equalizing terms of the same order. Then,

$$i\hbar \frac{\partial}{\partial t} a_n^{(0)}(t) = 0 \tag{C.50}$$

$$i\hbar \frac{\partial}{\partial t} a_n^{(1)}(t) = \sum_p a_p^{(0)}(t)\langle n|\mathcal{H}^{(1)}|p\rangle e^{(i/\hbar)(E_n - E_p)t} \tag{C.51}$$

$$i\hbar \frac{\partial}{\partial t} a_n^{(2)}(t) = \sum_p a_p^{(1)}(t)\langle n|\mathcal{H}^{(1)}|p\rangle e^{(i/\hbar)(E_n - E_p)t} \tag{C.52}$$

To establish the initial conditions for these equations, we take into account that we are calculating the probability to find the system in the eigenstate $|n\rangle$ at a time t, while it was in the state $|m\rangle$ at a time $t = 0$, while $n \neq m$. For this purpose it is sufficient to consider the case that the perturbation is switched on at exactly $t = 0$. Then, at $t = 0$, the perturbation has not yet created any higher order terms and all terms in the expansion of $a_n(0)$ are equal to zero, except for the zero-order term. So,

$$a_n^{(0)}(0) = \delta_{m,p} \tag{C.53}$$

$$a_n^{(1)}(0) = 0 \tag{C.54}$$

$$a_n^{(2)}(0) = 0 \tag{C.55}$$

etc.

The zero-order equation (C.50) simply yields

$$a_n^{(0)}(t) = a_n^{(0)}(0) = \delta_{m,p} = \text{constant}. \tag{C.56}$$

So, in zero order the system remains in the state $|m\rangle$, i.e. no transitions are induced by the perturbation.

In first-order transitions to the state $|n\rangle$ may occur. This is seen by inserting the zero order result in the first-order equation (C.51). We find,

$$i\hbar \frac{\partial}{\partial t} a_n^{(1)}(t) = \sum_p \delta_{m,p} \langle n|\mathcal{H}^{(1)}|p\rangle e^{(i/\hbar)(E_n-E_p)t}$$

$$= \langle n|\mathcal{H}^{(1)}|m\rangle e^{(i/\hbar)(E_n-E_m)t} \tag{C.57}$$

which can be integrated in a straightforward manner. Taking the initial condition (C.54) into account, we find

$$a_n^{(1)}(t) = a_n^{(1)}(0) + \langle n|\mathcal{H}^{(1)}|m\rangle \frac{1}{i\hbar} \int_0^t dt' e^{(i/\hbar)(E_n-E_m)t'}$$

$$= -\langle n|\mathcal{H}^{(1)}|m\rangle \frac{e^{(i/\hbar)(E_n-E_m)t} - 1}{(E_n - E_m)} \tag{C.58}$$

Thus, $a_n^{(1)}(t)$ and hence the probability (C.47) to find the system in the state $|n\rangle$ may become non-zero, provided the matrix element $\langle n|\mathcal{H}^{(1)}|m\rangle$ is non-zero. In other words, transitions from the state $|m\rangle$ to the state $|n\rangle$ are allowed in first order, provided this matrix element is non-zero.

In the case the matrix element $\langle n|\mathcal{H}^{(1)}|m\rangle$ is zero, we have to resort to second order. Then, we insert the first order result in the second order equation and we find,

$$i\hbar \frac{\partial}{\partial t} a_n^{(2)}(t) = -\sum_{\substack{p\neq m \\ p\neq n}} \left[\sum_q \delta_{m,q} \langle p|\mathcal{H}^{(1)}|q\rangle \frac{e^{(i/\hbar)(E_p-E_q)t} - 1}{(E_p - E_q)} \right] \times \langle n|\mathcal{H}^{(1)}|p\rangle e^{(i/\hbar)(E_n-E_p)t}$$

$$= -\sum_{\substack{p\neq m \\ p\neq n}} \frac{\langle n|\mathcal{H}^{(1)}|p\rangle \langle p|\mathcal{H}^{(1)}|m\rangle}{(E_p - E_m)} \times [e^{(i/\hbar)(E_n-E_m)t} - e^{(i/\hbar)(E_n-E_p)t}] \tag{C.59}$$

Here we explicitly exclude the terms $p = m$ and $p = n$ from the sum over p. This is possible, because we treat the case that $\langle n|\mathcal{H}^{(1)}|m\rangle = 0$, so such terms would be equal to zero anyhow. The exclusion of the terms $p = m$ and $p = n$ has the pleasant consequence that it kills any divergencies due to vanishing denominators in the above expression. Also eq. (C.59) is integrated in a straightforward manner. Inserting the initial condition (C.55), we find

$$a_n^{(2)}(t) = a_n^{(2)}(0) - \sum_{\substack{p \neq m \\ p \neq n}} \frac{\langle n | \mathcal{H}^{(1)} | p \rangle \langle p | \mathcal{H}^{(1)} | m \rangle}{(E_p - E_m)} \times \frac{1}{i\hbar} \int_0^t dt' [e^{(i/\hbar)(E_n - E_m)t'} - e^{(i/\hbar)(E_n - E_p)t'}]$$

$$= + \sum_{\substack{p \neq m \\ p \neq n}} \frac{\langle n | \mathcal{H}^{(1)} | p \rangle \langle p | \mathcal{H}^{(1)} | m \rangle}{(E_p - E_m)} \times \left[\frac{e^{(i/\hbar)(E_n - E_m)t} - 1}{(E_n - E_m)} - \frac{e^{(i/\hbar)(E_n - E_p)t} - 1}{(E_n - E_p)} \right] \quad (C.60)$$

FERMI'S GOLDEN RULE FOR FIRST-ORDER TRANSITIONS

We now continue calculating the transition probabilities. When

$$\langle n | \mathcal{H}^{(1)} | m \rangle \neq 0 \quad (C.61)$$

the first-order solution (C.58) suffices to calculate the probability to find the system in the eigenstate $|n\rangle$ at a time $t > 0$. We find

$$|a_n^{(1)}(t)|^2 = |\langle n | \mathcal{H}^{(1)} | m \rangle|^2 \left| \frac{e^{(i/\hbar)(E_n - E_m)t} - 1}{(E_n - E_m)} \right|^2$$

$$= |\langle n | \mathcal{H}^{(1)} | m \rangle|^2 \left| \frac{\cos(1/\hbar)(E_n - E_m)t - 1}{(E_n - E_m)} + i \frac{\sin(1/\hbar)(E_n - E_m)t}{(E_n - E_m)} \right|^2 \quad (C.62)$$

Noting that,

$$\frac{\partial}{\partial t}|a|^2 = \frac{\partial}{\partial t}(Re\{a\}^2 + Im\{a\}^2)$$

$$= 2Re\{a\} \frac{\partial}{\partial t} Re\{a\} + 2Im\{a\} \frac{\partial}{\partial t} Im\{a\} \quad (C.63)$$

the transition probability per unit of time is given by

$$W_{m,n} = \frac{\partial}{\partial t} |a_n^{(1)}(t)|^2 = |\langle n | \mathcal{H}^{(1)} | m \rangle|^2$$

$$\times \left[-2 \frac{\cos(1/\hbar)(E_n - E_m)t - 1}{(E_n - E_m)} \frac{\sin(1/\hbar)(E_n - E_m)t}{\hbar} \right.$$

$$\left. + 2 \frac{\sin(1/\hbar)(E_n - E_m)t}{(E_n - E_m)} \frac{\cos(1/\hbar)(E_n - E_m)t}{\hbar} \right]$$

$$= \frac{2\pi}{\hbar} |\langle n | \mathcal{H}^{(1)} | m \rangle|^2 \frac{\sin(1/\hbar)(E_n - E_m)t}{\pi(E_n - E_m)} \quad (C.64)$$

According to Heisenberg's uncertainty relation, a transition between two states $|m\rangle$ and $|n\rangle$ makes sense only on a timescale which is sufficiently long in order to obey

$$|E_n - E_m|t \gg \hbar \quad (C.65)$$

Under those circumstances we may approximate

$$\frac{\sin[(1/\hbar)(E_n - E_m)t]}{\pi(E_n - E_m)} \approx \delta(E_n - E_m) \tag{C.66}$$

and we get Fermi's golden rule for the first-order transition probability.

$$W_{m,n} = \frac{2\pi}{\hbar} |\langle n|\mathcal{H}^{(1)}|m\rangle|^2 \, \delta(E_n - E_m) \tag{C.67}$$

The latter Dirac δ-function can be interpreted as requiring conservation of energy. Only those transitions are allowed where the initial and final states of the system have the same energy.

SECOND-ORDER TRANSITION PROBABILITIES

A more complicated case arises when

$$\langle n|\mathcal{H}^{(1)}|m\rangle = 0 \tag{C.68}$$

so the second-order solution (C.60) is needed to calculate a transition probability between the states $|m\rangle$ and $|n\rangle$. As above we calculate

$$|a_n^{(2)}(t)|^2 = \left| \sum_{\substack{p \neq m \\ p \neq n}} \frac{\langle n|\mathcal{H}^{(1)}|p\rangle \langle p|\mathcal{H}^{(1)}|m\rangle}{(E_p - E_m)} \times \left[\frac{e^{(i/\hbar)(E_n - E_m)t} - 1}{(E_n - E_m)} - \frac{e^{(i/\hbar)(E_n - E_p)t} - 1}{(E_n - E_p)} \right] \right|^2$$

$$= \left| \sum_{\substack{p \neq m \\ p \neq n}} \frac{\langle n|\mathcal{H}^{(1)}|p\rangle \langle p|\mathcal{H}^{(1)}|m\rangle}{(E_p - E_m)} \right.$$

$$\left. \times \left\{ \left[\frac{\cos(1/\hbar)(E_n - E_m)t - 1}{(E_n - E_m)} - \frac{\cos(1/\hbar)(E_n - E_p)t - 1}{(E_n - E_p)} \right] \right. \right.$$

$$\left. \left. + i \left[\frac{\sin(1/\hbar)(E_n - E_m)t}{(E_n - E_m)} - \frac{\sin(1/\hbar)(E_n - E_p)t}{(E_n - E_p)} \right] \right\} \right|^2 \tag{C.69}$$

We note that the terms with $p = n$ and $p = m$ are excluded in the sum over p, because we assume eq. (C.68) to hold. Eq. (C.69) contains oscillating functions with frequencies

$$\omega_{nm} = (E_n - E_m)/\hbar \quad \text{and} \quad \omega_{np} = (E_n - E_p)/\hbar \tag{C.70}$$

Now, excluding degeneracy as well as the term $p = n$ ensures that the latter frequency is always non-zero. However, according to Heisenberg's uncertainty relation, transitions make sense only on a timescale which is sufficiently long in order to obey eq. (C.65). Clearly, the terms containing ω_{np} average out on this long timescale and

PERTURBATION THEORY

only the terms containing ω_{nm} survive. Hence,

$$|a_n^{(2)}(t)|^2 = \left| \sum_{\substack{p \neq m \\ p \neq n}} \frac{\langle n|\mathcal{H}^{(1)}|p\rangle \langle p|\mathcal{H}^{(1)}|m\rangle}{(E_p - E_m)} \right|^2$$

$$\times \left| \frac{\cos(1/\hbar)(E_n - E_m)t - 1}{(E_n - E_m)} + i \frac{\sin(1/\hbar)(E_n - E_m)t}{(E_n - E_m)} \right|^2. \quad \text{(C.71)}$$

As above the transition probability per unit time is obtained by differentiating this expression with respect to time. Remembering eqs (C.64) and (C.66), we find,

$$W_{m,n} = \left| \sum_{\substack{p \neq m \\ p \neq n}} \frac{\langle n|\mathcal{H}^{(1)}|p\rangle \langle p|\mathcal{H}^{(1)}|m\rangle}{(E_p - E_m)} \right|^2 \frac{2\sin(1/\hbar)(E_n - E_m)t}{\hbar(E_n - E_m)}$$

$$= \frac{2\pi}{\hbar} \left| \sum_{\substack{p \neq m \\ p \neq n}} \frac{\langle n|\mathcal{H}^{(1)}|p\rangle \langle p|\mathcal{H}^{(1)}|m\rangle}{(E_p - E_m)} \right|^2 \delta(E_n - E_m) \quad \text{(C.72)}$$

which is the second order equivalent of Fermi's golden rule. Note that the Dirac δ-function requires conservation of energy, just as in first order.

C.2.2 Time-dependent Perturbations
DIFFERENTIAL EQUATIONS OF INCREASING ORDER

In many relevant cases, e.g. when optical fields are concerned, the perturbation $\mathcal{H}^{(1)}$ may contain terms that oscillate in time. Therefore, we finish this appendix by extending the previous treatment to the most simple of these cases, where

$$\mathcal{H}^{(1)} = \mathcal{H}_0^{(1)} + \mathcal{H}_1^{(1)} e^{-i\omega t} \quad \text{(C.73)}$$

Here $\mathcal{H}_0^{(1)}$ represents the constant terms in the perturbation and $\mathcal{H}_1^{(1)}$ is the amplitude of the additional oscillating term. In a practical situation the former may be due to electron–phonon coupling and the latter to the interaction with an optical field. Note however, that, even for simple planar optical waves, the actual perturbation is more complicated as it then also contains a term proportional to $\exp(+i\omega t)$.

Now, nowhere in the derivation leading to the differential equations for $a_n^{(0)}$, $a_n^{(1)}$, $a_p^{(1)}$, etc., do we ever differentiate this perturbation with respect to time. Therefore, eqs (C.50)–(C.52) remain valid and, inserting eq. (C.73), they now read as

$$i\hbar \frac{\partial}{\partial t} a_n^{(0)}(t) = 0 \quad \text{(C.74)}$$

$$i\hbar \frac{\partial}{\partial t} a_n^{(1)}(t) = \sum_p a_p^{(0)}(t) \langle n| \mathcal{H}_0^{(1)} |p\rangle e^{(i/\hbar)(E_n-E_p)t}$$

$$+ \sum_p a_p^{(0)}(t) \langle n| \mathcal{H}_1^{(1)} |p\rangle e^{i((E_n-E_p)/\hbar-\omega)t} \qquad (C.75)$$

$$i\hbar \frac{\partial}{\partial t} a_n^{(2)}(t) = \sum_p a_p^{(1)}(t) \langle n| \mathcal{H}_0^{(1)} |p\rangle e^{(i/\hbar)(E_n-E_p)t}$$

$$+ \sum_p a_p^{(1)}(t) \langle n| \mathcal{H}_1^{(1)} |p\rangle e^{i((E_n-E_p)/\hbar-\omega)t} \qquad (C.76)$$

FIRST-ORDER SOLUTIONS

Subsequently, the solution of the first-order equation (C.75) is obtained by following the derivation from eq. (C.50) to eq. (C.58), where we simply replace

$$(E_n - E_m) \quad \text{by} \quad (E_n - E_m - \hbar\omega) \qquad (C.77)$$

wherever appropriate. Thus, eq. (C.58) changes to,

$$a_n^{(1)}(t) = -\langle n| \mathcal{H}_0^{(1)} |m\rangle \frac{e^{(i/\hbar)(E_n-E_m)t} - 1}{(E_n - E_m)} - \langle n| \mathcal{H}_1^{(1)} |m\rangle \frac{e^{i((E_n-E_m)/\hbar-\omega)t} - 1}{(E_n - E_m - \hbar\omega)} \qquad (C.78)$$

As long as we consider first-order transitions only, we confine ourselves to the separate effect of each individual term. For example, if the first term is due to electron–phonon coupling and the second term to coupling with the optical field, we consider the transition rates due to each of these couplings separately. Rates due to the time-independent perturbation $\mathcal{H}_0^{(1)}$ were treated in the previous section. So, here we only need to calculate the transition probability due to the oscillating perturbation with amplitude $\mathcal{H}_1^{(1)}$. We follow the derivation from eq. (C.62) to (C.67) and again we make the replacement (C.77) wherever appropriate. As a result, we find Fermi's golden rule for time-dependent perturbations,

$$W_{m,n} = \frac{2\pi}{\hbar} |\langle n| \mathcal{H}^{(1)} |m\rangle|^2 \, \delta(E_n - E_m - \hbar\omega) \qquad (C.79)$$

As in eq. (C.67) the Dirac δ-function can be interpreted as requiring conservation of energy. Now we need to take into account that the oscillating perturbation represents a field which is quantized in units $\hbar\omega$. For example, in the case of an optical field, these quanta are photons with an energy $\hbar\omega$. The δ-function allows only those transitions where the initial and final states of the system have an energy difference equal to the energy $\hbar\omega$ of these quanta, i.e. where the system absorbs such a quantum.

SECOND-ORDER SOLUTIONS

To obtain the equivalent of the second-order solution (C.60), we rewrite eq. (C.59) by inserting the time-dependent perturbation (C.73) and the first order solution (C.78).

PERTURBATION THEORY

We find a lengthy expression,

$$i\hbar \frac{\partial}{\partial t} a_n^{(2)}(t) = -\sum_{\substack{p \neq m \\ p \neq n}} \left[\sum_q \delta_{m,q} \langle p | \mathcal{H}_0^{(1)} | q \rangle \frac{e^{(i/\hbar)(E_p - E_q)t} - 1}{(E_p - E_q)} \right.$$

$$\left. + \sum_q \delta_{m,q} \langle p | \mathcal{H}_1^{(1)} | q \rangle \frac{e^{i((E_p - E_q)/\hbar - \omega)t} - 1}{(E_p - E_q - \hbar\omega)} \right]$$

$$\times \left[\langle n | \mathcal{H}_0^{(1)} | p \rangle e^{(i/\hbar)(E_n - E_p)t} + \langle n | \mathcal{H}_1^{(1)} | p \rangle e^{i((E_n - E_p)/\hbar - \omega)t} \right]$$

$$= -\sum_{\substack{p \neq m \\ p \neq n}} \frac{\langle n | \mathcal{H}_0^{(1)} | p \rangle \langle p | \mathcal{H}_0^{(1)} | m \rangle}{(E_p - E_m)} \times \left[e^{(i/\hbar)(E_n - E_m)t} - e^{(i/\hbar)(E_n - E_p)t} \right]$$

$$- \sum_{\substack{p \neq m \\ p \neq n}} \frac{\langle n | \mathcal{H}_1^{(1)} | p \rangle \langle p | \mathcal{H}_0^{(1)} | m \rangle}{(E_p - E_m)} \times \left[e^{i((E_n - E_m)/\hbar - \omega)t} - e^{(i/\hbar)(E_n - E_p)t} \right]$$

$$- \sum_{\substack{p \neq m \\ p \neq n}} \frac{\langle n | \mathcal{H}_0^{(1)} | p \rangle \langle p | \mathcal{H}_1^{(1)} | m \rangle}{(E_p - E_m - \hbar\omega)} \times \left[e^{i((E_n - E_m)/\hbar - \omega)t} - e^{(i/\hbar)(E_n - E_p)t} \right]$$

$$- \sum_{\substack{p \neq m \\ p \neq n}} \frac{\langle n | \mathcal{H}_1^{(1)} | p \rangle \langle p | \mathcal{H}_1^{(1)} | m \rangle}{(E_p - E_m - \hbar\omega)} \times \left[e^{i((E_n - E_m)/\hbar - 2\omega)t} - e^{i((E_n - E_p)/\hbar - \omega)t} \right]$$

(C.80)

Here the first term represents second-order transitions due to the time-independent part $\mathcal{H}_0^{(1)}$ of the perturbation. As this case was already treated in the previous section, we need not consider this term again. The second and the third term represent transitions where the constant and the oscillating terms in the perturbation work together to cause transitions, e.g. electron–phonon coupling and the optical field. These terms are important for indirect optical transitions as treated in chapter 8. The fourth and last term represent second order transitions caused by the oscillating part of the perturbation alone. It is one of the many terms that need to be considered when strong optical fields are applied, so non-linear optical effects take place. Here, we will not account for such non-linear optical effects and hence we will not consider this term any further. Finally we recall the discussion leading from eq. (C.69) to eq. (C.71), where it was argued that the second exponential in each of these four terms needs not be taken into account. As a result, we remain with

$$i\hbar \frac{\partial}{\partial t} a_n^{(2)}(t) = -\sum_{\substack{p \neq m \\ p \neq n}} \left[\frac{\langle n | \mathcal{H}_1^{(1)} | p \rangle \langle p | \mathcal{H}_0^{(1)} | m \rangle}{(E_p - E_m)} + \frac{\langle n | \mathcal{H}_0^{(1)} | p \rangle \langle p | \mathcal{H}_1^{(1)} | m \rangle}{(E_p - E_m - \hbar\omega)} \right]$$

$$\times e^{i((E_n - E_m)/\hbar - \omega)t}$$

(C.81)

As in eq. (C.60) integration is now straightforward. Inserting the initial condition (C.55), we find

$$a_n^{(2)}(t) = a_n^{(2)}(0) - \sum_{\substack{p \neq m \\ p \neq n}} \left[\frac{\langle n| \mathcal{H}_1^{(1)} |p\rangle \langle p| \mathcal{H}_0^{(1)} |m\rangle}{(E_p - E_m)} + \frac{\langle n| \mathcal{H}_0^{(1)} |p\rangle \langle p| \mathcal{H}_1^{(1)} |m\rangle}{(E_p - E_m - \hbar\omega)} \right]$$

$$\times \frac{1}{i\hbar} \int_0^t dt' \, e^{i((E_n - E_m)/\hbar - \omega)t'}$$

$$= + \sum_{\substack{p \neq m \\ p \neq n}} \left[\frac{\langle n| \mathcal{H}_1^{(1)} |p\rangle \langle p| \mathcal{H}_0^{(1)} |m\rangle}{(E_p - E_m)} + \frac{\langle n| \mathcal{H}_0^{(1)} |p\rangle \langle p| \mathcal{H}_1^{(1)} |m\rangle}{(E_p - E_m - \hbar\omega)} \right]$$

$$\times \frac{e^{i((E_n - E_m)/\hbar - \omega)t} - 1}{(E_n - E_m - \hbar\omega)} \tag{C.82}$$

SECOND-ORDER TRANSITION PROBABILITIES

The transition probability per unit time is obtained by following the procedure from eq. (C.69) to eq. (C.72). We find,

$$W_{m,n} = \frac{2\pi}{\hbar} tkl \left| \sum_{\substack{p \neq m \\ p \neq n}} \left[\frac{\langle n| \mathcal{H}_1^{(1)} |p\rangle \langle p| \mathcal{H}_0^{(1)} |m\rangle}{(E_p - E_m)} + \frac{\langle n| \mathcal{H}_0^{(1)} |p\rangle \langle p| \mathcal{H}_1^{(1)} |m\rangle}{(E_p - E_m - \hbar\omega)} \right] \right|^2$$

$$\times \delta(E_n - E_m - \hbar\omega) \tag{C.83}$$

where the Dirac δ-function requires conservation of energy in the same way as in eq. (C.79).

Often one of the terms in the sum over p is strongly dominant, because for the corresponding value of p either $E_p \approx E_m$ or $E_p \approx E_m + \hbar\omega$. In the former case we simply get

$$W_{m,n} = \frac{2\pi}{\hbar} \left| \frac{\langle n| \mathcal{H}_1^{(1)} |p\rangle \langle p| \mathcal{H}_0^{(1)} |m\rangle}{(E_p - E_m)} \right|^2 \delta(E_n - E_m - \hbar\omega) \tag{C.84}$$

while in the latter case we find in good approximation,

$$W_{m,n} = \frac{2\pi}{\hbar} \left| \frac{\langle n| \mathcal{H}_0^{(1)} |p\rangle \langle p| \mathcal{H}_1^{(1)} |m\rangle}{(E_p - E_m - \hbar\omega)} \right|^2 \delta(E_n - E_m - \hbar\omega)$$

$$= \frac{2\pi}{\hbar} \left| \frac{\langle n| \mathcal{H}_0^{(1)} |p\rangle \langle p| \mathcal{H}_1^{(1)} |m\rangle}{(E_p - E_n)} \right|^2 \delta(E_n - E_m - \hbar\omega) \tag{C.85}$$

where we insert the δ-function to rewrite the denominator. Clearly, the former expression is used when the intermediate state $|p\rangle$ lies close to the initial state $|m\rangle$, while the latter expression should be used when this intermediate state lies close to the final state $|n\rangle$.

Appendix D
Tensors in Cubic Crystals

TENSORS

Physical properties appear as scalars, vectors and tensors. A scalar S is a quantity which is invariant under rotation of the frame of reference. A vector **V** has three components V_x, V_y and V_z that transform in the same way as the three Cartesian components x, y and z of the position vector **r** in normal space. Thus, if the rotation operator **R** with components $R_{\mu'\nu}$ transforms from a frame where **r** has components (x, y, z) to a frame where it has components (x', y', z'), i.e.

$$\mu' = \sum_{\nu} R_{\mu'\nu} \nu \tag{D.1}$$

Then,

$$V_{\mu'} = \sum_{\nu} R_{\mu'\nu} V_{\nu} \tag{D.2}$$

where $\mu, \nu = x$, y or z. A tensor **T** of rank 2 has nine components $T_{\mu\nu} = T_{xx}$, $T_{xy}, \ldots, T_{zz}$ that transform in the same way as the products xx, xy, ..., zz of the Cartesian components of **r**. Hence, applying the same rotation,

$$T_{\sigma'\rho'} = \sum_{\mu,\nu} R^{-1}_{\sigma'\mu} T_{\mu\nu} R_{\nu\rho'} \tag{D.3}$$

where μ, ν, σ and ρ again denote x, y or z. Similarly, a tensor of rank 3 has 27 components $T_{\mu\nu\sigma} = T_{xxx}$, $T_{xxy}, \ldots, T_{zzz}$ which transform as the products $\mu\nu\sigma = xxx$, $xxy, \ldots, zzz$, while a tensor of rank 4 has 81 components $T_{\mu\nu\sigma\rho} = T_{xxxx}$, etc., that transform as the products $\mu\nu\sigma\rho = xxxx$, etc. Note that a vector is actually a tensor of rank 1 while a scalar is a tensor of rank 0.

CRYSTAL SYMMETRY

In crystals it is very profitable to invoke the crystal symmetry, which reduces the number of independent non-zero elements of a tensor. We illustrate this with the

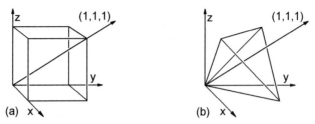

Figure D.1 Trigonal symmetry in a cube (a) and a tetrahedron (b).

cube and tetrahedron shown in Fig. D.1a and D.1b. In both cases we have chosen the Cartesian x, y and z axes in such a way that the so-called $(1, 1, 1)$ axis obeying $x = y = z$ is a trigonal symmetry axis. Here, if we rotate about this axis by $120°$ or $240°$, then the cube is transformed into itself. However, this rotation also implies a cyclic permutation of the coordinates from (x, y, z) to (z, x, y) or (y, z, x). Hence, the cube transforms into itself under such a cyclic permutation. Let us now consider a crystal which has cubic or tetrahedral symmetry. Then its properties should also be invariant under cyclic permutation of the coordinates. In particular, a vector property of the crystal should obey

$$V_x = V_y = V_z \tag{D.4}$$

while for a tensor property of rank 2 of the crystal, we should have

$$T_{xx} = T_{yy} = T_{zz} \tag{D.5}$$

$$T_{xy} = T_{yz} = T_{zx} \tag{D.6}$$

$$T_{yx} = T_{zy} = T_{xz} \tag{D.7}$$

Of course, similar equations can readily be written down for tensor properties of ranks 3 and 4.

CUBIC CRYSTALS

Clearly the interdependence of tensor components depends on crystal symmetry and a careful analysis should be made for each symmetry class. This can be done very elegantly by means of group theory [29], but such a treatment is beyond our present scope. Here we will restrict ourselves to giving the results for cubic and tetrahedral symmetry, corresponding to the two situations encountered at several places in this book. Then, for both symmetries real tensors of rank 2 have one independent non-zero component only,

$$T_{xx} = T_{yy} = T_{zz} \tag{D.8}$$

Hence, the other components, T_{xy}, etc., are equal to zero. For tetrahedral symmetry real tensors of rank 3 have also only one independent non-zero component,

$$T_{xyz} = T_{yzx} = T_{zxy} = T_{yxz} = T_{xzy} = T_{zyx} \tag{D.9}$$

Again all other components, T_{xxx}, etc., are equal to zero. However, for cubic symmetry $T_{xyz} = 0$, so all elements $T_{\mu\nu\sigma} = 0$. Finally, for both cubic and tetrahedral symmetry, real tensors of rank 4 have three independent non-zero components,

$$E_{xxxx} = E_{yyyy} = E_{zzzz} \tag{D.10}$$

$$E_{xxyy} = E_{yyzz} = E_{zzxx} = E_{yyxx} = E_{zzyy} = E_{xxzz} \tag{D.11}$$

$$E_{xyxy} = E_{yzyz} = E_{zxzx} = E_{yxyx} = E_{zyzy} = E_{xzxz} \tag{D.12}$$

so the other 66 components vanish.

Appendix E
The Classical Limit

E.1 The Correspondence Principle

In the limit where the extension of the wave function of a particle is much smaller than the distance covered by this particle, this motion should be classical, i.e. it should obey classical mechanics. In general textbooks on quantum mechanics describing this so-called correspondence principle, one always assumes an isotropic mass of the particle involved. However, one would expect that this principle should also be true for the motion of a band electron described by an anisotropic effective mass. This will be shown in the present appendix where we derive the classical equations of motion for a particle with an anisotropic mass $m_{\mu\nu}$ and exposed to externally applied electric and magnetic fields. We will assume the mass tensor to be symmetric, so $m_{\mu\nu} = m_{\nu\mu}$ and the externally applied magnetic field to be constant in space, so the vector potential can be written as

$$\bar{A}(\mathbf{r}, t) = \frac{1}{2}[\mathbf{r} \times \mathbf{B}(t)] \tag{E.1}$$

As discussed in section 5.2.3 the quantum mechanical Hamiltonian of such a particle is then given by eq. (5.64)

$$\mathcal{H} = \sum_{\mu,\nu} \frac{\left[\frac{\hbar}{i}\nabla_\mu - qA_\mu(\mathbf{r}, t)\right]\left[\frac{\hbar}{i}\nabla_\nu - qA_\nu(\mathbf{r}, t)\right]}{2m_{\mu\nu}} + qU(\mathbf{r}, t) \tag{E.2}$$

where the term (5.65) is equal to zero, while $\mu, \nu = x, y$ or z, $(\hbar/i)\nabla_\mu$ is the μ-component of the momentum operator, $A_\mu(\mathbf{r}, t)$ and $U(\mathbf{r}, t)$ the vector and scalar potentials and q the charge of the particle. The notation can be simplified by defining

$$\mathcal{P}_\mu \equiv \frac{\hbar}{i}\nabla_\mu - qA_\mu(\mathbf{r}, t) \tag{E.3}$$

THE CLASSICAL LIMIT

allowing us to write this Hamiltonian shortly as

$$\mathcal{H} = \sum_{\mu,\nu} \frac{P_\mu P_\nu}{2m_{\mu\nu}} + qU(\mathbf{r},t) \tag{E.4}$$

Its classical equivalent is given by

$$H = \sum_{\mu,\nu} \frac{P_\mu P_\nu}{2m_{\mu\nu}} + qU(\mathbf{r},t) \tag{E.5}$$

where

$$P_\mu \equiv p_\mu - qA_\mu(\mathbf{r},t) \tag{E.6}$$

In classical mechanics the motion of the particle is obtained by solving the Hamilton equations [11]. According to one of these equations the μ-component of the velocity is given by

$$v_\mu = \frac{\partial H}{\partial P_\mu} = \sum_\nu \left(\frac{1}{2m_{\mu\nu}} + \frac{1}{2m_{\nu\mu}}\right) P_\nu = \sum_\nu \frac{P_\nu}{m_{\mu\nu}} \tag{E.7}$$

because $m_{\mu\nu} = m_{\nu\mu}$. Thus, P_μ corresponds to *mass* × *velocity*, so it can be interpreted as mechanical momentum. The force on the particle should correspond to mass × acceleration, or

$$\frac{dP_\mu}{dt} = \frac{dp_\mu}{dt} - q\frac{dA_\mu(\mathbf{r},t)}{dt} = \frac{dp_\mu}{dt} - q\frac{\partial A_\mu(\mathbf{r},t)}{\partial t} - q(\mathbf{v}\cdot\nabla)A_\mu(\mathbf{r},t) \tag{E.8}$$

The time derivative of p_μ is obtained from the second Hamilton equation,

$$\frac{dp_\mu}{dt} = -\nabla_\mu H$$

$$= -\sum_{\rho,\nu} \left(\frac{1}{2m_{\rho\nu}} + \frac{1}{2m_{\nu\rho}}\right) P_\rho \nabla_\mu P_\nu - q\nabla_\mu U(\mathbf{r},t)$$

$$= -\sum_\nu v_\nu \nabla_\mu P_\nu - q\nabla_\mu U(\mathbf{r},t)$$

$$= +q\sum_\nu v_\nu \nabla_\mu A_\nu(\mathbf{r},t) - q\nabla_\mu U(\mathbf{r},t)$$

$$= +q\mathbf{v}\cdot[\nabla_\mu \mathbf{A}(\mathbf{r},t)] - q\nabla_\mu U(\mathbf{r},t)$$

$$= +q\{(\mathbf{v}\times[\nabla\times\mathbf{A}(\mathbf{r},t)]\}_\mu + q(\mathbf{v}\cdot\nabla)A_\mu(\mathbf{r},t) - q\nabla_\mu U(\mathbf{r},t) \tag{E.9}$$

where we inserted eq. (E.7) and the relation

$$\{\mathbf{v}\times[\nabla\times\mathbf{A}(\mathbf{r},t)]\}_\mu = \mathbf{v}\cdot[\nabla_\mu \mathbf{A}(\mathbf{r},t)] - (\mathbf{v}\cdot\nabla)A_\mu(\mathbf{r},t) \tag{E.10}$$

which is well known from vector analysis. Note furthermore that the partial derivatives were taken for constant p_ν. Hence, combining eqs (E.8) and (E.9)

$$\frac{dP_\mu}{dt} = +q\{\mathbf{v}\times[\nabla\times\mathbf{A}(\mathbf{r},t)]\}_\mu - q\nabla_\mu U(\mathbf{r},t) - q\frac{\partial A_\mu(\mathbf{r},t)}{\partial t} \tag{E.11}$$

Now according to classical electromagnetism [25],

$$\mathbf{B}(t) = \nabla \times \mathbf{A}(\mathbf{r}, t) \tag{E.12}$$

$$\mathbf{E}(\mathbf{r}, t) = -\nabla U(\mathbf{r}, t) - \frac{\partial \mathbf{A}(\mathbf{r}, t)}{\partial t} \tag{E.13}$$

so,

$$\frac{dP_\mu}{dt} = q[\mathbf{v} \times \mathbf{B}(t)]_\mu + qE_\mu(\mathbf{r}, t) \tag{E.14}$$

corresponding to the Lorentz force.

The next two sections are devoted to showing that these relations still hold for the quantum mechanical expectation values of the velocity and the time derivative of momentum.

E.2 The Velocity of a Particle

We first consider the velocity of the particle. It corresponds to the time derivative of the centre of gravity of the wave function $\phi(\mathbf{r})$ describing it, so its component in the μ direction is given by

$$v_\mu = \frac{d}{dt}\langle r_\mu \rangle$$

$$= \frac{d}{dt}\left[\int_\infty d\mathbf{r}\, \phi^*(\mathbf{r}) r_\mu \phi(\mathbf{r})\right]$$

$$= \int_\infty d\mathbf{r}\, \frac{\partial \phi^*(\mathbf{r})}{\partial t} r_\mu \phi(\mathbf{r}) + \int_\infty d\mathbf{r}\, \phi^*(\mathbf{r}) r_\mu \frac{\partial \phi(\mathbf{r})}{\partial t} + \int_\infty d\mathbf{r}\, \phi^*(\mathbf{r}) \frac{\partial r_\mu}{\partial t} \phi(\mathbf{r}) \tag{E.15}$$

The derivatives of the wave functions are calculated using the time dependent effective Schrödinger equation

$$\mathcal{H}\phi(\mathbf{r}) = i\hbar \frac{\partial \phi(\mathbf{r})}{\partial t} \tag{E.16}$$

and its complex conjugate

$$\mathcal{H}\phi^*(\mathbf{r}) = -i\hbar \frac{\partial \phi^*(\mathbf{r})}{\partial t} \tag{E.17}$$

Furthermore the position operator is intrinsically time independent, so

$$\frac{\partial r_\mu}{\partial t} = 0 \tag{E.18}$$

Then

$$v_\mu = \int_\infty d\mathbf{r}\, \left[-\frac{1}{i\hbar}\mathcal{H}\phi^*(\mathbf{r})\right] r_\mu \phi(\mathbf{r}) + \int_\infty d\mathbf{r}\, \phi^*(\mathbf{r}) r_\mu \left[\frac{1}{i\hbar}\mathcal{H}\phi(\mathbf{r})\right] \tag{E.19}$$

Now $\mathcal{H}$ is an Hermitian operator, so

$$\int_\infty d\mathbf{r}\, [\mathcal{H}\phi^*(\mathbf{r})] r_\mu \phi(\mathbf{r}) = \int_\infty d\mathbf{r}\, \phi^*(\mathbf{r})[\mathcal{H} r_\mu \phi(\mathbf{r})] \tag{E.20}$$

and

$$v_\mu = -\frac{1}{i\hbar} \int_\infty d\mathbf{r}\, \phi^*(\mathbf{r})(\mathcal{H} r_\mu - r_\mu \mathcal{H})\phi(\mathbf{r})$$
$$= \frac{i}{\hbar} \int_\infty d\mathbf{r}\, \phi^*(\mathbf{r})[\mathcal{H}, r_\mu]\phi(\mathbf{r}) \quad (E.21)$$

The commutator is readily calculated using eq. (E.2) for $\mathcal{H}$. As r_μ commutes with the scalar potential $U(\mathbf{r})$, only the kinetic energy term containing the vector potential needs to be taken into account. Then

$$[\mathcal{H}, r_\mu] = \sum_{\nu,\rho} \frac{1}{2m_{\nu\rho}} [\mathcal{P}_\nu \mathcal{P}_\rho, r_\mu] \quad (E.22)$$

Now

$$[\mathcal{P}_\nu \mathcal{P}_\rho, r_\mu] = \mathcal{P}_\nu \mathcal{P}_\rho r_\mu - r_\mu \mathcal{P}_\nu \mathcal{P}_\rho$$
$$= \mathcal{P}_\nu \mathcal{P}_\rho r_\mu - \mathcal{P}_\nu r_\mu \mathcal{P}_\rho + \mathcal{P}_\nu r_\mu \mathcal{P}_\rho - r_\mu \mathcal{P}_\nu \mathcal{P}_\rho$$
$$= \mathcal{P}_\nu [\mathcal{P}_\rho, r_\mu] + [\mathcal{P}_\nu, r_\mu]\mathcal{P}_\rho \quad (E.23)$$

while

$$[\mathcal{P}_\nu, r_\mu] = \left[\frac{\hbar}{i}\nabla_\nu - qA_\nu(\mathbf{r}, t), r_\mu\right] = \frac{\hbar}{i}\delta_{\nu,\mu} \quad (E.24)$$

so, replacing the sum over ρ by a sum over ν in the second term,

$$\frac{i}{\hbar}[\mathcal{H}, r_\mu] = \frac{i}{\hbar} \sum_{\nu,\rho} \frac{1}{2m_{\nu\rho}} \left(\mathcal{P}_\nu \frac{\hbar}{i}\delta_{\rho,\mu} + \frac{\hbar}{i}\delta_{\nu,\mu}\mathcal{P}_\rho\right)$$
$$= \sum_\nu \left(\frac{1}{2m_{\nu\mu}} + \frac{1}{2m_{\mu\nu}}\right)\mathcal{P}_\nu = \sum_\nu \frac{\mathcal{P}_\nu}{m_{\mu\nu}} \quad (E.25)$$

and

$$v_\mu = \int_\infty d\mathbf{r}\, \phi^*(\mathbf{r}) \left(\sum_\nu \frac{\mathcal{P}_\nu}{m_{\mu\nu}}\right)\phi(\mathbf{r})$$
$$= \sum_\nu \frac{1}{m_{\mu\nu}} \int_\infty d\mathbf{r}\, \phi^*(\mathbf{r})\mathcal{P}_\nu \phi(\mathbf{r}) = \sum_\nu \frac{\langle \mathcal{P}_\nu \rangle}{m_{\mu\nu}} \quad (E.26)$$

Thus we have recovered the classical shape (E.7) of the velocity, while the expectation value of the operator $\mathcal{P}_\mu$ appears to represent classical mechanical momentum.

E.3 The Force on a Particle

Next, we consider the force on a particle. For this purpose we pursue the quantum mechanical equivalent of the derivation given in eqs (E.8)–(E.14). Thus, we need to calculate the time derivative of the expectation value of $\mathcal{P}_\mu$. Equivalently to eqs (E.15)–(E.21) we find

$$\frac{\partial}{\partial t}\langle \mathcal{P}_\mu \rangle = \frac{i}{\hbar}\int_\infty d\mathbf{r}\phi^*(\mathbf{r})\left([\mathcal{H},\mathcal{P}_\mu] + \frac{\partial \mathcal{P}_\mu}{\partial t}\right)\phi(\mathbf{r}) \tag{E.27}$$

where we note that contrary to r_μ, $\mathcal{P}_\mu$ might be intrinsically time dependent through the time dependence of $A_\mu(\mathbf{r},t)$. Again we calculate the commutator using eq. (E.2) for $\mathcal{H}$. Then

$$[\mathcal{H},\mathcal{P}_\mu] = \sum_{\nu,\rho}\frac{1}{2m_{\nu\rho}}[\mathcal{P}_\nu\mathcal{P}_\rho,\mathcal{P}_\mu] + q[U(\mathbf{r},t),\mathcal{P}_\mu] \tag{E.28}$$

The second commutator is most easily calculated. We find

$$[U(\mathbf{r},t),\mathcal{P}_\mu] = \left[U(\mathbf{r},t),\frac{\hbar}{i}\nabla_\mu - qA_\mu(\mathbf{r},t)\right] = -\frac{\hbar}{i}\nabla_\mu U(\mathbf{r},t) \tag{E.29}$$

Now we first consider the first commutator. Similar to eq. (E.23) we can expand it into

$$[\mathcal{P}_\nu\mathcal{P}_\rho,\mathcal{P}_\mu] = \mathcal{P}_\nu[\mathcal{P}_\rho,\mathcal{P}_\mu] + [\mathcal{P}_\nu,\mathcal{P}_\mu]\mathcal{P}_\rho \tag{E.30}$$

For $\nu = \mu$,

$$[\mathcal{P}_\nu,\mathcal{P}_\mu] = 0 \tag{E.31}$$

while eq. (E.3) yields for $\nu \neq \mu$,

$$[\mathcal{P}_\nu,\mathcal{P}_\mu] = \left[\frac{\hbar}{i}\nabla_\nu - qA_\nu(\mathbf{r},t), \frac{\hbar}{i}\nabla_\mu - qA_\mu(\mathbf{r},t)\right]$$

$$= \left[\frac{\hbar}{i}\nabla_\nu, \frac{\hbar}{i}\nabla_\mu\right] - \left[\frac{\hbar}{i}\nabla_\nu, qA_\mu(\mathbf{r},t)\right] - \left[qA_\nu(\mathbf{r},t), \frac{\hbar}{i}\nabla_\mu\right] + [qA_\nu(\mathbf{r},t), qA_\mu(\mathbf{r},t)]$$

$$= -q\frac{\hbar}{i}\nabla_\nu A_\mu(\mathbf{r},t) + q\frac{\hbar}{i}\nabla_\mu A_\nu(\mathbf{r},t)$$

$$= -q\frac{\hbar}{i}(\nabla \times \mathbf{A}(\mathbf{r},t))_{\nu\times\mu} = -q\frac{\hbar}{i}B_{\nu\times\mu}(t) \tag{E.32}$$

where we use eq. (E.12) and furthermore introduce the notation $\nu \times \mu$ which is defined as follows:

$$B_{x\times y} = +B_z \quad \text{and} \quad B_{y\times x} = -B_z \tag{E.33}$$

$$B_{y\times z} = +B_x \quad \text{and} \quad B_{z\times y} = -B_x \tag{E.34}$$

$$B_{z\times x} = +B_y \quad \text{and} \quad B_{x\times z} = -B_y \tag{E.35}$$

Then

$$\sum_{\nu,\rho} \frac{1}{2m_{\nu\rho}} [\mathcal{P}_\nu \mathcal{P}_\rho, \mathcal{P}_\mu] = \sum_{\nu,\rho} \frac{1}{2m_{\nu\rho}} (\mathcal{P}_\nu[\mathcal{P}_\rho, \mathcal{P}_\mu] + [\mathcal{P}_\nu, \mathcal{P}_\mu]\mathcal{P}_\rho)$$

$$= \sum_{\nu,\rho} \frac{1}{2m_{\nu\rho}} \mathcal{P}_\nu[\mathcal{P}_\rho, \mathcal{P}_\mu] + \sum_{\nu,\rho} \frac{1}{2m_{\rho\nu}} [\mathcal{P}_\rho, \mathcal{P}_\mu]\mathcal{P}_\nu$$

$$= -q\frac{\hbar}{i} \left[\sum_{\nu,\rho} \frac{1}{2m_{\nu\rho}} \mathcal{P}_\nu B_{\rho\times\mu}(t) + \sum_{\rho,\nu} \frac{1}{2m_{\rho\nu}} B_{\rho\times\mu}(t)\mathcal{P}_\nu \right] \quad \text{(E.36)}$$

where we interchanged the indices ν and ρ in the second sum. Combining this result with eq. (E.29), we find the commutator in eq. (E.27) to be given by

$$[\mathcal{H}, \mathcal{P}_\mu] = -q\frac{\hbar}{i} \left[\sum_{\nu,\rho} \frac{1}{2m_{\nu\rho}} \mathcal{P}_\nu B_{\rho\times\mu}(t) + \sum_{\rho,\nu} \frac{1}{2m_{\rho\nu}} B_{\rho\times\mu}(\mathbf{r},t)\mathcal{P}_\nu \right] - q\frac{\hbar}{i} \nabla_\mu U(t) \quad \text{(E.37)}$$

We insert this commutator in eq. (E.27) to obtain the time derivative of the expectation value of the generalized momentum. We assume the externally applied fields and their associated potentials to be long range, so they vary little over the extension of the wave function and we take their value at the centre of gravity of the wave packet,

$$U(\mathbf{r}, t) \approx U(\langle \mathbf{r} \rangle, t) \quad \text{(E.38)}$$

$$\mathbf{A}(\mathbf{r}, t) \approx \mathbf{A}(\langle \mathbf{r} \rangle, t) \quad \text{(E.39)}$$

$$\mathbf{E}(\mathbf{r}, t) \approx \mathbf{E}(\langle \mathbf{r} \rangle, t) \quad \text{(E.40)}$$

while of course,

$$\mathbf{B}(\mathbf{r}, t) = \mathbf{B}(\langle \mathbf{r} \rangle, t) = \mathbf{B}(t) \quad \text{(E.41)}$$

Then,

$$\frac{\partial}{\partial t} \langle \mathcal{P}_\mu \rangle = -q \left[\sum_{\nu,\rho} \frac{1}{2m_{\nu\rho}} \langle \mathcal{P}_\nu \rangle B_{\rho\times\mu}(t) + \sum_{\rho,\nu} \frac{1}{2m_{\rho\nu}} B_{\rho\times\mu}(t) \langle \mathcal{P}_\nu \rangle \right]$$

$$- \left\langle q\nabla_\mu U(\langle \mathbf{r} \rangle, t) - q\frac{\partial A_\mu(\langle \mathbf{r} \rangle, t)}{\partial t} \right\rangle$$

$$= -\frac{q}{2} \left[\sum_\rho v_\rho B_{\rho\times\mu}(t) + \sum_\rho B_{\rho\times\mu}(t)v_\rho \right] - qE_\mu(\langle \mathbf{r} \rangle, t) \quad \text{(E.42)}$$

where we insert eq. (E.26) for the velocity v_ρ and eq. (E.13) for E_μ. We extend our notation further by writing $\rho = \mu$, $(\mu+1)$ and $(\mu+2)$ for the three possible values it may take. Thus for $\mu = y$, we have $(\mu+1) = z$ and $(\mu+2) = x$, etc. Then, expanding the sums and noting that $B_{\mu \times \mu}(t) = 0$, we obtain

$$\frac{\partial}{\partial t} \langle \mathcal{P}_\mu \rangle = -\frac{q}{2} [v_{(\mu+1)} B_{(\mu+1) \times \mu}(t) + v_{(\mu+2)} B_{(\mu+2) \times \mu}(t)$$
$$+ B_{(\mu+1) \times \mu}(t) v_{(\mu+1)} + B_{(\mu+2) \times \mu}(t) v_{(\mu+2)}] - q E_\mu(\langle \mathbf{r} \rangle, t)$$
$$= -\frac{q}{2} \left[-v_{(\mu+1)} B_{(\mu+2)}(t) + v_{(\mu+2)} B_{(\mu+1)}(t) - B_{(\mu+2)}(t) v_{(\mu+1)} + B_{(\mu+1)}(t) v_{(\mu+2)} \right]$$
$$- q E_\mu(\langle \mathbf{r} \rangle, t)$$
$$= - q [v_{(\mu+2)} B_{(\mu+1)}(t) - v_{(\mu+1)} B_{(\mu+2)}(t)] - q E_\mu(\langle \mathbf{r} \rangle, t) \quad \text{(E.43)}$$

In the final step we recover the classical Lorentz force (E.14),

$$\frac{\partial}{\partial t} \langle \mathcal{P}_\mu \rangle = q [\mathbf{v} \times \mathbf{B}(t)]_\mu + q E_\mu(\langle \mathbf{r} \rangle, t) \quad \text{(E.44)}$$

Appendix F
Some Fourier Transforms

F.1 The Coulomb Potential

The Coulomb potential of a charge Q at position $\mathbf{r}$

$$\frac{Q}{4\pi\varepsilon_0\varepsilon_r |\mathbf{r}|} \tag{F.1}$$

is long range, because it diminishes slowly with the distance $|\mathbf{r}|$. Therefore the calculation of its Fourier transform is rather tricky. In particular the integral

$$I_1 = \frac{1}{V}\int_V d\mathbf{r}\, e^{i\mathbf{k}\cdot\mathbf{r}} \frac{1}{|\mathbf{r}|} \tag{F.2}$$

converges badly, because the integrant diminishes too slowly for $\mathbf{r} \to \infty$.

This problem may be solved by inferring some physical considerations, in particular the fact that charges never occur fully isolated. In the real world they are surrounded by other charges of the same and of the opposite sign. On a large space scale the average charge is even expected to be zero, so charges of opposite sign compensate the initial charge Q and moreover, also compensate the potential of Q. To account for such so-called shielding, we may add an exponential factor $\exp(-\alpha|\mathbf{r}|)$ to the Coulomb potential. Here $1/\alpha$ is the characteristic distance at which shielding by other charges dominates the potential of the charge Q. Note that this provides a very rough model for shielding only, as neither the exponential decay, nor the characteristic length $1/\alpha$ are based on a detailed picture of how shielding actually works.

The existence of shielding precisely provides the mathematical method by which the integral (F.2) can be evaluated. First, we multiply the integrant by an exponential factor $\exp(-\alpha|\mathbf{r}|)$. As a result an analytic solution for the integral can be found, provided α is so large that the integrant vanishes at the integration limits defined by the integration volume V. This assumption is correct if we assume the integration volume as a whole to be neutral, i.e. when the integration volume represents a semiconductor crystal, this crystal as a whole is assumed to be neutral. At the end of the calculation, we may want to study the case that shielding is very small. We will see that this case can be handled by simply taking the limit $\alpha \to 0$ corresponding to the case that $\exp(-\alpha|\mathbf{r}|) = 1$.

So, first we calculate,

$$I'_1 = \frac{1}{V} \int_V d\mathbf{r}\, e^{i\mathbf{k}\cdot\mathbf{r}} \frac{1}{|\mathbf{r}|} e^{-\alpha|\mathbf{r}|} \tag{F.3}$$

We have assumed shielding to be so important that the integrant vanishes at the integration limits defined by the integration volume V. This implies that we may approximate the integral by extending it to infinite space. Then,

$$I'_1 \approx \frac{1}{V} \int_\infty d\mathbf{r}\, e^{i\mathbf{k}\cdot\mathbf{r}} \frac{1}{|\mathbf{r}|} e^{-\alpha|\mathbf{r}|} \tag{F.4}$$

The integrant is spherically symmetric. This allows us a free choice of the direction of the z-axis. We choose it parallel to $\mathbf{k}$. We furthermore change to spherical coordinates, so, defining $r \equiv |\mathbf{r}|$, we write $x = r\sin\theta\cos\phi$, $y = r\sin\theta\sin\phi$ and $z = r\cos\theta$. Defining furthermore $k \equiv |\mathbf{k}|$, the integral becomes

$$I'_1 = \frac{1}{V} \int_0^\infty dr\, r^2 \int_0^\pi d\theta\, \sin\theta \int_0^{2\pi} d\phi\, e^{ikr\cos\theta} \frac{1}{r} e^{-\alpha r} \tag{F.5}$$

Our first integration is over ϕ. If we also rewrite

$$\int_0^\pi d\theta\, \sin\theta = \int_{-1}^{+1} d\cos\theta \tag{F.6}$$

then the integral becomes

$$\begin{aligned}
I'_1 &= \frac{2\pi}{V} \int_0^\infty dr\, r^2 \frac{1}{r} e^{-\alpha r} \int_{-1}^{+1} d\cos\theta\, e^{ikr\cos\theta} \\
&= \frac{2\pi}{V} \int_0^\infty dr\, r\, e^{-\alpha r} \int_{-1}^{+1} d\cos\theta\, e^{ikr\cos\theta}
\end{aligned} \tag{F.7}$$

The next integral over θ is performed by defining $x = kr\cos\theta$ and replacing it by an integral over x. Then

$$\begin{aligned}
I'_1 &= \frac{2\pi}{V} \int_0^\infty dr\, r\, e^{-\alpha r} \frac{1}{kr} \int_{-kr}^{+kr} dx\, e^{ix} \\
&= \frac{2\pi}{V} \int_0^\infty dr\, e^{-\alpha r} \frac{1}{k} \int_{-kr}^{+kr} dx\, e^{ix} \\
&= \frac{2\pi}{V} \int_0^\infty dr\, e^{-\alpha r} \frac{1}{ik} [e^{ix}]_{-kr}^{+kr} \\
&= \frac{2\pi}{V} \frac{1}{ik} \int_0^\infty dr\, e^{-\alpha r} [e^{ikr} - e^{-ikr}] \\
&= \frac{2\pi}{V} \frac{1}{ik} \int_0^\infty dr\, [e^{-(\alpha-ik)r} - e^{-(\alpha+ik)r}] \\
&= \frac{2\pi}{V} \frac{1}{ik} \left[-\frac{1}{\alpha-ik} e^{-(\alpha-ik)r} + \frac{1}{\alpha+ik} e^{-(\alpha+ik)r} \right]_0^\infty
\end{aligned} \tag{F.8}$$

Now, for $r \to \infty$ the exponential functions tend to zero, while for $r \to 0$ they become equal to one. Hence,

$$\begin{aligned} I'_1 &= \frac{2\pi}{V} \frac{1}{ik} \left[0 - \left(-\frac{1}{\alpha - ik} + \frac{1}{\alpha + ik} \right) \right] \\ &= \frac{2\pi}{V} \frac{1}{ik} \left[\frac{-(\alpha - ik) + (\alpha + ik)}{(\alpha + ik)(\alpha - ik)} \right] \\ &= \frac{2\pi}{V} \frac{1}{ik} \frac{2ik}{\alpha^2 + k^2} \\ &= \frac{4\pi}{V} \frac{1}{\alpha^2 + k^2} \end{aligned}$$ (F.9)

It is now easy to take the limit $\alpha \to 0$. Then,

$$I_1 = \lim_{\alpha \to 0} I'_1 = \lim_{\alpha \to 0} \frac{4\pi}{V} \frac{1}{\alpha^2 + k^2} = \frac{4\pi}{V} \frac{1}{k^2}$$ (F.10)

This last step also yields a good criterion to decide whether shielding should be taken into account: when shielding is so large that $\alpha \approx k$, then shielding effects are important. We recall however that we have not provided a detailed model for shielding. We have only invoked shielding to make a proper choice for the method to evaluate the Fourier transform of the Coulomb potential. Hence, when shielding is small, eq. (F.10) may be considered to give a correct result. However, when shielding becomes important, a detailed model for shielding should be provided first.

F.2 The Potential of an Electric Dipole

The potential of an electric dipole $\mathbf{p} = (p_x, p_y, p_z)$ at position $\mathbf{r}$

$$\frac{\mathbf{p}}{4\pi\varepsilon_0\varepsilon_r} \cdot \nabla \frac{1}{|\mathbf{r}|}$$ (F.11)

is still long range, because it still diminishes rather slowly with the distance $|\mathbf{r}|$. Therefore, also the calculation of the Fourier transform of the potential of an electric dipole should be handled with care. Now we need to calculate the integrals,

$$I_{2\mu} = \frac{1}{V} \int_V d\mathbf{r}\, e^{i\mathbf{k}\cdot\mathbf{r}} \nabla_\mu \frac{1}{|\mathbf{r}|}$$ (F.12)

where $\mu = x$, y or z. However, it proves to be possible to reduce these integrals to the one treated in the previous section. Now, shielding applies just as well for the potential of an electric dipole as it does for an electric charge. Therefore, once we have reduced eq. (F.12) to eq. (F.4), we may simply insert the result (F.10).

The present problem has two well defined directions: one is determined by the electric dipole **p**, the other by the **k**-vector. This time it is advantageous to use Cartesian coordinates (x,y,z) and to choose the z-axis along the direction of the **k**-vector. Retaining the definitions $r \equiv |\mathbf{r}|$ and $k \equiv |\mathbf{k}|$, we hence need to calculate

$$I_{2x} = \frac{1}{V}\int_{-\infty}^{+\infty} dx \int_{-\infty}^{+\infty} dy \int_{-\infty}^{+\infty} dz\, e^{ikz} \frac{\partial}{\partial x} \frac{1}{r} \tag{F.13}$$

$$I_{2y} = \frac{1}{V}\int_{-\infty}^{+\infty} dx \int_{-\infty}^{+\infty} dy \int_{-\infty}^{+\infty} dz\, e^{ikz} \frac{\partial}{\partial y} \frac{1}{r} \tag{F.14}$$

$$I_{2z} = \frac{1}{V}\int_{-\infty}^{+\infty} dx \int_{-\infty}^{+\infty} dy \int_{-\infty}^{+\infty} dz\, e^{ikz} \frac{\partial}{\partial z} \frac{1}{r} \tag{F.15}$$

We start by considering the integral I_{2x}. It can be rewritten as

$$\begin{aligned} I_{2x} &= \frac{1}{V}\int_{-\infty}^{+\infty} dx \int_{-\infty}^{+\infty} dy \int_{-\infty}^{+\infty} dz\, e^{ikz} \frac{\partial}{\partial x} \frac{1}{r} \\ &= \frac{1}{V}\int_{-\infty}^{+\infty} dy \int_{-\infty}^{+\infty} dz \int_{-\infty}^{+\infty} dx \left(\frac{\partial}{\partial x} e^{ikz} \frac{1}{r} \right) \\ &= \frac{1}{V}\int_{-\infty}^{+\infty} dy \int_{-\infty}^{+\infty} dz \left[e^{ikz} \frac{1}{r} \right]_{-\infty}^{+\infty} \\ &= 0 \end{aligned} \tag{F.16}$$

because the function $1/r \to 0$ for $z \to \infty$. Similarly,

$$I_{2y} = 0 \tag{F.17}$$

However,

$$\begin{aligned} I_{2z} &= \frac{1}{V}\int_{-\infty}^{+\infty} dx \int_{-\infty}^{+\infty} dy \int_{-\infty}^{+\infty} dz\, e^{ikz} \frac{\partial}{\partial z} \frac{1}{r} \\ &= \frac{1}{V}\int_{-\infty}^{+\infty} dx \int_{-\infty}^{+\infty} dy \int_{-\infty}^{+\infty} dz \left(\frac{\partial}{\partial z} e^{ikz} \frac{1}{r} \right) \\ &\quad - \frac{1}{V}\int_{-\infty}^{+\infty} dx \int_{-\infty}^{+\infty} dy \int_{-\infty}^{+\infty} dz\, \frac{1}{r} \frac{\partial}{\partial z} e^{ikz} \end{aligned} \tag{F.18}$$

Here the first integral is equivalent to expression (F.16) for I_{2x}, so it is equal to zero. Performing the differentiation the second integral, we find,

$$I_{2z} = -ik \frac{1}{V}\int_{-\infty}^{+\infty} dx \int_{-\infty}^{+\infty} dy \int_{-\infty}^{+\infty} dz\, \frac{1}{r} e^{ikz} = -ikI_1 \tag{F.19}$$

Thus we have reduced I_{2z} to I_1. As discussed above, we may use the same method to calculate this integral as we used in the previous section. Hence,

$$I_{2z} = -ikI_1 = -ik \frac{4\pi}{V} \frac{1}{k^2} = -i \frac{4\pi}{V} \frac{1}{k} \tag{F.20}$$

EXERCISES

EXERCISE 2.1

In order to understand homopolar binding we try to reproduce Fig. 2.4. For this purpose we consider the more simple case of the hydrogen ion. It consists of two nuclei with charge $+e$ and one electron with charge $-e$.

(a) First we try a simple classical model. An electron with charge $-e$ is positioned in the origin and two nuclei, each with charge $+e$, are placed at positions $+(0,0,r_0)$ and $-(0,0,r_0)$. Express the total potential energy of the ion in r_0/a, where a is the Bohr radius. Express the result in Rydbergs and plot it as a function of r_0/a. (See Appendix A for the definitions of the Rydberg and the Bohr radius.)

(b) What happens if $r_0 \to 0$? Suppose the model of the previous question is correct, what does this mean for our hydrogen ion?

(c) Apparently a simple classical model is not able to describe our hydrogen ion. We now try a quantum mechanical model to reproduce Fig. 2.4. We still simplify the model as much as possible. We suppose that the nuclei are again located at positions $+(0,0,r_0)$ and $-(0,0,r_0)$. The electron is described by a wave function

$$\phi(r) = \frac{2}{\sqrt{4\pi a^3}} \exp\left(-\frac{r}{a}\right)$$

where a is the Bohr radius, so it is precisely a 1s-orbital centred on the origin. Hence, the charge inside a sphere with radius r_0 is given by

$$q = -e\, 4\pi \int_0^{r_0} dr\, r^2\, |\phi(r)|^2$$

As is well known from electrostatics, the attraction between a nucleus in position $\pm(0,0,r_0)$ and this spherically symmetric electron cloud is obtained by replacing this cloud by a charge q in the origin. This observation allows us to redo the calculation of the potential energy of the hydrogen ion as a function of r_0. First, express q/e as a function of r_0/a. Hint, use partial integration to evaluate the integral.

(d) Next, express the total potential energy of the ion in r_0. Plot the result in the same figure as the result of the first part of the question.
(e) Indicate the equilibrium positions of the nuclei in the plot.
(f) Finally, we consider the anti-bonding state. Again we start with a classical model. The nuclei, each with charge $+e$, are again positioned at $+(0,0,r_0)$ and $-(0,0,r_0)$. Now we assume the electron to be 'divided' over two positions, so there are charges $-e/2$ at $+(0,0,r_0+a)$ and $-(0,0,r_0+a)$, respectively. Redo the calculation of the total potential energy and plot the result in the same figure as the results obtained above.
(g) What happens if $r_0 \to 0$? Do you think that the problem can be avoided by invoking a quantum mechanical description? Why?

EXERCISE 2.2

The separation of core and valence electrons is based on the shell model for atoms. In this model electrons are grouped in shells, each shell corresponding to all electrons with a given principal quantum number n. Thus in diamond we have two electrons in the inner shell with $n = 1$ and four electrons in the outer shell with $n = 2$. The inner shell electrons are supposed to 'see' the full nuclear charge $Ze = 6e$, while the outer shell 'sees' a core with charge $(Z-2)e = 4e$.

(a) Use the results of appendix A to calculate the effective Bohr radius for each of the two shells.
(b) Plot the formula found in exercise 2.1c to find graphically the radius r_0 of the sphere containing 50% of the charge of the two electrons in the inner shell. Compare r_0 with the effective Bohr radius for the outer shell.
(c) In silicon we have three shells, corresponding to $n = 1, 2$ and 3, containing two, eight and four electrons. Calculate the effective Bohr radius for each shell.
(d) Find the radius r_0 of the sphere containing 50% of the charge of the two electrons in the innermost shell. Compare r_0 with the effective Bohr radius for the outer shell.
(e) Finally repeat the procedure for germanium, which has four shells with 2, 8, 18 and 4 electrons, respectively.
(f) Compare the three cases. What happens with the 'size' of the atom when moving from diamond to germanium? What happens with the 'size' of the inner shells ($n = 1$)? Now, to calculate the effective Bohr radius in the outer shells, we approximated the inner shell ($n = 1$) by a charge $-2e$ at the nucleus. Was this acceptable in all three cases? Why? Consider the size of the unit cell of diamond, silicon and germanium as given in Table 2.2. Discuss this size in view of the results of this exercise.

EXERCISE 2.3

The derivation of the lattice Hamiltonian is cluttered by various indices because one has to account for many nuclei and electrons. Also the derivation in section 2.3 is rather formal because e.g. the quantities $\Phi^{nm}_{l\mu,l'\mu'}$ are not evaluated explicitly. To clarify the subject it is worthwhile to consider the simple case of the H_2^+ ion treated in exercise

2.1. It consists of two nuclei with charge $+e$ at positions $\mathbf{R}_1$ and $\mathbf{R}_2$ and one electron with charge $-e$ at position $\mathbf{r}$. For further simplicity we assume the nuclei to remain on the z-axis, i.e. we assume that $X_1 = X_2 = Y_1 = Y_2 = 0$, so $R_{12} = Z_2 - Z_1$. As there is no nuclear motion in the x- and y-directions we also neglect the corresponding kinetic energy terms in the Hamiltonian. As a result the Hamiltonian (2.30) reads

$$\mathcal{H} = -\frac{\hbar^2}{2m_e}\left(\frac{\partial^2}{\partial x^2} + \frac{\partial^2}{\partial y^2} + \frac{\partial^2}{\partial z^2}\right) - \frac{\hbar^2}{2M}\left(\frac{\partial^2}{\partial Z_1^2} + \frac{\partial^2}{\partial Z_2^2}\right)$$
$$-\frac{e^2}{4\pi\varepsilon_0}\left(\frac{1}{|\mathbf{r}-\mathbf{R}_1|} + \frac{1}{|\mathbf{r}-\mathbf{R}_2|} + \frac{1}{Z_2-Z_1}\right)$$

where M is the nuclear mass.

(a) Expand the nuclear positions around equilibrium values $\mathbf{R}_1^0 = -(0, 0, Z_0)$ and $\mathbf{R}_2^0 = +(0, 0, Z_0)$. Note that there are no displacements in the x- and y-directions, so we only need to account for the nuclear displacements $u_1 = Z_1 - Z_1^0 = Z_1 + Z_0$ and $u_2 = Z_2 - Z_2^0 = Z_2 - Z_0$. Express $\mathcal{H}_e$ and $\mathcal{H}_n$, defined in eqs (2.36) and (2.37), in the electron position $\mathbf{r} = (x, y, z)$, the nuclear displacements u_1 and u_2 and Z_0.

(b) We now suppose the electron ground state to be given by $\Psi_0(\mathbf{r})$. Rewrite eq. (2.42) for the simple case of one electron and two nuclei which may only move in the z-direction. Use the results obtained above to obtain explicit expressions for Φ_{1z}^{00}, Φ_{2z}^{00}, $\Phi_{1z,1z}^{00}$, $\Phi_{2z,2z}^{00}$ and $\Phi_{1z,2z}^{00}$. We assume $\Psi_0(\mathbf{r})$ to be normalized so we may set

$$\int d\mathbf{r}\, \Psi_0^*(\mathbf{r})\, \Psi_0(\mathbf{r}) = 1$$

whenever possible.

(c) If we wish to use a quantum mechanical model, we need to insert a wave function like the one used in exercise 2.1c. Unfortunately the integrals then become quite complicated. Therefore we restrict ourselves to the 'classical' model of exercise 2.1a, where we assume the electron to be positioned exactly at $\mathbf{r} = (0, 0, 0)$. Mathematically this implies that we write $\Psi_0^*(\mathbf{r})\Psi_0(\mathbf{r}) = \delta(\mathbf{r})$. In the sequel we will see that this approximation again leads to impossible results. Insert these Dirac δ-functions in Φ_{1z}^{00}, Φ_{2z}^{00}, $\Phi_{1z,1z}^{00}$, $\Phi_{2z,2z}^{00}$ and $\Phi_{1z,2z}^{00}$ and show that

$$\mathcal{H}_L^{00} = -\frac{\hbar^2}{2M}\left(\frac{\partial^2}{\partial u_1^2} + \frac{\partial^2}{\partial u_2^2}\right)$$
$$+ \frac{e^2}{4\pi\varepsilon_0}\left[\frac{3}{4Z_0^2}(u_2 - u_1) - \frac{1}{2Z_0^3}(u_2 + u_1)^2 - \frac{3}{8Z_0^3}(u_2 - u_1)^2\right]$$

We clearly see a problem: the linear term cannot be made to vanish as eq. (2.43) requires for the equilibrium position of the nuclei. Discuss this result by comparing it with the result of exercise 2.1a.

(d) To be able to continue we simply skip the term linear in $(u_2 - u_1)$. Rewrite the previous result using reduced displacements $w_1 = \sqrt{M/\hbar}\, u_1$ and

(g) Next, consider the two lowest levels only, so this matrix is a 2×2 matrix. Convince yourself that it has the shape

$$\begin{pmatrix} \dfrac{\hbar^2 k_0^2}{2m_e} & W_1 \\ W_1^* & \dfrac{\hbar^2 (k_0 - b)^2}{2m_e} \end{pmatrix}$$

Next calculate its eigenvalues as a function of k_0. Hint: first introduce the variable $\kappa = k_0 - b/2$. Next, rewrite the matrix as

$$\begin{pmatrix} H_{11} & H_{12} \\ H_{21} & H_{22} \end{pmatrix} = \frac{1}{2} \begin{pmatrix} H_{11} + H_{22} & 0 \\ 0 & H_{11} + H_{22} \end{pmatrix} + \frac{1}{2} \begin{pmatrix} H_{11} - H_{22} & 2H_{12} \\ 2H_{21} & -H_{11} + H_{22} \end{pmatrix}$$

and subsequently calculate the eigenvalues of the second term.

(h) Assume that $W_1/e = W_{-1}/e = 0.1$ eV. Plot the result of the previous question in the same graph as the empty lattice structure.

(i) At $k_0 = b/2$ the upper band will show a minimum and the lower band a maximum. What is the splitting between the two bands at this point? How many electrons do you need in each primitive cell to fill the lower band completely, leaving the upper band empty? Take account of spin degeneracy.

(j) Finally expand the eigenvalues for the energy as a function of $\kappa = k_0 - b/2$ around $k_0 = b/2$ up to second order in κ. In particular show that the energy of the lower two levels can be written as

$$E_1 = \frac{\hbar^2}{2m_e} \left(\frac{b}{2}\right)^2 - |W_1| + \frac{\hbar^2}{2m_e} \left[1 - \frac{1}{|W_1|} \frac{\hbar^2}{m_e} \left(\frac{b}{2}\right)^2 \right] \kappa^2$$

$$E_2 = \frac{\hbar^2}{2m_e} \left(\frac{b}{2}\right)^2 + |W_1| + \frac{\hbar^2}{2m_e} \left[1 + \frac{1}{|W_1|} \frac{\hbar^2}{m_e} \left(\frac{b}{2}\right)^2 \right] \kappa^2$$

EXERCISE 3.2

At first sight aluminium is quite similar to silicon. The cores of the two atoms are the same. Only, silicon has four valence electrons while aluminium has only three. Moreover, the shape of the primitive cells of aluminium and silicon are the same, though in aluminium there is only one atom per primitive cell, while in silicon there are two.

(a) Assume that aluminium and silicon have the same band structure. Argue why aluminium is a metal and silicon a semiconductor. Draw the estimated position of the Fermi level in the band structure given in Fig. 3.7.

(b) Provide an argument why the band structure of aluminium should differ from the band structure of silicon.

EXERCISE 4.1

The hypothetical case of the one-dimensional crystal defined in exercise 3.1 offers an interesting possibility to connect the nearly free electron model presented in chapter 3

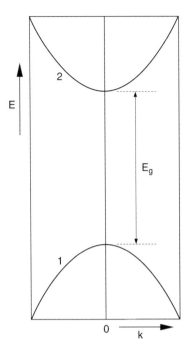

Figure Ex. 4.1 Band structure of a hypothetical one-dimensional crystal.

to the **k·p**-approximation as treated in chapter 4. Consider the band structure of a hypothetical one-dimensional crystal as shown in Fig. Ex. 4.1. Neglect spin, so none of the bands is degenerate and it is possible to use the **k·p**-approximation for non-degenerate bands. We define $|1k\rangle$ and $|2k\rangle$ to be the periodical parts of the Bloch states corresponding to the lower and upper bands, respectively. Moreover, E_g is the splitting between $|10\rangle$ and $|20\rangle$.

(a) Provide the one-dimensional equivalents of eqs (4.34)–(4.36). Restrict any sums to the two bands shown in Fig. Ex. 4.1.
(b) Compare the results with the expressions obtained in exercise 3.1. Note that one may arbitrarily choose the origin of reciprocal space. Express E_g and $P^{12} = -(P^{21})^*$ in terms of the parameters $|W_1|$ and b used in exercise 3.1. Use this comparison to explain the physical origin of E_g and P.

EXERCISE 4.2

In sections 4.2.4 and 4.2.5 spin-orbit coupling was treated as a perturbation. We consider the validity of this treatment and the problems arising when a perturbation treatment is not allowed.

(a) Consider the band parameters given in Table 4.1. Indicate the cases where a perturbation treatment of the spin-orbit coupling may prove to be quantitatively wrong.

(b) In the cases indicated in the previous question we need to incorporate spin-orbit coupling in the Hamiltonian. Consider the treatment of the spin-orbit coupling in section 4.2.4 and the matrix elements of the electron spin given in section 4.2.5. Next try to redo the argument of section 4.1.2 proving that the unperturbed states in the $\mathbf{k} \cdot \mathbf{p}$-approximation can always be chosen to be real. Indicate why this argument fails and where.

(c) Show that eq. (4.32) does not hold if spin-orbit coupling is incorporated in the unperturbed Hamiltonian. Consider subsequently eqs (4.33)–(4.36) and show that the energy may contain an extra term

$$\frac{\hbar^2}{im_e} \sum_\mu k_\mu P_\mu^{nn}$$

Note: whether such a term really exists depends also on crystal symmetry. In the diamond structure this term is excluded. However, in the zinc blende structure it may occur.

(d) Determine the position in reciprocal space where the band extrema occur. We can still define an effective mass as

$$\frac{1}{m^*_{\mu\nu}} = \frac{2m_e}{\hbar^2} \frac{\partial^2 E_k^n}{\partial k_\mu \partial k_\nu}$$

Does the value of this effective mass change because of the extra term in E_k^n?

EXERCISE 4.3

In silicon the spin-orbit coupling is quite small. let us assume that it can be neglected, so the valence band structure can be calculated via a straightforward diagonalization of the matrix (4.105). Consider the special case that $k_z = 0$, so it reads as

$$\frac{\hbar^2}{2m_e} \begin{pmatrix} Lk_x^2 + Mk_y^2 & Nk_x k_y & 0 \\ Nk_x k_y & Lk_y^2 + Mk_x^2 & 0 \\ 0 & 0 & M(k_x^2 + k_y^2) \end{pmatrix}$$

It is convenient to define $k^2 = k_x^2 + k_y^2$ and the angle ϕ in the (k_x, k_y)-plane such that $\cos\phi = k_x/k$ and $\sin\phi = k_y/k$.

(a) Show that the eigenvalues of this matrix are given by

$$E_1 = \frac{\hbar^2 k^2}{2m_e m_1^*}$$

$$E_2 = \frac{\hbar^2 k^2}{2m_e m_2^*}$$

$$E_3 = \frac{\hbar^2 k^2}{2m_e m_3^*}$$

where

$$\frac{1}{m_1^*} = M$$

$$\frac{1}{m_2^*} = \frac{1}{2}(L+M) + \frac{1}{2}\sqrt{(L-M)^2 \cos^2 2\phi + N^2 \sin^2 2\phi}$$

$$\frac{1}{m_3^*} = \frac{1}{2}(L+M) - \frac{1}{2}\sqrt{(L-M)^2 \cos^2 2\phi + N^2 \sin^2 2\phi}$$

Hint: note that we only need to diagonalize the upper left 2×2 matrix. Next use the trick of exercise 3.1g.

(b) Plot the angular dependence of $1/m_1^*$, $1/m_2^*$ and $1/m_3^*$ as a function of ϕ. Take the values for silicon given in Table 4.2, i.e. $L = -3.6$, $M = -6.5$ and $N = -8.7$.

(c) Is it still possible to talk about light and heavy holes? Why is it fortunate that $|L+M| > |N|$? Consider eqs (4.113) and (4.114) and verify whether this fortunate condition persists in the case of the extremely simple model presented in sections 4.2.1–4.2.5.

EXERCISE 4.4

In section 4.2.5 the valence band structure was obtained in a somewhat unorthodox way. The more standard treatment starts with the Hamiltonian (4.73), where we replace the first two terms by their more general form (4.101) and (4.105). For simplicity we restrict ourselves to the special case that $k_x = k_y = 0$.

(a) Insert the Pauli matrices (4.77) and and show that the Hamiltonian is written as the following 6×6 matrix

$$E_0^v \mathbf{1} + \frac{\hbar^2}{2m_e} \begin{pmatrix} Mk_z^2 & 0 & -i\lambda & 0 & 0 & \lambda \\ 0 & Mk_z^2 & 0 & i\lambda & -\lambda & 0 \\ i\lambda & 0 & Mk_z^2 & 0 & 0 & -i\lambda \\ 0 & -i\lambda & 0 & Mk_z^2 & -i\lambda & 0 \\ 0 & -\lambda & 0 & i\lambda & Lk_z^2 & 0 \\ \lambda & 0 & i\lambda & 0 & 0 & Lk_z^2 \end{pmatrix}$$

on the basis

$$|vx0\rangle |\alpha\rangle$$
$$|vx0\rangle |\beta\rangle$$
$$|vy0\rangle |\alpha\rangle$$
$$|vy0\rangle |\beta\rangle$$
$$|vz0\rangle |\alpha\rangle$$
$$|vz0\rangle |\beta\rangle$$

where
$$|\alpha\rangle = \begin{pmatrix} 1 \\ 0 \end{pmatrix}$$
$$|\beta\rangle = \begin{pmatrix} 0 \\ 1 \end{pmatrix}$$

(b) it is now claimed that the following states are eigenvectors of this matrix:

$$|hh+\rangle = -\frac{1}{\sqrt{2}}(|vx0\rangle + i|vy0\rangle)|\alpha\rangle$$

$$|lh+\rangle = -\frac{1}{\sqrt{6}}(|vx0\rangle + i|vy0\rangle)|\beta\rangle + \sqrt{\frac{2}{3}}|vz0\rangle|\alpha\rangle$$

$$|lh-\rangle = \frac{1}{\sqrt{6}}(|vx0\rangle - i|vy0\rangle)|\alpha\rangle + \sqrt{\frac{2}{3}}|vz0\rangle|\beta\rangle$$

$$|hh-\rangle = \frac{1}{\sqrt{2}}(|vx0\rangle - i|vy0\rangle)|\beta\rangle$$

$$|\lambda h+\rangle = \frac{1}{\sqrt{3}}|vz0\rangle|\alpha\rangle + \frac{1}{\sqrt{3}}(|vx0\rangle + i|vy0\rangle)|\beta\rangle$$

$$|\lambda h-\rangle = \frac{1}{\sqrt{3}}|vz0\rangle|\beta\rangle - \frac{1}{\sqrt{3}}(|vx0\rangle - i|vy0\rangle)|\alpha\rangle$$

These states are orthonormal. Check this for two cases, i.e. check that
$$\langle hh+|hh+\rangle = 1 \quad \text{and that} \quad \langle hh+|lh+\rangle = 0$$

(c) Show by insertion that

$$\langle hh+|\mathcal{H}|hh+\rangle = \langle hh-|\mathcal{H}|hh-\rangle = E_0^v + M\frac{\hbar^2 k^2}{2m_e} + \frac{1}{2}\lambda$$

$$\langle lh+|\mathcal{H}|lh+\rangle = \langle lh-|\mathcal{H}|lh-\rangle = \frac{1}{3}(M+2L)\frac{\hbar^2 k^2}{2m_e} + \frac{1}{2}\lambda$$

$$\langle \lambda h+|\mathcal{H}|\lambda h+\rangle = \langle \lambda h-|\mathcal{H}|\lambda h-\rangle = \frac{1}{3}(L+2M)\frac{\hbar^2 k^2}{2m_e} - \lambda$$

(d) Identify the heavy holes, light holes and spin-orbit bands. Compare the present results with eqs (4.106)–(4.112).

EXERCISE 5.1

The treatment in section 5.2 leading to the effective Schrödinger equations is mainly complicated because it takes potentials into account. To single out the essential steps it is instructive to repeat the derivation in the absence of such potentials.

(a) Repeat the derivation leading from eqs (5.36) and (5.37) to eqs (5.62), (5.63) and (5.64) for the case that the potentials $\mathbf{A}(\mathbf{r},t)$ and $V(\mathbf{r},t)$ are equal to zero. While doing so, consider carefully which steps can be shortened or omitted.

(b) In section 5.2.4 we studied the limit that the external fields vanish. Then the envelope function in normal space $F^n(\mathbf{r})$ is a simple plane wave. Here we also assume this plane wave to be time-dependent, so we put

$$e^{i(\mathbf{k}_0 \cdot \mathbf{r} - \omega t)}$$

Solve the effective Schrödinger equations in order to obtain E^n and ω as a function of $\mathbf{k}_0$.

(c) Use eqs (5.35) and (5.31) to obtain explicit expressions for the envelop functions in reciprocal space $\tilde{G}^n(\mathbf{k}, t)$ and $G^n(\mathbf{k}, t)$. Note: assume that $\langle_0 n 0 | \exp(i\mathbf{k} \cdot \mathbf{r}) | \widehat{nk} \rangle \approx 1$. Hint: recall the Fourier transforms in chapter 3. Subsequently insert the result in eq. (5.25) to obtain the full electron wave function. Comment on the result.

EXERCISE 5.2

In section 5.3.2 we treated shallow donors. However, we were not very careful with respect to the conditions that effective mass theory may be used for such a treatment. Here we will consider some of these conditions. We consider GaAs. Let

$$e = 1.602 \cdot 10^{-19} \text{ C}$$
$$4\pi\varepsilon_0 = 10^7 c^{-2}$$
$$c = 2.998 \cdot 10^8 \text{ m s}^{-1}$$
$$\varepsilon_r = 12.8$$
$$\hbar = 1.055 \cdot 10^{-34} \text{ J s}$$
$$m_e = 9.110 \cdot 10^{-31} \text{ kg}$$
$$m^* = 0.0665$$
$$|\mathbf{c}| = 0.5653 \text{ nm}$$

where $|\mathbf{c}|$ is the length of an edge of the cubic unit cell. Using effective mass theory and assuming the electron to be in the ground state, the envelope function is given by

$$F(r) = \frac{2}{\sqrt{4\pi a^{*3}}} \exp\left(-\frac{r}{a^*}\right)$$

where a^* is the effective Bohr radius given by eq. (5.128). Hence, the probability to find the electron in a sphere with radius r_0 is given by

$$P = 4\pi \int_0^{r_0} dr\, r^2\, |F(r)|^2$$

In exercise 2.1c we obtained an analytic expression for P as a function of r_0/a^*. In exercise 2.2b we plotted this expression and determine graphically the value of r_0/a^* where $P = 0.5$. We will call r_0 the 'radius' of the envelope function.

(a) Determine a^* and this radius for four cases: (a) a donor in GaAs; (b) the same donor, but assuming $\varepsilon_r = 1.28$; (c) the same donor but now assuming the effective mass to be $m^* = 0.665$; and (d) the same donor but assuming both $\varepsilon_r = 1.28$ and $m^* = 0.665$.

(b) Use Fig. 3.1 and Table 2.2 to determine the distance between a donor at a Ga position and its nearest neighbour As atom. We call this distance d_0. Determine a^*/d_0 for each of the four cases mentioned above and indicate where you expect the effective mass approximation to be wrong.

EXERCISE 5.3

As we saw in section 5.3.3 heterostructures offer clean realizations of textbook examples of basic quantum mechanics. Fig. Ex. 5.3 shows a heterojunction between GaAs and $Al_{0.2}Ga_{0.8}As$ where the doping is so low that space charge effects can be ignored. Only the conduction band energy at the Γ-point is shown. As in Fig. 5.3 the height of the barrier is $V_B = 140$ meV. In the present exercise we choose the zero level of energy at the bottom of the conduction band of GaAs and $z = 0$ at the interface between GaAs and $Al_{0.2}Ga_{0.8}As$. For simplicity we assume the parameters for GaAs and $Al_{0.2}Ga_{0.8}As$ to be the same. They are given in the previous exercise.

We consider the case that an electron arrives from the negative z-direction and is either reflected by the barrier or continues through it. We restrict ourselves to wave functions that are independent of x and y, i.e. wave functions that are functions of z only.

(a) For each of the two layers, solve the time-independent Schrödinger equation for an electron with an energy $E > V_B$. In particular, show that for $z \leq 0$,

$$F_1(z) = e^{ikz} + A_1 e^{-ikz}$$

and for $z \geq 0$,

$$F_2(z) = A_2 e^{ik'z}$$

obey the time-independent Schrödinger equation. Express k and k' in E, V_B, $\hbar$, m_e and m^*. Give a physical interpretation of the two terms in $F_1(z)$ as well as of $F_3(z)$.

(b) At the interface the following boundary conditions apply:

$$F_1(0) = F_2(0)$$

$$\left(\frac{\partial F_1(z)}{\partial z}\right)_{z=0} = \left(\frac{\partial F_2(z)}{\partial z}\right)_{z=0}$$

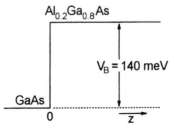

Figure Ex. 5.3 Figure with exercise 5.3. Heterojunction between GaAs and $Al_{0.2}Ga_{0.8}As$.

EXERCISES

Use these boundary conditions to express A_1 and A_2 in k and k'. Subsequently, express A_1 and A_2 in E, V_B, $\hbar$, m_e and m^*.

(c) Finally assume that the electron arrives at the first interface with an energy of 280 meV. Determine the probability that the electron is reflected by the barrier.

EXERCISE 5.4

Fig. Ex. 5.4 provides another example of quantum effects in effective mass theory. A thin layer of GaAs has been grown between two thick layers of $Al_{0.2}Ga_{0.8}As$. Thus, the square quantum well of section 5.3.3 is obtained. Next a large part of the crystal has been etched away, so the electrons are not only confined in the z-direction perpendicular to the layer but also in the x- and y-directions. We call the 'box' confining these electrons a quantum dot. In the present exercise we assume this box to be rectangular and to have a thickness $z_0 = 3$ nm, a width $x_0 = 30$ nm and a length $y_0 = 10$ nm. We assume the potentials at the boundaries of the quantum dot to be infinitely high. All other parameters are the same as in the previous exercises.

(a) Assume that the quantum dot contains one electron in the conduction band. Give expressions for its energy levels and the corresponding envelop functions. Note that these envelop functions can be separated as follows,

$$F(\mathbf{r}) = F_x^n(x) F_y^m(y) F_z^l(z)$$

where n, m and l are independent quantum numbers and use the results of section 5.3.3.

(b) Plot the resulting energy level scheme. Express the vertical axis in eV.

(c) In the actual case of a GaAs quantum dot embedded in GaAs to $Al_{0.2}Ga_{0.8}As$ the height of the barriers in the z-direction is 140 meV. Because electrons have spin $s = \frac{1}{2}$, each level may contain two electrons. Clearly something goes wrong. Indicate ways to make a more useful quantum dot.

EXERCISE 6.1

Not only the derivation of the band structure is best understood by considering the simplified case of a one dimensional crystal. Also the treatment of the crystal lattice profits from a reduction to one dimension. If we also limit ourselves to one atom per

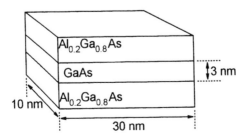

Figure Ex. 5.4 Example of a quantum dot.

primitive cell, the lattice Hamiltonian (6.1) lost most of its summations and indices and reads as

$$\mathcal{H}_L = -\sum_{l}^{N} \frac{\hbar^2}{2M} \left(\frac{\partial}{\partial u^l}\right)^2 + \frac{1}{2} \sum_{ll'}^{N} \Phi^{ll'} u^l u^{l'}$$

where N is the number of primitive cells in our one dimensional crystal, $l = 0, 1, \ldots, N$ numbers these cells, M is the nuclear mass, u^l is the displacement of the nucleus in the lth primitive cell and $\Phi^{ll'}$ are the corresponding interaction constants. We use the results of exercise 3.1a, and let $R_l = la$. Whenever applicable we omit the indices μ, ν, s and s'.

(a) First reproduce eqs (6.2)–(6.9) for our one dimensional crystal.
(b) Next rewrite eqs. (6.15)–(6.18) for our one dimensional crystal and obtain the transform of the potential to reciprocal space. Do the same with eqs (6.22)–(6.23) and combine the results to obtain the lattice Hamiltonian in reciprocal space.
(c) Consider the arguments leading to eqs (6.38), (6.41) and (6.43). Reformulate these arguments to show that $\Psi^0 = 0$. Thus, we only get an acoustical branch in the phonon spectrum. Why don't we obtain an optical branch?
(d) Rewrite eqs (6.45)–(6.47) and (6.50)–(6.52) for our one-dimensional crystal. Henceforward we omit the index A as there exists an acoustical branch only so there is no need to distinguish it from an optical branch.
(e) Introduce reduced displacements and interaction constants

$$w^q = \sqrt{\frac{M}{\hbar}} v^q \quad \text{and} \quad D^q = \frac{1}{M} \Psi^q$$

and insert these in the lattice Hamiltonian. Explain why we may write $w^q = Q^q$ in the next step. Prove that $Q^{-q} = Q^{q*}$. Argue why we may write $D^q = (\omega^q)^2$, so

$$\mathcal{H}_L = \frac{\hbar}{2} \sum_{q}^{N} \left[\frac{\partial}{\partial Q^q} \frac{\partial}{\partial Q^{q*}} + (\omega^q)^2 Q^q Q^{q*} \right]$$

(f) We finally define $Q^{q1} = \sqrt{2}\, \Re\{Q^q\}$ and $Q^{q2} = \sqrt{2}\, \Im\{Q^q\}$. Rewrite eqs (6.80)–(6.83) to obtain a sum of real one-dimensional harmonic oscillators.
(g) Note that—except for a phase difference of the imaginary part—standing waves with wave vectors k_q and k_{-q} are indistinguishable. Express ω^q in k_q and k_{-q}, use this expression to show that $\omega^q = \omega^{-q}$ and finally plot the dispersion curve for our one-dimensional crystal.

EXERCISE 6.2

(a) Earlier in exercise 3.2 we discussed the aluminium crystal structure which has the same primitive cell as silicon except that that it has only one atom in each primitive cell instead of two. Assume that the elastic tensor $c_{\mu\rho\nu\sigma}$ is the same for silicon and aluminium. Use Fig. 6.4 to construct the phonon dispersion curves for aluminium.

(b) Also the rocksalt (NaCl) structure has the primitive cell shown in Fig. 3.1a. However, while the first atom site is at the corners 1,2 to 8 of this cell, the second atomic site is at the centre of the octahedron formed by the corners 2,3 to 7. Again assume the elastic tensor $c_{\mu\rho\nu\sigma}$ to be the same as for silicon. Use Fig. 6.4 to construct as much of the phonon dispersion curves as is possible with this information. Explain the nature of the parts you cannot draw.

EXERCISE 7.1

The derivation leading to expressions (7.54), (7.55), (7.96) and (7.107) for $(\Xi_j^{DO})_{ni,nf}$, $(\Xi_j^{DA})_{ni,nf}$, $(\Xi_j^{PLO})_{ni,nf}$ and $(\Xi_j^{PA})_{ni,nf}$ is long. But these results are essential to calculate scattering rates in chapter 8 and optical absorption coefficients in chapter 9. For many purposes it is important to be able to estimate the relative importance of the four different types of electron–phonon coupling. We consider the case of the Γ-point minimum of the conduction band in GaAs. As we saw in section 7.2.1, $(\Xi_j^{DO})_{ni,nf}$ vanishes for this special case, so we will ignore it. Also we use the parameters for GaAs given in exercise 5, while we set

$$k_B = 1.381 \cdot 10^{-23} \text{ J K}^{-1}$$

$$\rho = 5.318 \cdot 10^3 \text{ kg m}^{-3}$$

$$E/e = 7.0 \text{ eV (see ref. [32])}$$

$$\varepsilon_r(\infty) = 10.63 \text{ (see ref. [6])}$$

$$\varepsilon_r(0) = 12.44 \text{ (see ref. [6])}$$

$$\epsilon^{pe} = 0.160 \text{ C m}^{-2} \text{ (see ref. [6])}$$

where k_B is Boltzman's constant and ρ the mass density of GaAs.

(a) Therefore we start by considering eq. (7.117) for $(\Xi^{DLA})_{ci,cf}$. First consider the phonon dispersion curves 6.5 for GaAs. From the slope of the LA phonon branch at the Γ-point in the Δ-direction determine the ration $\hbar\omega^{qLA}/|\mathbf{k}_q|$. Remember Table 3.1 to infer the horizontal scale of Fig. 6.5. Next, estimate k_q itself by assuming it to be equal to the value of $|\mathbf{k}_q|$ of an electron in the Γ-valley of the conduction band with an energy k_BT, where $T = 300$ K. Subsequently, calculate $\hbar\omega^{qLA}$ for this value of $|\mathbf{k}_q|$. Finally, refer to Fig. 1.3 to count the number of Ga and As atoms in a single cubic unit cell and use the result to obtain $M_1 + M_2$ and

$$\frac{1}{\sqrt{\hbar\omega^{qLA}(M_1 + M_2)}}$$

Finally determine

$$(\Xi^{DLA})_{ci,cf} = \frac{k_q E}{\sqrt{\hbar\omega^{qLA}(M_1 + M_2)}}.$$

(b) We continue by considering eq. (7.118) for $(\Xi^{PLO})_{ci,cf}$. Again we use the phonon dispersion curves 6.5 for GaAs, now to obtain ω^{OLO} and $\hbar\omega^{OLO}$, i.e. near the Γ-point. According to ref. [32],

$$\frac{(e^*)^2}{(\omega^{OLO})^2 \varepsilon_0 V_C} \frac{M_1 + M_2}{M_1 M_2} = \left[\frac{1}{\varepsilon_r(\infty)} - \frac{1}{\varepsilon_r(0)}\right]$$

allowing us to calculate $|\mathbf{k}_q|(\Xi^{PLO})_{ci,cf}$. Finally use the same estimate for $|\mathbf{k}_q|$ as in the previous part of the exercise in order to obtain $(\Xi^{PLO})_{ci,cf}$ itself.

(c) We finish this exercise by considering eq. (7.119) for $(\Xi_j^{PA})_{ci,cf}$. We restrict ourselves to coupling with LA phonons. We approximate the sum over μ in eq. (7.119) by $1/3$, so we need to calculate

$$(\Xi^{PLA})_{ci,cf} \approx \frac{i}{3} \frac{ee^{pe}}{\varepsilon_0 \varepsilon_r} \frac{1}{\sqrt{\hbar\omega^{qLA}(M_1 + M_2)}}$$

As the frequency of LA phonons is below ω^{LO}, we need to insert $\varepsilon_r(0)$ from the table above. Calculate $(\Xi^{PLA})_{ci,cf}$ using the value of

$$\frac{1}{\sqrt{\hbar\omega^{qLA}(M_1 + M_2)}}$$

obtained in the first part of this exercise.

EXERCISE 8.1

As mentioned in section 8.1.2, eq. (8.17) for the classical velocity of an electron has a more general validity than suggested by the derivation given in that section. In particular it applies to holes in the valence band, even though the energy of that band has a complicated dependence on $\mathbf{k}$. We consider the motion of a hole in such a complicated band.

(a) Use eqs (4.107) and (4.109) to express the energy of light holes as a function of spherical coordinates k, θ and ϕ. Next simplify the result by choosing $k_z = 0$. Note: be aware that the energy of holes has opposite sign compared to the energy of electrons.
(b) Start again with eqs (4.107) and (4.109) and use eq. (8.17) to calculate the components v_x and v_y of the velocity of the light holes, first as a function of k_x and k_y, subsequently as a function of k and ϕ. Assume right from the beginning that $k_z = 0$.
(c) Are $\mathbf{v}$ and $\mathbf{k}$ generally parallel? Are there preferred crystallographic directions where they remain parallel? Consider Table 4.2, calculate A, B and C for each of the semiconductors given in that table and arrange the various semiconductors according to the importance of this effect.

EXERCISE 8.2

As long as an analytical expression for the band energy has been obtained it is also possible to obtain an analytical expression for the differential density of states.

(a) Use eqs (4.106)–(4.109) to express the energy of heavy and light holes in spherical coordinates k, θ and ϕ. Next show that the differential density of states for the heavy and light hole bands can be written in the shape (8.57), provided we replace m^* by $1/f_{hh}$ or $1/f_{lh}$.

(b) Consider Fig. 4.6 and indicate the directions with the highest density of states for the light and heavy hole bands, respectively. Next, consider the various possible transitions that may be induced by electron–phonon coupling: heavy hole to heavy hole, heavy hole to light hole, light hole to heavy hole and light hole to light hole. Indicate the preferred final direction of the hole after each type of scattering.

EXERCISE 8.3

The result (8.57) for the differential density of states is strongly dependent on the number of dimensions in reciprocal space. As an exercise we repeat the derivation for one and two dimensions. This exercise is relevant for e.g. 2-DEGs in quantum wells.

(a) First, rewrite eqs (8.38)–(8.42) for two instead of three dimensions in order to derive a general expression for the differential density of states in two dimensions. For this purpose, first replace the integral over the spherical coordinates k_f, θ_f and ϕ_f by

$$\frac{N}{V_B} \int_{V_B} d\phi_f \, dk_f \, k_f (W^{ij}_{ni,nf})^{em/abs}$$

and define

$$(W^{ij}_i)^{em/abs}(\phi_f) = \frac{N}{V_B} \int_0^\infty dk_f \, k_f \left(W^{ij}_{ni,nf} \right)^{em/abs}$$

Next, introduce the differential density of states in a similar way as in eqs (8.41) and (8.42).

(b) Give a physical interpretation of the density of states, just as done in the text for three dimensions.

(c) Consider an isotropic band, so

$$E^n_f = \frac{\hbar^2 k_f^2}{2m_e m^*}$$

and obtain the equivalent of eq. (8.57) for two dimensions.

(d) Show that in one dimension the density of states is defined as

$$N_d(E^n_f) = \frac{N}{V_B} \frac{1}{\left(\dfrac{\partial E^n_f}{\partial k_f} \right)}$$

(e) Give a physical interpretation of the result. In particular explain why we may omit the qualification 'differential'.

(f) Finally, use the band energy of (c) to obtain the equivalent of eq. (8.57) for one dimension.

EXERCISE 9.1

In chapter 9 only a few of all possible optical transitions are treated. An important type of optical transition consists of indirect intraband transitions where e.g., a photon is absorbed and an optical phonon is emitted as illustrated in Fig. Ex. 9.1. In the present exercise we consider the conduction band valley near the Γ-point.

(a) To be able to treat this type of optical transition we wish to use eq. (C.83). For this purpose Fig. Ex. 9.1 is not adequate. Consider Fig. 9.3 and depict an indirect intraband transition in such a way that eq. (C.83) may be applied. Denote the initial, intermediate and final conduction band states by $|ck_i\rangle$, $|ck_p\rangle$ and $|ck_f\rangle$.

(b) As in section 9.2.2 we need to determine the optical transition matrix element between the states $|ck_i\rangle$ and $|ck_p\rangle$ and the matrix element of the electron-phonon coupling between the states $|ck_p\rangle$ and $|ck_f\rangle$. Note that we consider emission of an optical phonon. Argue why this type of indirect intraband optical transition is forbidden near the Γ-point minimum of the conduction band of germanium. In the remainder of this exercise we consider the case of GaAs and we assume polar optical phonon coupling to be responsible for the emission of the optical phonon. Then the matrix element between the states $|ck_p\rangle$ and $|ck_f\rangle$ is given by eq. (9.90) where we replace $(\Xi^{DLO})_{cp,cf}$ by $(\Xi^{PLO})_{cp,cf}$ as given in eq. (7.118). Why may we approximate $k_p \approx k_i$ and $\omega^{q_{pf}LO} \approx \omega^{0LO}$? Insert these approximations when writing down eqs. (9.90) and (7.118) for our particular case. Also write down the equivalent of eq. (9.94) for the present case.

(c) The optical transition matrix element between $|ck_i\rangle$ and $|ck_p\rangle$ is given by eq. (9.61). Investigate how eq. (9.59) enters in the transition probability (9.95) and derive a similar equation for the transition probability of indirect intraband optical transitions. Also derive the equivalent of eq. (9.102). In both cases insert the expression for $(\Xi^{PLO})_{ci,cf}$.

(d) Note that $\mathbf{k}_f$ is absent in our previous results. Moreover, $\mathbf{k}_i$ is present in a factor $|\mathbf{A}_0 \cdot \mathbf{k}_i|/|\mathbf{k}_0|^2$ only. If we choose the z-axis parallel to $\mathbf{A}_0$ then this factor reduces

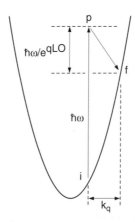

Figure Ex. 9.1 Indirect intraband optical transition.

to $|A_0|^2 \cos^2 \theta_i$. Perform the integrations over the angles θ_i, ϕ_i, θ_f, ϕ_f. As the conduction band near the Γ-point is isotropic, we may use eq. (8.57) for the differential density of states. Insert this expression in the equivalent of eq. (9.102) and perform the integration over the energy as in eq. (9.106). Discuss the resulting dependence on the optical frequency ω.

EXERCISE 9.2

We return to the heterojunction treated in exercise exercise 5.3. We investigate whether an intraband optical transition is possible near this heterojunction. As in exercise 5.3 we assume the parameters for GaAs and $Al_{0.2}Ga_{0.8}As$ to be the same. We consider an electron with an energy E_i^n which is larger than the potential barrier. Follow the arguments of section 9.2.3 to show that an optical field is not able to act on this electron. Note: this exercise shows that one really needs quantum confinement to enable an optical field to induce transitions without the help of phonons.

Bibliography

[1] N. W. Ashcroft and N. D. Mermin, *Solid State Physics*, Saunders College, Philadelphia, 1976.
[2] G. Bastard, *Wave Mechanics Applied to Semiconductors and Semiconductor Structures*, Les Editions de Physique, Les Ulis, 1992.
[3] H. A. Bethe and E. E. Salpeter, Quantum Mechanics of One- and Two-electron Systems. In *Handbuch der Physik* vol. 35, pp. 88–436, Springer Verlag, Berlin, 1957.
[4] G. L. Bir and G. E. Pikus, *Symmetry and Strain-induced Effects in Semiconductors*, John Wiley & Sons, New York, 1974.
[5] J. S. Blakemore, *Semiconductor Statistics*, Dover Publications, 1987.
[6] J. S. Blakemore, *Semiconducting and other Major Properties of Gallium Arsenide*, J. Appl. Phys. **53**, (1983) R123–R181.
[7] K. W. Böer, *Survey of Semiconductor Physics*, Van Nostrand Reinhold, New York, 1990.
[8] M. Borne and K. Huang, *Dynamical Theory of Crystal lattices*, Clarendon Press, Oxford, 1954.
[9] F. Capasso, *Physics of Quantum Electron Devices*, Springer Verlag, 1990.
[10] M. L. Cohen and J. R. Chelikowsky, *Electronic Structure and Optical Properties of Semi-Conductors*, Springer Series in Solid-State Sciences, 2nd Edition, Springer Verlag, Berlin, 1989.
[11] C. Cohen-Tanoudji, B. Diu and F. Laloë, *Quantum Mechanics*, John Wiley & Sons, New York, 1977.
[12] J. F. Cornwell, *Group Theory and Electronic Energy Bands in Solids*, North-Holland, Amsterdam, 1969.
[13] J. F. Cornwell, *Group Theory in Physics*, Vol. I, Academic Press, London, 1984.
[14] J. F. Cornwell, *Group Theory in Physics; An Introduction*, Academic Press, London, 1997.
[15] P. A. M. Dirac, *The Principles of Quantum Mechanics*, revised 4th Edition, The Clarendon Press, Oxford, 1967.
[16] W. A. Harrison, *Solid State Theory*, Dover Publications, New York, 1980.
[17] W. A. Harrison, *Electronic Structure and the Properties of Solids, The Physics of the Chemical Bond*, Dover Publications, New York, 1989.
[18] T. Manku and A. Nathan, Valence Energy-band Structure for Strained Group-IV Semiconductors, *J. Appl. Phys.* **73** (1993) 1205–1213; J. Dijkstra and W. Th. Wenckebach, Hole transport in strained silicon, *J. Appl. Phys.* **81** (1997) 1259–1262.
[19] E. O. Kane, Energy Band Theory. In *Handbook on Semiconductors*, Vol. 1, p. 193, ed. T. S. Moss, North Holland, Amsterdam, 1982.
[20] C. Kittel, *Introduction to Solid State Physics*, John Wiley, New York, 1956.
[21] C. Kittel, *Quantum Theory of Solids*, John Wiley, New York, 1963.
[22] C. F. Klingshirn, *Semiconductor Optics*, Springer Verlag, Berlin, 1995.
[23] L. D. Landau and E. M. Lifshitz, *Quantum Mechanics*, Pergamon Press, London–Paris, 1959.

[24] *Landolt-Börnstein, Numerical Data and Functional Relationships in Science and Technology,* New Series, Vol. 17a, *Physics of Group IV Elements and III–V Compounds,* ed. O. Madelung, Springer Verlag, Berlin, 1982.
[25] P. Lorain, D. R. Corson and F. Lorain, *Electromagnetic Fields and Waves,* 3rd edition, W. H. Freeman, New York, 1988.
[26] H. Margenau and G. M. Murphy, *The Mathematics of Physics and Chemistry,* D. van Nostrand, Princeton, 1964.
[27] C. Moglestue, *Monte Carlo Simulation of Semiconductor Devices,* Chapman and Hall, London, 1993.
[28] *Handbook on Semiconductors,* ed. T. S. Moss, *Vol. 2, Optical Properties of Solids,* ed. M. Balkanski, North Holland, Amsterdam, 1982.
[29] J. F. Nye, *Physical Properties of Crystals,* Clarendon Press, Oxford, 1985.
[30] J. Pankove, *Optical Processes in Semiconductors,* Dover Publications, Inc., New York, 1975.
[31] T. Brudevoll, T. A. Fjeldly, J. Baek, M. S. Shur, *Scattering Rates for Holes near the Valence-Band Edge in Semiconductors, J. Appl. Phys.* **67** (1990) 7373–7382 and references therein.
[32] B. K. Ridley, *Quantum Processes in Semiconductors,* Clarendon Press, Oxford, 1988.
[33] B. K. Ridley, *Electrons and Phonons in Semiconductor Multilayers,* Cambridge Studies in Semiconductor Physics and Microelectronic Engineering, Vol. 5, Cambridge University Press, Cambridge, 1997.
[34] L. M. Roth, Dynamics of Electrons in Semiconductors in Electric and Magnetic Fields. In *Handbook on Semiconductors,* Vol. 1, pp. 451–485, ed. T. S. Moss, North Holland, Amsterdam, 1982.
[35] K. Seeger, *Semiconductor Physics,* Springer Verlag, Berlin, 1985.
[36] J. Singh, *Physics of Semiconductors and their Heterostructures,* McGraw-Hill, New York, 1993.
[37] M. D. Sturge, *Optical Absorption of Gallium Arsenide between 0.6 and 2.75 eV, Phys. Rev.* **127** (1962) 768.
[38] S. M. Sze, *Physics of Semiconductor Devices,* John Wiley, New York, 1981.
[39] M. Tinkham, *Group Theory and Quantum Mechanics,* McGraw-Hill, New York, 1964.
[40] C. Weisbuch and B. Vinter, *Quantum Semiconductor Structures,* Academic Press, Boston, 1991.
[41] G. K. Woodgate, *Elementary Atomic Structure,* Clarendon Press, Oxford, 1980.
[42] B. G. Wybourne, *Classical Groups for Physicists,* John Wiley, New York, 1974.
[43] P. Y. Yu and M. Cardona, *Fundamentals of Semiconductors,* Springer Verlag, Berlin, 1995.
[44] H. J. Zeiger and G. W. Pratt, *Magnetic Interactions in Solids,* Clarendon Press, Oxford, 1973.

Index

absorption coefficient, optical, 236
acceptor, 18, 120, 210
acoustical branch, 136
acoustical deformation potential, 159, 163
acoustical waves, 143
angular momentum, 115
annihilation operators, 29, 116, 260
antibonding state, 17
approximation, $\mathbf{k} \cdot \mathbf{p}$, 69
approximation, $\mathbf{P} \cdot \pi$, 105
atomic monolayer, 66, 123
atomic orbitals, 11
atomic orbitals, angular part, 12, 79
atomic orbitals, radial part, 11, 79
atomic structure, 9
average, of applied fields, 106

band gap, 62
band structure, free electrons, 52, 53
band structure, near $\mathbf{k} = 0$, 89
band structure, of GaAs, 58
band structure, of Ge, 58
band structure, of Si, 58
band structure, of ZnSe, 58
basis, 35
basis states, completeness, 73, 74
basis states, for $\mathbf{k} \cdot \mathbf{p}$-method, 71
basis states, for $\mathbf{k} \cdot \mathbf{p}$-method, simple model, 79
basis states, orthonormality, 73, 74
Bloch function, 56
Bloch function, finite crystal, 56
Bloch function, normalization, 56
Bloch function, translationally symmetric part, 56
Bloch state, final, 156
Bloch state, initial, 156
Bohr magneton, 85
bonding state, 17

Born–Oppenheimer approximation, 24
boundary conditions, periodic, 48, 56, 130, 156
Brillouin zone, 38
Brillouin zone, diamond structure, 40
Brillouin zone, zinc blende structure, 40

central force model, 10, 79
central force model, corrections, 80
chemical binding, 14
chemical bonds, bending and stretching, 151
chemical bonds, elastic constants, 152
chemical vapour deposition, 66
classical electron position, 189
classical equations of motion, 188, 282
compensation, 210
conduction band, 62
conduction band, $\mathbf{k} \cdot \mathbf{p}$-treatment, 82
conduction band, effective mass, 83
conduction band, non-degeneracy, 82
conduction band, s-character, 82
conductivity, 186
core electrons, 13, 22
creation operators, 29, 116, 260
crystal momentum, 52, 96, 190
crystal symmetry, 279
cubic symmetry, 280
cubic symmetry, in primitive cell, 36
current density, 186
cyclotron resonance frequency, 115

deformation potential, 155
deformation potential acoustical phonon coupling, 176, 184, 194, 204
deformation potential acoustical phonon coupling, at the Γ-point, 180
deformation potential and crystal size, 158

INDEX

deformation potential optical phonon coupling, 173, 182, 193, 204
deformation potential optical phonon coupling, at the Γ-point, 179
deformation potential, translational symmetry, 158
degeneracy, in central force model, 12
degeneracy, of p-type states, 80
density of states, 57
density, electron, 186
density, hole, 186
diamond structure, 18, 35, 60
differential density of states, 199, 212, 246
differential phonon absorption rate, 198
differential phonon emission rate, 198
differential scattering rate, 197, 199
differential transition rate, 235
Dirac δ-function, 43
Dirac brackets, 73
direct band gap, 63
direct optical transitions, 230, 232
direct semiconductor, 63, 230
discrete wave function, 44, 47
donor, 18, 120, 209

effective Bohr radius, 122
effective deformation potential, 171, 172
effective deformation potential, in reciprocal space, 171
effective mass, 111
effective mass tensor, 75
effective mass theory, 96, 103, 170, 200, 210
effective mass theory, quantization effects, 113
effective mass, anisotropy, 76, 111, 282
effective mass, isotropic, 83, 201, 203, 213, 237, 247, 249, 282
effective Rydberg, 122
elastic scattering process, 210
elastic tensor, 140
electric dipole moment, 173
electric field, externally applied, 95, 282
electromagnetic plane wave, 218
electron density, GaAs, 62
electron density, Ge, 62
electron density, Si, 62
electron density, ZnSe, 62
electron gas, two-dimensional, 125
electron, g-value, 85
electron, dipole moment, 84
electron, spin, 85, 199
electron, wave function, 64
electron–acoustical phonon coupling, 160
electron–lattice coupling, 26, 154
electron–optical phonon coupling, 160

electro phonon coupling, $\mathbf{k}\cdot\mathbf{p}$-approximation, 161, 179
electron–phonon coupling, in effective mass theory, 170
electron–phonon coupling, in reciprocal space, 157
electronic phonon coupling, simplified, 170
electronic eigenstate, final, 156
electronic eigenstate, initial, 156
electronic eigenstate, normalization, 156
empty bands, 61
empty lattice structure, 53, 59
envelope function, 103
envelope function, in normal space, 103, 104, 190
envelope function, in reciprocal space, 104, 190
envelope function, normalization, 104, 190
excitation, of electron, 62

Fermi's golden rule, 195, 211, 233, 273
finite crystal size, 48
force on a particle, 285
Fourier sum, 45
Fourier transform, 41, 42, 289
Fourier transform, Coulomb potential, 289
Fourier transform, dipole field, 291
Fourier transform, discrete, 49
Fourier transform, unitary, 42, 43
full bands, 61

gauge, 105, 111, 114, 189, 218
gerade wave function, 17
group theory, 79, 183, 280

Hamiltonian, $\mathbf{k}\cdot\mathbf{p}$-, 70
Hamiltonian, $\mathbf{k}\cdot\mathbf{p}$-, matrix representation, 74
Hamiltonian, effective, 105, 110, 188
Hamiltonian, effective, free electron, 112, 115
Hamiltonian, effective, with optical field, 220, 226
Hamiltonian, electron–phonon coupling, 155
Hamiltonian, in optical field, 220
Hamiltonian, Landau, 115
Hamiltonian, of atom, 10
Hamiltonian, of crystal, 24
Hamiltonian, of hydrogen, 11, 254
Hamiltonian, of the electrons, 19, 20
Hamiltonian, of the lattice, 26, 129
Hamiltonian, of the lattice, acoustical branch, 143
Hamiltonian, of the lattice, eigenvalues, 147
Hamiltonian, of the lattice, in reciprocal space, 135

Hamiltonian, of the lattice, optical branch, 143
Hamiltonian, single electron, 11, 20, 50, 70, 98
Hamiltonian, single electron, with external fields, 99
harmonic oscillator, one-dimensional, 27, 28, 29, 115, 130, 146, 166, 167, 259
harmonic oscillator, multi-dimensional, 27, 129
Hartree–Fock approximation, 21
heavy holes band, 89
heterojunction, 123
heterostructure, 123
hole, 63
hole, wave function, 65
hole-phonon coupling, at the Γ-point, 182
holes, positive effective mass, 90
homopolar binding, 14
Hooke's law, 140
hybridization, 13
hydrogen atom, 254
hydrogen, eigenstates, 255
hydrogen, energy levels, 255

impurities, acceptor, 120
impurities, deep, 121
impurities, donor, 120
impurities, hydrogenic, 121
impurities, ionization, 122
impurities, isoelectronic, 120
impurities, isotropic effective mass, 121
impurities, shallow, 121, 209
indirect band gap, 62
indirect optical transitions, 232, 240
indirect semiconductor, 63, 232
indirect transition; final state, 242
indirect transitions, initial state, 242
indirect transitions, intermediate state, 242
indirect transitions, matrix elements, 242
indirect transitions, total rate, 245
indirect transitions, transition rate, 244
interband coupling, 229
interband mixing, 102
interband transition, 102
intraband coupling, 230
intraband mixing, 102
intraband transition, 102, 249
inversion symmetry, 191
ionized impurities, 210
ionized impurity scattering, 211
ionized impurity scattering rate, 211
ionized impurity scattering, matrix element, 211

Jahn–Teller effect, 180, 181, 182
joint differential density of states, 235

ket, 73
kinetic energy, in reciprocal space, 51, 134
Kronecker δ-function, 46

LA phonon modes, 143, 144
Landau levels, 115, 119
Landau subbands, 115
lattice eigenstates, 167
lattice modes, 29
lattice vibrations, 23
LCAO, 15
light holes band, 89
light intensity, 219
linear combination of atomic orbitals, 15
LO phonon modes, 144
longitudinal acoustical phonon modes, 143, 164
longitudinal optical phonon modes, 144, 164, 175

magnetic field, constant, 111
magnetic field, externally applied, 95, 282
magnetization, 97
MBE, 123
minibands, 68
mobility, electron, 186
mobility, hole, 186
molecular beam epitaxy, 66, 123
momentum space, 52
Monte Carlo simulations, 187, 188
motion, of band electron, 95

nearly free electron model, 58
NFEM, 58
non-polar semiconductors, 33
normalization, in reciprocal space, 43
normalization, of periodic functions, 46, 47
nuclear displacements, 24, 129
nuclear displacements, centre of mass, 136, 160
nuclear displacements, centre of mass, in reciprocal space, 136
nuclear displacements, complex normal, 144, 163, 175, 178
nuclear displacements, generalized, 28, 31
nuclear displacements, in reciprocal space, 130
nuclear displacements, real normal, 145, 166
nuclear displacements, reduced, 27, 31, 141, 163, 175, 178
nuclear displacements, relative, 136, 160
nuclear displacements, relative, in reciprocal space, 136, 174
nuclear mass, 129

optical absorption in GaAs, 239

INDEX

optical absorption in Si, 249
optical branch, 136
optical deformation potential, 160, 162
optical deformation potential, for holes, 183
optical devices, 217
optical field, 218
optical field, effective mass treatment, 226, 249
optical spectroscopy, 217
orbitals, hybrid, 13

Pauli exclusion principle, 12, 21
periodic potential, in reciprocal space, 54
permittivity, 97, 98, 219
perturbation theory, 262
perturbation theory, degenerate, 76, 265
perturbation theory, first-order, 195, 264, 268, 273, 276
perturbation theory, non-degenerate, 75, 262
perturbation theory, second-order, 240, 264, 269, 274, 276
perturbation theory, time-dependent, 269
perturbation theory, time-independent, 107
perturbation, time-dependent, 275
phonon absorption, 32
phonon annihilation operator, 30, 167, 194
phonon creation operator, 30, 167, 194
phonon dispersion curves, 147
phonon dispersion curves for GaAs, 147
phonon dispersion curves for Si, 147
phonon emission, 32
phonon scattering rate, 202
phonon transitions, allowed, 169
phonon transitions, forbidden, 169
phonons, 30
piezoelectric effect, 176
piezoelectric electron–phonon coupling, 176, 194, 208
piezoelectric electron–phonon coupling, at the Γ-point, 182
piezoelectric tensor, 176
plane wave, 113, 171
plane wave, final, 227, 249
plane wave, initial, 227, 249
plane waves, acoustical phonon, 137
plane waves, amplitude, 132
plane waves, mixing, 250
plane waves, optical phonon, 137
plane waves, phase, 132
plane waves, propagating, 132
plane waves, standing, 132
plane waves, standing, acoustical, 149
polar electron–optical phonon coupling, 173, 194, 206
polar electron-phonon coupling, at the Γ-point, 182

polar hole-phonon coupling, 185
polar semiconductors, 34, 172
polarization, 97, 98
polarization of phonon modes, 143, 144
potential energy, in reciprocal space, 133
potential energy, translational symmetry, 133
potential, scalar, 97, 218
potential, vector, 97, 218, 250
Poynting vector, 219
primitive cell, 35

quantum well, 124, 250
quantum well, square, 124, 125, 250
quantum well, triangular, 124, 126
quasi-classical motion, 187

reciprocal lattice, 38
reciprocal space, 38
reciprocal space lattice, 39
reciprocal space lattice vectors, 39
reduced coordinate, 259

scattering processes, 192
scattering rates in GaAs, 206, 207, 209, 216
scattering, final state, 197
scattering, initial state, 197
Schrödinger equation, in reciprocal space, 52, 54
Schrödinger equations, effective, 105, 110
screening, 176, 214, 289, 291
selection rules, 195, 253
self consistent methods, 20
silicon, atomic structure, 13
space lattice, 35, 37
space lattice vectors, 38
spin-orbit band, 89
spin-orbit coupling, 85
spin-orbit coupling, matrix elements, 85, 86
spontaneous emission, 217
strain tensor, 139, 162, 176
supercell, 66
superlattice, 66, 152
susceptibility, magnetic, 97

TA phonon modes, 143
TBM, 58
tensor, 279
tetrahedral environment, 18, 78, 79
tetrahedral symmetry, 280
tetrahedral symmetry, in primitive cell, 35
tight binding model, 58
TO phonon modes, 144
total scattering rate, 197, 213
total transition rate, 234, 236
transition matrix element,
  $\mathbf{k} \cdot \mathbf{p}$-approximation, 223

transition matrix elements, 100, 167, 221
transition matrix elements, optical, 221, 250
transitions, across the band gap, 229, 230
transitions, interband, 224
transitions, intraband, 224
translational invariance, 38
translational symmetry, 35, 44, 47
transverse acoustical phonon modes, 143, 164
transverse optical phonon modes, 44, 164

umklapp processes, 198
ungerade wave function, 17
unit cell, 35

valence band, 62
valence band, $\mathbf{k}\cdot\mathbf{p}$ treatment, 84
valence band, three-fold degeneracy, 84
valence band, eigenvalues, 87
valence band, p-character, 81
valence band, warping, 93
valence electrons, 13, 14, 22
velocity of a wave packet, 190, 284
velocity of sound, 150

wave packet, 95, 103
wave packet, normalization, 96
Wigner–Seitz cell, 37, 78
wurtzite structure, 18

Zeeman interaction, 115
zinc blende structure, 18, 35, 60, 176